Proceedings in Adaptation, Learning and Optimization

Volume 9

About this Series

The role of adaptation, learning and optimization are becoming increasingly essential and intertwined. The capability of a system to adapt either through modification of its physiological structure or via some revalidation process of internal mechanisms that directly dictate the response or behavior is crucial in many real world applications. Optimization lies at the heart of most machine learning approaches while learning and optimization are two primary means to effect adaptation in various forms. They usually involve computational processes incorporated within the system that trigger parametric updating and knowledge or model enhancement, giving rise to progressive improvement. This book series serves as a channel to consolidate work related to topics linked to adaptation, learning and optimization in systems and structures. Topics covered under this series include:

- complex adaptive systems including evolutionary computation, memetic computing, swarm intelligence, neural networks, fuzzy systems, tabu search, simulated annealing, etc.
- machine learning, data mining & mathematical programming
- hybridization of techniques that span across artificial intelligence and computational intelligence for synergistic alliance of strategies for problem-solving
- aspects of adaptation in robotics
- agent-based computing
- autonomic/pervasive computing
- dynamic optimization/learning in noisy and uncertain environment
- systemic alliance of stochastic and conventional search techniques
- all aspects of adaptations in man-machine systems.

This book series bridges the dichotomy of modern and conventional mathematical and heuristic/meta-heuristics approaches to bring about effective adaptation, learning and optimization. It propels the maxim that the old and the new can come together and be combined synergistically to scale new heights in problem-solving. To reach such a level, numerous research issues will emerge and researchers will find the book series a convenient medium to track the progresses made.

More information about this series at http://www.springer.com/series/13543

Jiuwen Cao · Erik Cambria
Amaury Lendasse · Yoan Miche
Chi Man Vong
Editors

Proceedings of ELM-2016

Editors
Jiuwen Cao
Institute of Information and Control
Hangzhou Dianzi University
Zhejiang
China

Erik Cambria
School of Computer Science and Engineering
Nanyang Technological University
Singapore
Singapore

Amaury Lendasse
Department of Mechanical and Industrial Engineering
University of Iowa
Iowa City, IA
USA

Yoan Miche
Department of Information and Computer Science
Aalto University school of science
Aalto
Finland

Chi Man Vong
Department of Computer and Information Science
University of Macau
Macau
China

ISSN 2363-6084 ISSN 2363-6092 (electronic)
Proceedings in Adaptation, Learning and Optimization
ISBN 978-3-319-86157-9 ISBN 978-3-319-57421-9 (eBook)
DOI 10.1007/978-3-319-57421-9

Printed on acid-free paper

This Springer imprint is published by Springer Nature
The registered company is Springer International Publishing AG
The registered company address is: Gewerbestrasse 11, 6330 Cham, Switzerland

Contents

Contributors

Anton Akusok Arcada University of Applied Sciences, Helsinki, Finland

Buse Atli Bell Labs, Nokia, Finland; Aalto University, Espoo, Finland

Tuomas Aura Aalto University, Espoo, Finland

Stephen Baek Department of Mechanical and Industrial Engineering, The University of Iowa, Iowa City, USA

Peng Bai Department of Electrical and Electronic Engineering, Shijiazhuang Tiedao University, Shijiazhuang, China; State Key Laboratory of Intelligent Technology and Systems, Tsinghua University TNLIST, Beijing, China

Peng Bian College of Mechanical and Material Engineering, North China University of Technology, Beijing, People's Republic of China

Kaj-Mikael Bjork Arcada University of Applied Sciences, Helsinki, Finland

Kaj-Mikael Björk Risklab at Arcada University of Applied Sciences, Helsinki, Finland

Ali R. Butt Department of Computer Science, Virginia Polytechnic Institute and State University, Blacksburg, VA, USA

Canhui Cai School of Engineering, Huaqiao University, Quanzhou, China

Lei Cai School of Information Science and Engineering, Huaqiao University, Xiamen, China

Jiuwen Cao Institute of Information and Control, Hangzhou Dianzi University, Zhejiang, China

Yajie Chang Department of Computer of Information Science, University of Macau, Macau, China

Jing Chen School of Information Science and Engineering, Huaqiao University, Xiamen, China

Jie Du Department of Computer of Information Science, University of Macau, Macau, China

Emil Eirola Arcada University of Applied Sciences, Helsinki, Finland

Fei Fu School of Cultural Heritage, Northwest University, Xi'an, China

Meng Gao Department of Electrical and Electronic Engineering, Shijiazhuang Tiedao University, Shijiazhuang, China

Hongjie Geng School of Electrical Engineering, Hebei University of Technology, Tianjin, China

Francesco Grasso Department of Information Engineering (DINFO), University of Florence, Florence, Italy

Zixiao Guan School of Automation, Beijing Institute of Technology, Beijing, China

Jia Guo School of Information Science and Engineering, Ocean University of China, Qingdao, China

Andrey Gritsenko Department of Mechanical and Industrial Engineering, The Iowa Informatics Initiative, The University of Iowa, Iowa City, USA

Abdul Hafeez Computer Systems Engineering, University of Engineering & Tech, Peshawar, Pakistan

Bo He School of Information Science and Engineering, Ocean University of China, Qingdao, China

Silke Holtmanns Bell Labs, Nokia, Finland

Guang-Bin Huang School of Electrical & Electronic Engineering, Nanyang Technological University, Singapore, Singapore

Samir M. Iqbal Department of Electrical Engineering, Nano-Bio Lab, Nanotechnology Research Center, Department of Bioengineering, University of Texas at Arlington, Arlington, TX, USA; Department of Urology, University of Texas Southwestern Medical Center at Dallas, Dallas, TX, USA

Wei Jiang Department of Control Science and Engineering, Zhejiang University, Hangzhou, People's Republic of China

Yang Jiao Department of Computer of Information Science, University of Macau, Macau, China

Yi Jin School of Computer and Information Technology, Beijing Jiaotong University, Beijing, People's Republic of China; Beijing Key Lab of Traffic Data Analysis and Mining, Beijing, People's Republic of China

Xu Jingting School of Information Science and Technology, Northwest University, Xi'an, China

Hans Johnson Department of Electrical Engineering, The University of Iowa, Iowa City, USA

Feng Jun School of Information Science and Technology, Northwest University, Xi'an, China

Aapo Kalliola Bell Labs, Nokia, Finland; Aalto University, Espoo, Finland

Zhang Lei School of Information Science and Technology, Northwest University, Xi'an, China

Amaury Lendasse Department of Mechanical and Industrial Engineering and the Iowa Informatics Initiative, The University of Iowa, Iowa City, IA, USA

Xiaodong Li School of Computer Science and Techonogy, Hangzhou Dianzi University, Hangzhou, People's Republic of China

Zhigang Liao Monash Business School, Monash University, Melbourne, VIC, Australia

Fangye Lin The College of Mathematics and Computer Science, Fuzhou University, Fuzhou, Fujian, China

Huaping Liu Department of Computer Science and Technology, State Key Laboratory of Intelligent Technology and Systems, Tsinghua University, TNLIST, Beijing, People's Republic of China

Antonio Luchetta Department of Information Engineering (DINFO), University of Florence, Florence, Italy

Hui Lv Department of Mathematics, Dalian Maritime University, Dalian, China

Pengfei Lv School of Information Science and Engineering, Ocean University of China, Qingdao, China

Stefano Manetti Department of Information Engineering (DINFO), University of Florence, Florence, Italy

Shangbo Mao Rolls-Royce@NTU Corporate Lab, Nanyang Technological University, Singapore, Singapore

Weijie Mao Department of Control Science and Engineering, Zhejiang University, Hangzhou, People's Republic of China

Yoan Miche Bell Labs, Nokia, Finland; Nokia Solutions and Networks Group, Espoo, Finland; Department of Computer Science, Aalto University, Espoo, Finland

Ian Oliver Nokia Solutions and Networks Group, Espoo, Finland; Bell Labs, Nokia, Finland

Yogesh Parth Space Applications Centre (SAC), ISRO, Ahmedabad, India

Yue Qi The Ministry Key Laboratory of Electronic Information Countermeasure and Simulation, School of Electronic Engineering, Xidian University, Xi'an, China

Lily Rachmawati Computational Engineering Team, Advanced Technology Centre, Rolls-Royce Singapore Pte Ltd, Singapore, Singapore

Eshan Rajabally Strategic Research Center, Rolls-Royce Plc, London, UK

Muhammad Rizwan School of Electrical and Computer Engineering, Georgia Institute of Technology, Atlanta, GA, USA

Peng Song Department of Computer Science and Technology, School of Computer and Communication Engineering, University of Science and Technology Beijing, Beijing, China

Ly-Fie Sugianto Monash Business School, Monash University, Melbourne, VIC, Australia

Fuchun Sun Department of Computer Science and Technology, State Key Laboratory of Intelligent Technology and Systems, Tsinghua University, TNLIST, Beijing, People's Republic of China

Enmei Tu Rolls-Royce@NTU Corporate Lab, Nanyang Technological University, Singapore, Singapore

Xuyan Tu Department of Computer Science and Technology, School of Computer and Communication Engineering, University of Science and Technology Beijing, Beijing, People's Republic of China

Chi-Man Vong Department of Computer of Information Science, University of Macau, Macau, China

Bowen Wang School of Electrical Engineering, Hebei University of Technology, Tianjin, China

Chengyao Wang Department of Computer Science and Technology, School of Computer and Communication Engineering, University of Science and Technology Beijing, Beijing, China

Hongbo Wang Department of Computer Science and Technology, School of Computer and Communication Engineering, University of Science and Technology Beijing, Beijing, People's Republic of China

Kezhen Wang Department of Computer Science and Technology, School of Computer and Communication Engineering, University of Science and Technology Beijing, Beijing, People's Republic of China

Sun Xia School of Information Science and Technology, Northwest University, Xi'an, China

Liu Xiaoning School of Information Science and Technology, Northwest University, Xi'an, China

Tianhong Yan School of Mechanical and Electrical Engineering, China Jiliang University, Hangzhou, China

Fenxi Yao School of Automation, Beijing Institute of Technology, Beijing, China

Ye Yao School of Computer Science and Techonogy, Hangzhou Dianzi University, Hangzhou, People's Republic of China

Yuanlong Yu The College of Mathematics and Computer Science, Fuzhou University, Fuzhou, Fujian, China

Huanqiang Zeng School of Information Science and Engineering, Huaqiao University, Xiamen, China

Baihai Zhang School of Automation, Beijing Institute of Technology, Beijing, China

Guanghao Zhang School of Electrical & Electronic Engineering, Nanyang Technological University, Singapore, Singapore

Huisheng Zhang Department of Mathematics, Dalian Maritime University, Dalian, China

Xinliang Zhang The Ministry Key Laboratory of Electronic Information Countermeasure and Simulation, School of Electronic Engineering, Xidian University, Xi'an, China

Wang Zhaoxia Institute of High Performance Computing (IHPC), A*STAR, Singapore, Singapore

Jianqing Zhu School of Engineering, Huaqiao University, Quanzhou, China

Mingzhe Zhu The Ministry Key Laboratory of Electronic Information Countermeasure and Simulation, School of Electronic Engineering, Xidian University, Xi'an, China

Weidong Zou School of Automation, Beijing Institute of Technology, Beijing, China

Earthen Archaeological Site Monitoring Data Analysis Using Kernel-based ELM and Non-uniform Sampling TFR

Yue Qi, Mingzhe Zhu, Xinliang Zhang and Fei Fu

Abstract Known as an ancient civilization, there exists a large amount of earthen archaeological sites in China. Various types of environment monitoring data have been accumulated waiting to be analyzed for the aim of future protection. In this paper, a non-stationary data processing strategy is proposed for the better understanding of such monitoring data. The kernel-based extreme learning machine (ELM) is utilized to preprocess the original data and restore the missing parts. Then a new non-uniform sampling time-frequency representation (TFR) is proposed to analyze the non-stationary characteristic of restored data from a signal processing perspective. The test data is the real environment monitoring data of the burial pit at the Yang Mausoleum of the Han dynasty. The experimental result shows that the proposed scheme can extract different information from the original data.

Keywords Data prediction · Monitoring data analysis · Extreme learning machine · Time-frequency representation

1 Introduction

Electronic-based earthen archaeological sites protection is a multidisciplinary research field and the studies of it are full of opportunities and challenges. The preliminary work has been carried out for many years. For example, the environment monitoring data series of the burial pit at the Yang Mausoleum of the Han dynasty, known as the first enclosed earthen site museum in China, have been accumulated more than 7 million. However, the systemic analysis of such data is rare because of the complicated characteristics and the relatively poor quality of the

Y. Qi · M. Zhu · X. Zhang (✉)
The Ministry Key Laboratory of Electronic Information Countermeasure and Simulation, School of Electronic Engineering, Xidian University, Xi'an, China
e-mail: zxl_xddc@foxmail.com

F. Fu
School of Cultural Heritage, Northwest University, Xi'an, China

J. Cao et al. (eds.), *Proceedings of ELM-2016*, Proceedings in Adaptation, Learning and Optimization 9, DOI 10.1007/978-3-319-57421-9_1

data. The monitoring data is usually time series with non-uniform interval, which makes it difficult to be analyzed by traditional time-varying processing methods. As a result, some powerful TFRs can't be directly used such as the short-time Fourier transform (STFT) [1], wavelet transform (WT) [2] and the S-transform (ST) [3]. Moreover, there are a lot of interrupt parts and abrupt changes in the data which may degrade the subsequent processing. Therefore, a high performance preprocessing is desired to restore the original data before the characteristic analysis.

Extreme learning machine (ELM) is a kind of single hidden layer feedforward networks (SLFNs), which is suitable for various applications including forecast, regression, classification and so on [4–6]. Compared with another two of the most popular methods, i.e., back propagation neural networks (BPNN) [7] and support vector machine (SVM) [8], ELM achieve the faster learning speed owning to the random generation of the hidden layer parameters [9]. In our work, the kernel-based ELM is employed for preprocessing because of its better performance compared to the traditional ELM [10]. After data restoration, a new time-frequency representation (TFR) is proposed to deal with the non-uniform sampling problem. Combining the kernel-based ELM with non-uniform sampling TFR, we could extract non-stationary information form monitoring data in the view of signal processing other than only data mining.

2 Data Preprocessing Using Kernel ELM Algorithm

In this part, kernel ELM is employed to predict and restore the original monitoring data. The ELM and kernel ELM are briefly introduced here. Then the restoration performance is demonstrated by the real monitoring data of the burial pit at the Yang Mausoleum of the Han dynasty (BPYMHD). After the preprocessing, the non-stationary data information will be extracted with less error.

Given N arbitrary samples $(\mathbf{X}, \mathbf{t})$, where $\mathbf{X} = [\mathbf{x}_1, \mathbf{x}_2, \ldots, \mathbf{x}_N]^{\mathrm{T}}$ represents the feature vector, and $\mathbf{t} = [t_1, t_2, \ldots, t_N]^{\mathrm{T}}$ is target data vector. The output weight vector $\boldsymbol{\beta} = [\beta_1, \beta_2, \ldots, \beta_N]^{\mathrm{T}}$ is from the hidden nodes to the output layer and $g(x)$ is the activation function. The standard SLFN output with M hidden layer nodes neural networks is defined as follows:

$$\sum_{i=1}^{M} \beta_i g(\mathbf{w}_i \cdot \mathbf{x}_j + b_i) = y(\mathbf{x}_j), j = 1, 2, \ldots, N \tag{1}$$

which can be rewritten as:

$$\mathbf{H}\boldsymbol{\beta} = \mathbf{y} \tag{2}$$

where $\boldsymbol{\beta} = [\beta_1, \beta_2, \ldots, \beta_N]^T$.

To find the least square solution by using Moore-Penrose generalized inverse:

$$\hat{\boldsymbol{\beta}} = (\mathbf{H}^T\mathbf{H})^{-1}\mathbf{H}^T\mathbf{y} \tag{3}$$

Aiming at a better performance, the classical ELM can be improved by the kernel ELM. In a newly developed kernel ELM, the output function of ELM can be written as:

$$\mathbf{f}(\mathbf{x}) = \mathbf{h}(\mathbf{x})\mathbf{H}^T(\frac{\mathbf{I}}{\lambda} + \mathbf{H}\mathbf{H}^T)^{-1}\mathbf{y} \tag{4}$$

where λ is a coefficient used to revise the diagonal matrix $\mathbf{HH}^{\mathrm{T}}$ in order to value the weight vector $\boldsymbol{\beta}$. The advanced kernel ELM makes the learning system more stable. The kernel matrix for ELM is defined as:

$$\boldsymbol{\Omega}_{kernel} = \mathbf{HH}^T\boldsymbol{\Omega}_{kernel}(i, j) = \mathbf{h}(\mathbf{x}_i)\mathbf{h}(\mathbf{x}_j) = \mathbf{K}(\mathbf{x}_i, \mathbf{x}_j) \tag{5}$$

$$\mathbf{f}(\mathbf{x})\begin{bmatrix} \mathbf{K}(\mathbf{x}, \mathbf{x}_1) \\ \vdots \\ \mathbf{K}(\mathbf{x}, \mathbf{x}_N) \end{bmatrix}^T (\frac{\mathbf{I}}{\lambda} + \boldsymbol{\Omega}_{kernel})^{-1}\mathbf{y} \tag{6}$$

The hidden layer feature using kernel function h(x) here the Gaussian kernel is:

$$K(u, v) = e^{-\gamma\|u - v\|^2} \tag{7}$$

The test data is real monitoring environment data offered by BPYMHD museum. We choose multi-features including time moment, temperature and frost point information as inputs for testing and predicting. Continuous data of ten days are used for experiment and randomly sampling points are used for testing. The weight function and results are showed in Fig. 1. The mean square error is 0.0321, which meets the requirement for predicting.

The weight parameters learned by kernel ELM then can be used in prediction. The result is showed in Fig. 2. Red points represent the reality data, blue points represent prediction results. The mean square error of the prediction is 0.0838. According to the prediction curve, the day model trend can be learned well by kernel ELM algorithm.

3 Non-stationary Information Analysis by ST-LS Based TFR

In our previous research, it was found that both the interrupt parts and abrupt changes could lead spectrum spread in TF plane. The former is often caused by hardware problem and network congestion such as sensor faults, power failure,

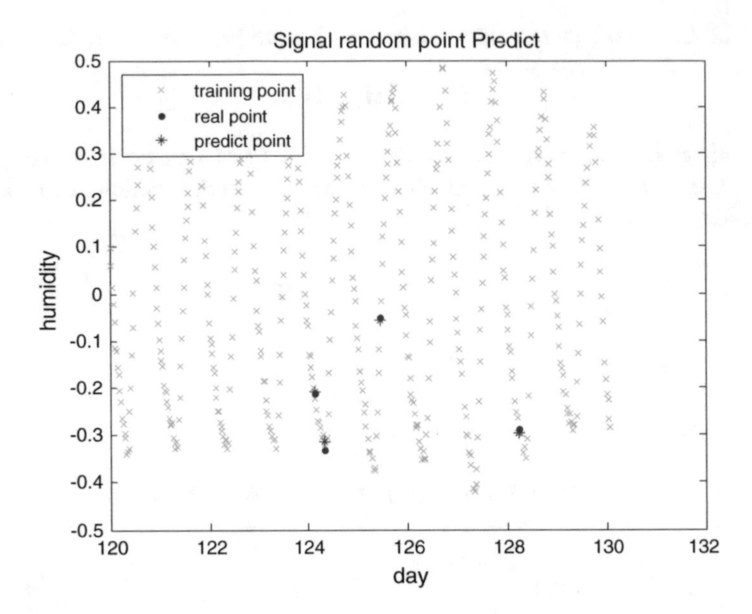

Fig. 1 Random points prediction

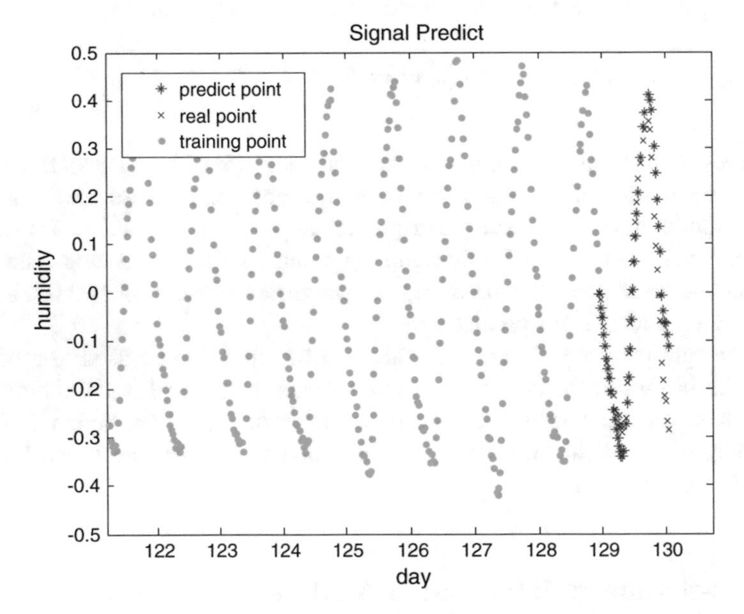

Fig. 2 Continuous point prediction

network, delay and so on, which may have the less significance for protection scheme decision. But the later often indicates environment change which may be harmful to the earthen archaeological site. After ELM-based preprocessing, the effect of missing data is suppressed. But the restored data is a kind of time series with non-uniform interval. It means that the traditional TFRs can't be directly used. In this part, the relationship between discrete Fourier transform (DFT) and least squares (LS) is discussed. Then a new short-time-LS (ST-LS) based TFR is proposed to solve the non-uniform sampling TF analysis problem.

3.1 Comparison of DFT and LS

The definition of DFT of N-points signal $x(\mathrm{n})$ can be written as

$$x(k) = \sum_{n=0}^{N-1} x(n)e^{-j2\pi nk/N} \tag{8}$$

And the M-points inverse DFT of $X(\mathrm{k})$ is formed as bellow:

$$x(n) = \frac{1}{M}\sum_{k=0}^{M-1} X(k)e^{j2\pi nk/M} \tag{9}$$

which the matrix representation is:

$$\frac{1}{M}\exp(j2\pi\frac{1}{M}\begin{bmatrix} 0 & 0 & \cdots & \cdots & 0 \\ 0 & 1 & & & M-1 \\ \vdots & & nk & & \vdots \\ \vdots & & & \ddots & \vdots \\ 0 & N-1 & \cdots & \cdots & (N-1)(M-1) \end{bmatrix}) \cdot \mathbf{X} = \mathbf{x} \tag{10}$$

From the perspective of solving linear equations, if $N = M$, the $\mathbf{X}$ has unique solution. In the situation that $M < N$, the Eq. (10) is overdetermined. The Fourier coefficient $\mathbf{X}$ can be obtained using the Least squares (LS) method. The LS provides a criterion of overdetermined equations:

$$\mathbf{\Phi\theta} = \mathrm{y} \tag{11}$$

By minimize the error function (12), the estimation of θ can be confirm by (14).

$$J(\theta) = \frac{1}{2}\|\mathbf{\Phi\theta} - \mathrm{y}\| \tag{12}$$

$$\nabla_{\theta}J(\theta) = (\frac{\partial J(\theta)}{\theta_1} \cdots \frac{\partial J(\theta)}{\theta_{end}})^{\mathrm{T}} = \mathbf{\Phi}^{\mathrm{T}}\mathbf{\Phi}\boldsymbol{\theta} - \mathbf{\Phi}^{\mathrm{T}}\mathbf{y} \tag{13}$$

$$\hat{\boldsymbol{\theta}} = (\mathbf{\Phi}^{\mathrm{T}}\mathbf{\Phi})^{-1}\mathbf{\Phi}^{\mathrm{T}}\mathrm{y} \tag{14}$$

where $(\mathbf{\Phi}^{\mathrm{T}}\mathbf{\Phi})^{-1}\mathbf{\Phi}^{\mathrm{T}}$ is the generalized inverse of coefficient matrix $\mathbf{\Phi}$. The DFT interlinked with LS method.

3.2 ST-LS and Non-stationary Information Analysis

The STFT is a useful tool for non-stationary signal analyzing, which is formed by:

$$STFT(m,k) = \sum_{n=0}^{N-1} x(n)w(n-m)e^{-j2\pi nk/N} \tag{15}$$

If the analyzing signal is unequal interval sampled, the traditional DFT-based TFR algorithms are disabled. The ST-LS is proposed using the Fourier basis LS to determine replace the DFT is the normal STFT. The Fourier basis function is set as bellow:

$$\phi(\mathbf{t},p) = \begin{cases} 1, & p=0 \\ \sin(2\pi \cdot pfs/M \cdot \mathbf{t}), & p=2r-1, r=1,2,3,\ldots,M/2 \\ \cos(2\pi \cdot pfs/M \cdot \mathbf{t}), & p=2r, r=1,2,3,\ldots,M/2 \end{cases} \tag{16}$$

where the $\mathbf{t}$ is a vector that represent time of each signal point. The basis function $\phi(t,p)$ ensures the feasibility of unequal interval sampled signal. The ST-LS can be established by (17):

$$STLS(m,k) = \sum_{n=0}^{N-1} x(n)w(n-m)L_s(m,k) \tag{17}$$

where $L_s(n,k)$ is the Fourier Spectrum Coefficient generated by LS through (18)–(20).

$$\begin{bmatrix} \phi(t,0) \\ \phi(t,1) \\ \vdots \\ \phi(t,2M) \end{bmatrix} \boldsymbol{\theta}_m = \begin{bmatrix} x(0)w(0-m) \\ x(1)w(1-m) \\ \vdots \\ x(N-1)w(N-1-m) \end{bmatrix} \tag{18}$$

$$\hat{\boldsymbol{\theta}}_m = (\begin{bmatrix} \phi(t,0) \\ \phi(t,1) \\ \vdots \\ \phi(t,2M) \end{bmatrix} \begin{bmatrix} \phi(t,0) \\ \phi(t,1) \\ \vdots \\ \phi(t,2M) \end{bmatrix}^{\mathrm{T}})^{-1} \begin{bmatrix} \phi(t,0) \\ \phi(t,1) \\ \vdots \\ \phi(t,2M) \end{bmatrix} \begin{bmatrix} x(0)w(0-m) \\ x(1)w(1-m) \\ \vdots \\ x(N-1)w(N-1-m) \end{bmatrix} \quad (19)$$

$$L_s(m,k) = \begin{cases} \hat{\boldsymbol{\theta}}_m(0), & k=0 \\ \sqrt{\hat{\boldsymbol{\theta}}_m^2(2k-1) + \hat{\boldsymbol{\theta}}_m^2(2k)}, & k=1,2,\ldots,M \end{cases} \quad (20)$$

Experiment chooses 110116th pit data and using the same feature that as in part 2 for humidity restore. The analysis range is from 01-01-2011 till now. By tabbing every day from the start, this experiment chooses 1600th–1800th days for analysis. The humidity is recovered by kernel ELM method.

The monitoring data of humidity is illustrated in Fig. 3. The data missing is around the 388th day. The sample time interval is about 30 min randomly with sudden data point missing. The time-frequency spectrum generated by ST-LS method is presented in Fig. 4, which the window width is equal to 10 days. The high spectrum amplitude appeared when the missing data segment begin enter into and shift out of the analyzing window. The spectrum information is covered by the data missing.

Using the restored data, to ensure data appear in every 30–40 min. The restore data is shown in Fig. 5, and the time-frequency spectrum is given in Fig. 6. The time varying information of humidity is clearly illustrate in time-frequency spec-

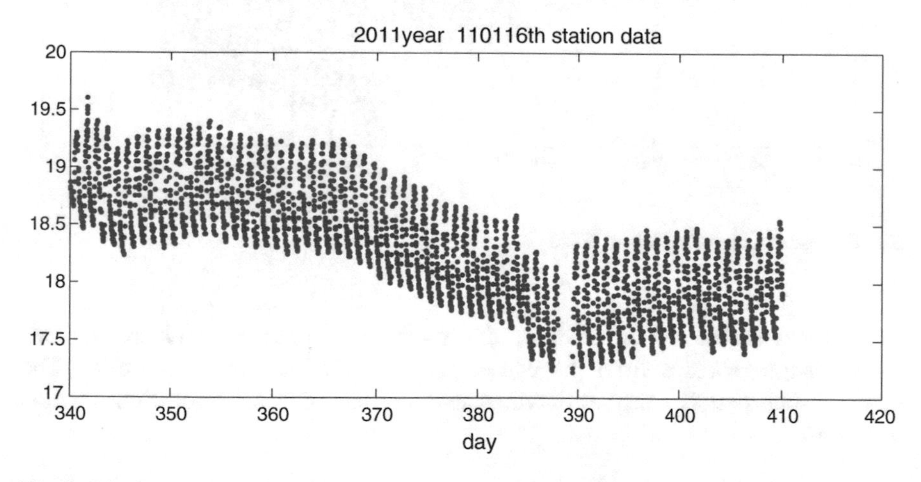

Fig. 3 Monitoring data of humidity

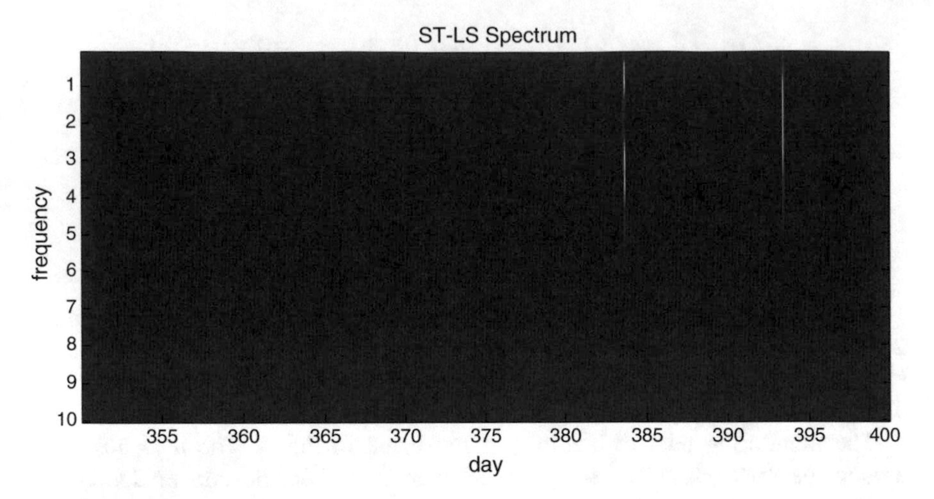

Fig. 4 Time-frequency spectrum of monitoring data

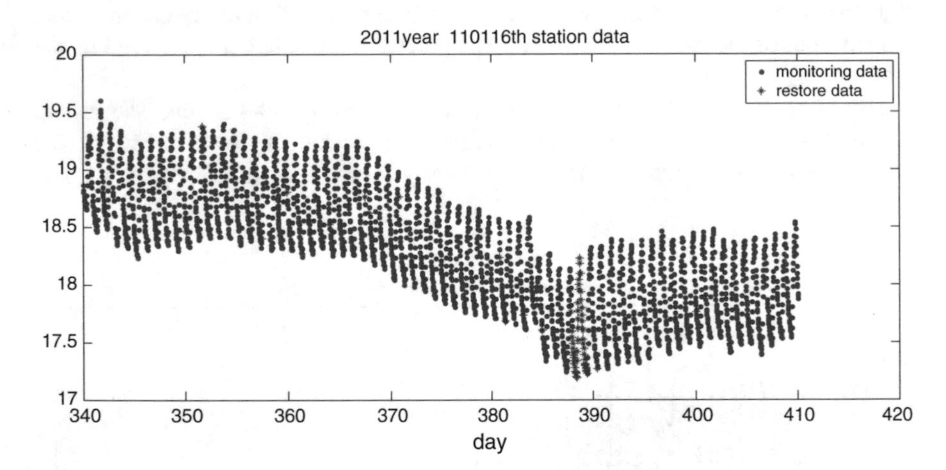

Fig. 5 Restore data using kernel ELM

trum. By interference elimination, the spectrum caused by sudden environment changes spreads from the carrier frequency to the whole spectrum. The mark of such part is essential for preventive conservation and protection scheme decision.

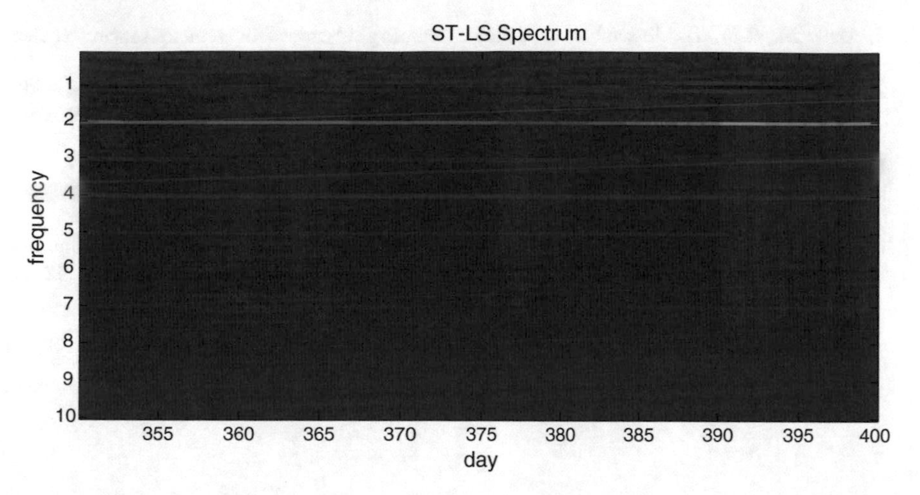

Fig. 6 Time-frequency spectrum of restore data

4 Conclusion

In this paper, we try to analyze monitoring data of earthen archaeological site in the view of non-stationary signal processing. Combining with kernel-based ELM and non-uniform sampling TFR, we are able to extract non-stationary information hidden in original records. The real data from BPYMHD museum verifies the validity of proposed method.

Acknowledgement This research was supported in part by the National Natural Science Foundation of China (61301286) and the Fundamental Research Funds for the Central Universities (JB160210).

References

1. Zhu, M.Z., Zhang, X.L., Qi, Y.: An adaptive STFT using energy concentration optimization. In: 10th International Conference on Information, Communications and Signal Processing, Singapore, Dec 2015
2. Sejdic, E., Djurovic, I., Jiang, J.: Time–frequency feature representation using energy concentration: an overview of recent advances. Digit. Sig. Proc. **19**(1), 153–183 (2009)
3. Zhang, S.Q., Li, P., Zhang, L.G., Li, H.J., Jiang, W.L., Hu, Y.T.: Modified S transform and ELM algorithms and their applications in power quality analysis. Neurocomputing **185**, 231–241 (2016)
4. Bai, Z., Huang, G.B., Wang, D.W., Wang, H.: Sparse extreme learning machine for classification. IEEE Trans. Cybern. **44**(10), 1858–1870 (2014)
5. Silvestre, L.J., Lemos, A.P., Braga, J.P., Braga, A.P.: Dataset structure as prior information for parameter-free regularization of extreme learning machines. Neurocomputing **169**, 288–294 (2015)

6. Han, M., Liu, B.: Ensemble of extreme learning machine for remote sensing image classification. Neurocomputing **149**, 65–70 (2015)
7. Hu, J.G., Zhou, G.M., Xu, X.J.: Using an improved back propagation neural network to study spatial distribution of sunshine illumination from sensor network data. Ecol. Model. **266**, 86–96 (2013)
8. Cortes, C., Vapnik, V.: Support vector networks. Mach. Learn. **20**(3), 273–297 (1995)
9. Feng, G.R., Lan, Y., Zhang, X.P., Qian, Z.X.: Dynamic adjustment of hidden node parameters for extreme learning machine. IEEE Trans. Cybern. **45**(2), 279–288 (2015)
10. Huang, G.B., Zhou, H., Ding, X., Zhang, R.: Extreme learning machine for regression and multiclass classification. IEEE Trans ON systems, Man, and Cybernetics-Part B **42**(2), 513–529 (2012)

A Multi-valued Neuron ELM with Complex-Valued Inputs for System Identification Using FRA

Francesco Grasso, Antonio Luchetta and Stefano Manetti

Abstract In the paper a new kind of ELM network is presented, which uses a MVN (multivalued neuron) with complex weights and complex inputs and that seems to be particularly suitable for fault diagnosis and identification in the frequency domain with very simple structures, given their high generalization performance. The presented network has high potentiality with a very low number of neurons. The ELM architecture is then designed with general approach, following the philosophy of this class of neural techniques, and then applied to some specific example.

Keywords Extreme learning machines · Multi-valued neuron · Frequency response analysis (FRA) · Fault diagnosis · Lumped model identification

1 Introduction

A complex-valued neural network (CVNN) is naturally predisposed to the proper elaboration of the complete information contained in a frequency response (module and, mainly, phase). A summary of techniques and applications in the CVNN area is included in [1]. CVNNs have been used in approaching and solving many real-world problems. Among them can be recalled the landmine detection [2], the forecasting of wind profiles [3], and medical image analysis [4].

In the present work, a particular kind of complex-valued neural network, the Multi-Layer Multi-Valued Neuron Networks MLMVNN, is integrated in an ELM

F. Grasso · A. Luchetta (✉) · S. Manetti
Department of Information Engineering (DINFO), University of Florence, Florence, Italy
e-mail: antonio.luchetta@unifi.it

F. Grasso
e-mail: francesco.grasso@unifi.it

S. Manetti
e-mail: stefano.manetti@unifi.it

J. Cao et al. (eds.), *Proceedings of ELM-2016*, Proceedings in Adaptation, Learning and Optimization 9, DOI 10.1007/978-3-319-57421-9_2

architecture, with a further modification to make it able to treat complex input coming from Frequency Response Analysis FRA. This approach can be extremely useful in all those cases when frequency response can be elaborated to the aim of circuital model identification or of parametric fault diagnosis. In this work we will focus over lumped model identification, but all the proposed structure can be easily adapted, with minimal changes, to parametric fault diagnosis issues.

Researchers who operate in this area well know that in many operative situations related to naturally distributed system, the identification of the lumped model still constitutes an important challenge that can help the designer in many tasks. The motivation can be of various nature: approximation of microwave filters to lumped models, extraction or parasitic parameters of analog circuit sensitive to parasitic effects, design centering, parametric analysis. The goal can be difficult to be reached, due to many problems: (i) the huge number of parameters, (ii) the non-linear nature of the system to solve (nonlinear expressions appear for linear systems either), (iii) the location of the most adequate parameters to be extracted, dealing with the testability concept, "solvability" and sensitivity of them. On the other hand, this scenery can be present in several application areas, as, for instance, the study of the time response, the evaluation of the EM compatibility, the estimate of the harmonic content, the detection and localization of faults, the complete description of a more complex structure. In the last few years, soft computing techniques have been applied in some case in order to solve this problem. We should mention artificial neural networks (ANNs) [5], genetic algorithms (GAs) [6], and particle swarm optimizers (PSOs) [7, 8]. However, it should be underlined that most of these techniques do not take direct advantage of the complex domain of the frequency response data and do not take into full consideration testability, ambiguity groups, and/or sensitivity of the model to be identified.

In this paper, a very lean neural structure with great performance is proposed for face this family of applications that comes from the convergence of MLMVNN [9, 10] and ELM [11] and that will be called CMVN-ELM (Complex MultiValued Neuron-ELM) from now on. Moreover, the multi-valued neuron is modified to receive arbitrary complex valued inputs. The kind of neuron used in this work appears to be particularly useful in that kind of problems where input data are directly represented by the frequency response of the device or system under exam, that is a number intrinsically formed by a module and a phase, or a real and an imaginary part. In fact, in the solution of system identification it is an important advantage that no conversion or normalization of the input data is needed and a low number of network parameters (of neurons) is usually enough to achieve an excellent performance.

The proposed technique uses a set of simulations or measurements made on the device/system, evaluated over different values of electrical parameters and at different frequencies, to train a CMVN-ELM, in order to estimate the electrical parameters of the lumped model, or in other words to "invert" the circuital model. This operation requires a preliminary evaluation of the testability of the circuit which is modeled, in order to determine the solvability degree with respect to the circuit parameters, following the classic definition given in [12].

In this work then CMVN paradigm has been included in an ELM architecture, because of some considerations which has been proved by experiments; the combination increases the tendency to a good generalization, which can be an important advantage in a modeling/inversion problem; moreover a particular version of CMVN algorithm is naturally predisposed to be inserted in a ELM, for the reasons that will be exposed in the next Section. Then, it is easy to enhance the common aspects and harmonize the complementary ones, as demonstrated by the good results.

2 The Techniques

2.1 Extreme Learning Machine

ELM is formalized in [12]. The network is trained over a dataset of N distinct samples $(\mathbf{x}_i, d_i)$, where $\mathbf{x}_i$ is a $n \times 1$ input vector $\mathbf{x}_i = (x_{i,1}, x_{i,2}, \ldots, x_{i,n})$ and d_i is the desired output, and the ELM output is given by:

$$\mathbf{o}_j = \sum_{i=1}^{M} \beta_i \psi_i(\mathbf{x}_j) = \sum_{i=1}^{M} \beta_i \psi(\mathbf{w}_i \cdot \mathbf{x}_j + w_{0i}); \quad j = 1, \ldots, N \tag{1}$$

where M is the number of neurons in the hidden layer, β_i is the output weight of the ith hidden node, $\mathbf{w}_i \cdot \mathbf{x}_j$ is the inner product of the weight vector $\mathbf{w}_i$ and $\mathbf{x}_j$ and w_{0i} is the threshold of the ith hidden node. In the original version of the network [12] the output nodes are linear. The equations in (1) can be written in the compact form:

$$\mathbf{O} = \mathbf{H}\beta \tag{2}$$

where

$$\mathbf{H} = \begin{bmatrix} \psi(\mathbf{w}_1 \cdot \mathbf{x}_1 + w_{01}) & \cdots & \psi(\mathbf{w}_M \cdot \mathbf{x}_1 + w_{0M}) \\ \vdots & \cdots & \vdots \\ \psi(\mathbf{w}_1 \cdot \mathbf{x}_N + w_{01}) & \cdots & \psi(\mathbf{w}_M \cdot \mathbf{x}_N + w_{0M}) \end{bmatrix} \tag{3}$$

is the output matrix of the hidden layer and includes the activation functions $\psi(\mathbf{w}_i \cdot \mathbf{x}_j + b_i)$ of the ith hidden neuron, for any jth sample of the training dataset and $\mathbf{O} = [o_1, o_2, \ldots, o_N]^{\mathrm{T}}$.

While the values of the input weights $\mathbf{w}_i$ are kept fixed, the N samples are approximated by training an ELM to find a solution to the equation:

$$\mathrm{Min}\left(||\mathbf{O} - \mathbf{D}||^2\right) = \mathrm{Min}\left(||\mathbf{H}\beta - \mathbf{D}||^2\right) \tag{4}$$

where $\mathbf{D} = [d_1, d_2, \ldots, d_N]^T$ is the vector of training target (desired output). The output weights β_i are the only parameters to be determined, finding the optimal solution of

$$\beta^* = \mathbf{H}^{\dagger}\mathbf{D} \tag{5}$$

where $\mathbf{H}^{\dagger}$ is the Moore-Penrose generalized inverse of the matrix $\mathbf{H}$. Several approaches have been proposed to estimate them [13].

2.2 *Multi-valued Neuron*

The multi-valued neuron MVN is a neurocomputing paradigm operating in complex domain algebra, formalized for both discrete and continuous data. The continuous version, used in this work, was introduced in [10]. The mapping between n inputs and the output is described by a multi-valued function of n variables $f(x_1, \ldots, x_n): \mathbb{C}_O \rightarrow \mathbb{C}_O^n$ where $\mathbb{C}_O$ is a set of points $e^{i\varphi}$ located on the unit circle of the complex number field. The continuous MVN activation function is:

$$P(z) = e^{i\arg z} = z/|z| \tag{6}$$

where $z = w_0 + w_1x_1 + \ldots w_nx_n$ is the weighted sum of the inputs, and $\arg(z)$ is the main value of the argument (phase) of the complex number z. The output is then, for the continuous MVN, the projection of the weighted sum on the unit circle, determined by (6) and shown in Fig. 1. MVNs are comprehensively reviewed in [14]. In a Neural Network based on the MVN, the learning algorithm is given by an error-correction rule, rather than a minimization iterative algorithm. In general, the rule for the adjustment of weights is something like:

$$W_{r+1} = W_r + \frac{C_r}{(n+1)|z_r|}\delta\bar{X} \tag{7}$$

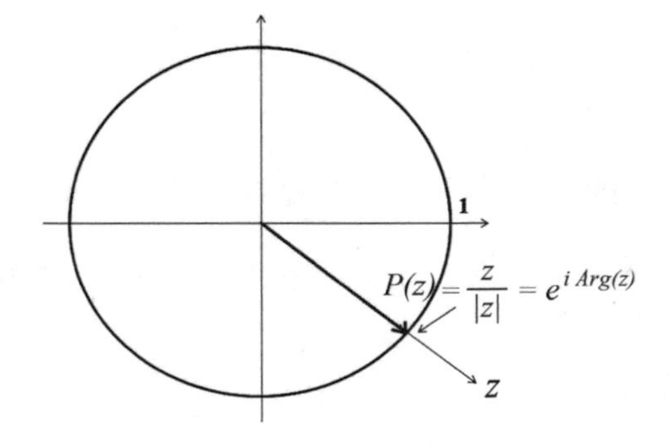

Fig. 1 Geometrical interpretation of the continuous MVN activation function

where δ is the error, $\bar{X}$ is the complex-conjugated vector of the input, r is the index of the learning epoch, n is the number of neuron inputs, W_r and W_{r+1} are the weighting vectors before and after correction, respectively, C_r is the learning rate.

In a multilayer neural network based on this kind of neuron, with a feedforward topology, the neurons are integrated into layers, and the output of each neuron from the current layer is connected to the corresponding inputs of neurons from the following layer. The use of MVN as a basic neuron determines important distinctions and advantages of this kind of network with respect to a classical multilayer perceptron MLP, that are described in detail in [14]. The canonical learning algorithm for this network uses the same error-correction learning rule as the one of a single MVN, backpropagated to the hidden layers as described in [14]. Assuming, for instance, that network is formed by M neurons in the hidden layer, and the error is:

$$\delta_k^* = D_k - Y_k \tag{8}$$

where D_k and Y_k are, respectively, the target output and the calculated output of the network. To train the input layer neurons, the local error is used:

$$\delta_k = \frac{1}{M}\delta_k^* \tag{9}$$

backpropagated by means of the algebraic complex rule:

$$\tilde{w}_i^k = w_i^k + \frac{C_r}{(N_{in}+1)|z_k|}\delta_k \bar{\tilde{Y}}_i; \quad i = 1, \ldots, N_{in} \tag{10}$$

where $\tilde{w}_i^k$ is a corrected weight, N_{in} is the number of input lines, C_r is the learning rate, Y_i is the actual output of the ith neuron of the hidden layer (corrected when the $\sim$ superscript is used and conjugated when the “bar” superscript is used). The convergence of the learning process based on the learning rule (7) is again proven in [14].

In a two layers network with a single output, with inputs and output in the real domain, the real inputs are converted into complex inputs using (1). Finally, the value given by the output neuron is converted to a real value using the arg() function.

The most severe limitation of MVN based networks is to be “slow” in the learning time with respect to the real value NN [15], and to reach same or better performance they require a very high number of epochs. This problem has been completely overcome by a modification in the algorithm introduced in [16]. The core of this modification consists in using a least squares solution for the weights of the output neuron. In this case, the use of a least squares solution is possible also without resorting to a linear output neuron, because the hidden layer output matrix is of the form:

$$\mathbf{H} = \begin{bmatrix} \mathbf{w}_1 \cdot \mathbf{x}_1 + w_{01} & \cdots & \mathbf{w}_M \cdot \mathbf{x}_1 + w_{0M} \\ \vdots & \cdots & \vdots \\ \mathbf{w}_1 \cdot \mathbf{x}_N + w_{01}) & \cdots & \mathbf{w}_M \cdot \mathbf{x}_N + w_{0M} \end{bmatrix} \quad (11)$$

where $\mathbf{x}_i$, b_i and $\mathbf{w}_i \in \mathbb{C}$.

The complex output of the network is of the form:

$$\mathbf{o}_j = \sum_{i=1}^{M} \beta_i(\mathbf{x}_j) = \sum_{i=1}^{M} \beta_i(\mathbf{w}_i \cdot \mathbf{x}_j + w_{0i}); \quad j = 1, \ldots, N \quad (12)$$

that is a linear combination in the complex domain of the outputs of the first layer.

2.3 CMVN-ELM

It is worth to point out that the learning technique exposed in the previous paragraph, as regards the adjustment of the output neuron weights, follows, substantially, the same procedure used by ELM. For the MVN based network, the output complex weights β_{i} are determined, calculating the optimal solution of

$$\beta^* = \mathbf{H}^{\dagger}\mathbf{D} \quad (13)$$

in the complex domain.

In order to exploit the acclaimed advantages of both recalled paradigms, and to apply it to FRA problems aimed to model identification or fault diagnosis, as described in the introduction, the authors have implemented a new architecture that can be summarized in the following steps:

1. Create a network with n multi-valued complex neurons in the hidden layer; these neurons are extension of the canonical MVN, because their inputs are constituted by complex (and not real) values;
2. Connect the output lines of the hidden layer neurons to the neurons of output layer (the number of output neurons is determined by the nature of the process to model); these neurons have a linear activation function in the real or complex domain;
3. Initialize to random values the weights of the hidden layer;
4. Submit to the network the set of the examples to learn;
5. Initiate the procedure to solve the Eq. (13), using QR decomposition method.

The minimum number of needed neurons in the hidden layer suitable for a given application can be obtained repeating the steps 3–5 for different numbers of hidden neurons.

The extension introduced in the point 1 requires just a slight change in the error-correction learning rule, which is explained and demonstrated to be mathematically consistent in [17].

The initialization of the weights required at point 3 is done following some consideration on the complex-valued nature of the network. The random complex weights are located in a ±0.5 band around the unitary circle of the complex plane. From initialization area is excluded a slide around the discontinuity point $\pm\pi$, that can be empirically set in a range $\pi/2$.

It is noteworthy to consider that one of the benefit of this new architecture is the possibility to keeping low the number of neurons in the hidden layer, much lower in general of a real values network and of a complex (MVN) network with complex weights but with real inputs. This aspect furtherly improves the quality of generalization, as verified in the tests, and generalization is an important point of an inversion problem (both if aimed to identification and to fault diagnosis), where the valued to be associated to the correct response often are not present in the training dataset, or very far from them.

3 CMVN-ELM for Parameter Extraction

This developed architecture is able to directly elaborate a set of frequency response values of the circuit, given in a complex form (magnitude and phase, or real and imaginary parts), using a minimal amount of hidden neurons and a very fast convergence, just training the output layer. Then, this particular ELM can be a great tool to use in any system that requires a complex-valued input evaluation.

3.1 Testability

The first step in any system designed to parameter extraction starts from the network function of the lumped model, where $\mathbf{p} = [p_1, p_2, \ldots.p_R]^t$ is the vector of the circuit parameters, (a_i are the coefficients of numerator terms and b_j are the coefficients of denominator terms):

$$H(\mathbf{p},\omega) = H(\mathbf{p},s)|_{s=j\omega} = \left.\frac{N(\mathbf{p},s)}{D(\mathbf{p},s)}\right|_{s=j\omega} = \left.\frac{\sum_{i=0}^{n} \frac{a_i(\mathbf{p})}{b_m(\mathbf{p})} \cdot s^i}{s^m + \sum_{j=0}^{m-1} \frac{b_j(\mathbf{p})}{b_m(\mathbf{p})} \cdot s^j}\right|_{s=j\omega} \tag{14}$$

The circuit parameter values can be determined from the knowledge of a set of network function values measured, in phase and magnitude, at selected frequencies. The system of nonlinear equations to solve to do that is obtained from (14), where the model parameters $\mathbf{p} = [p_1, p_2, \ldots, p_R]^t$ are the unknowns [18]. If a unique solution does not exist for the unknowns, testability gives a measure of how many parameters cannot be identified with the set of chosen measurements. The testability T is equal to the rank of a matrix $\mathbf{B}$, whose elements are the derivatives of the coefficients of the network function with respect to the circuit parameters [19], as reported in (15).

$$B = \begin{bmatrix} \frac{\partial \frac{a_0}{b_m}}{\partial p_1} & \frac{\partial \frac{a_0}{b_m}}{\partial p_2} & \cdots & \frac{\partial \frac{a_0}{b_m}}{\partial p_R} \\ \cdots & \cdots & & \cdots \\ \frac{\partial \frac{a_n}{b_m}}{\partial p_1} & \frac{\partial \frac{a_n}{b_m}}{\partial p_2} & & \frac{\partial \frac{a_n}{b_m}}{\partial p_R} \\ \cdots & \cdots & & \cdots \end{bmatrix} \tag{15}$$

If $T = \mathrm{rank}(\mathbf{B})$ is equal to the number of unknown parameters R, their values can be uniquely calculated. However, if $T < R$, a locally unique solution cannot be determined, unless $R - T$ parameters are assumed to be known (or fixed). Seen that each column of $\mathbf{B}$ is associated to a specific circuit parameter, then each set of linearly dependent columns of $\mathbf{B}$ localizes an ambiguity group constituted by the circuit parameters corresponding to these columns [20]. This is a group of parameters where it is not possible to uniquely identify the value of each of them starting from the measurement data. In general, both testability value and ambiguity groups do not depend on component values [18], so they can be evaluated by assigning some random values to the parameters (to avoid algebraic varieties). In a problem of model identification, the testability provides the solvability degree that can be obtained with the considered network function, i.e. the number of parameters which must be fixed a priori and the number of parameters to consider as the unknowns. Furthermore, the knowledge of testability and ambiguity groups allows to determine the parameters that can be considered as unknowns. In [21], a further algorithm development avoids the problem of pole/zero cancellations in network functions.

3.2 Parameter Extraction Procedure

The CMVN-ELM introduced in this work is used to associate measurement or simulations to an equivalent lumped circuit of the structure to identify. The block diagram is shown in Fig. 2.

After that architecture is chosen, the measured network function and the unknown parameters are fixed, based on testability evaluation, and the unknown parameters become the CMVN-ELM outputs. The network function is in fact used for determining testability and ambiguity groups, suitable to identify the unknown parameters giving a unique solution when inverted by the CMVN-ELM.

The parameter identification process can be outlined as follows:

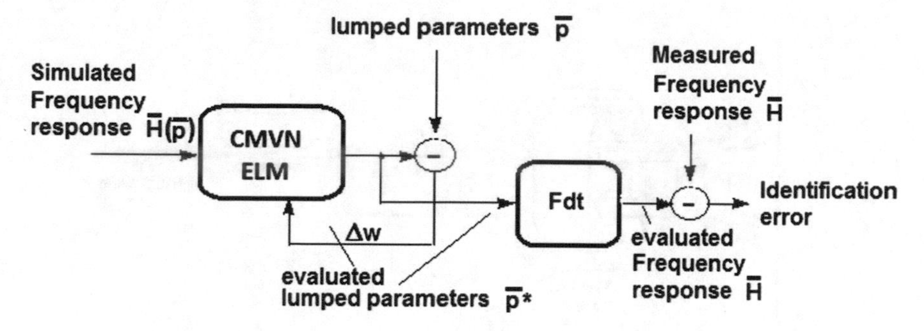

Fig. 2 General scheme of the neural system CMVN-ELM

1. *choose the lumped model equivalent circuit of the structure;*
2. *based on testability and ambiguity groups evaluation, determine which parameters should be assumed as unknowns;*
3. *generate an adequate number of samples to be used in the training phase;*
4. *train the CMVN-ELM part of the neural identification system (see Fig. 2);*
5. *extract the parameters, as the output of the CMVN-ELM part of the whole system;*
6. *evaluate the quality of identification, by comparing the measured or simulated frequency response with the one calculated with extracted parameters.*

The lumped model equivalent circuit is designed matching the collected data (measured or simulated) representing the relation $\boldsymbol{H}(\mathbf{p},\omega)$. The number of samples used in the training phase depends on the problem under exam.

4 Applications

In this section, two examples are given of the application of method to specific configuration. Anyway, it should be taken into account that the proposed approach is very general, just in the spirit of ELM systems, and not restricted to any particular form of the model to be identified.

4.1 Antenna Balun

As a first example, let us use the balun module of a Schwarzbeck half-wave dipole antenna shown in Fig. 3 [22]. An equivalent lumped circuit of the balun is obtained by physical/electrical considerations, together with imposing symmetry between the two conductors.

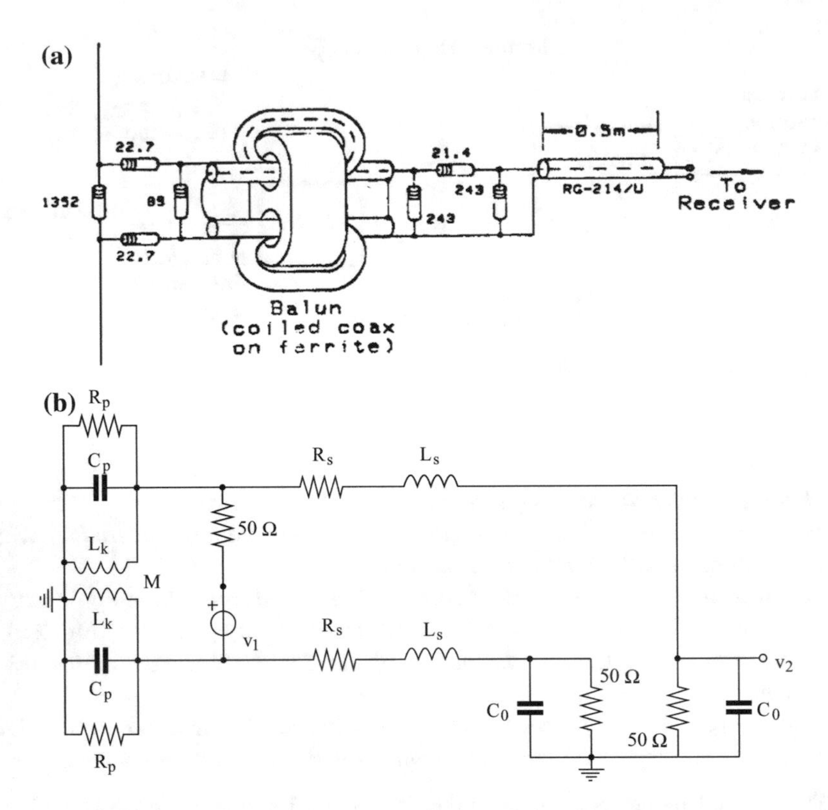

Fig. 3 Balun construction details (**a**) and equivalent circuit (**b**)

The reference values are taken to be the initial values of the components and are obtained using an empirical approach over the physical model. The CMVN-ELM is therefore used to extract and tune the component values, which give a frequency response as close as possible to the measured one. To do that, the following steps are made:

(i) the testability of the circuit is evaluated, seen that its value is maximum and equal to 7, that means that all the elements can be potentially extracted;
(ii) a set of 2000 samples is generated; to do that the circuit simulator SapWin [23, 24] is used, where the component values are varying in a random range of ±10% with respect to their nominal values;
(iii) a CMVN-ELM is trained over the set of generated samples, using a part of that (1400 samples) for the training, and the other portion of 600 samples for the validation.
(iv) the measurements are finally used for extracting the identified components of the model and to test the quality of the global approximation (as reported in the Fig. 4).

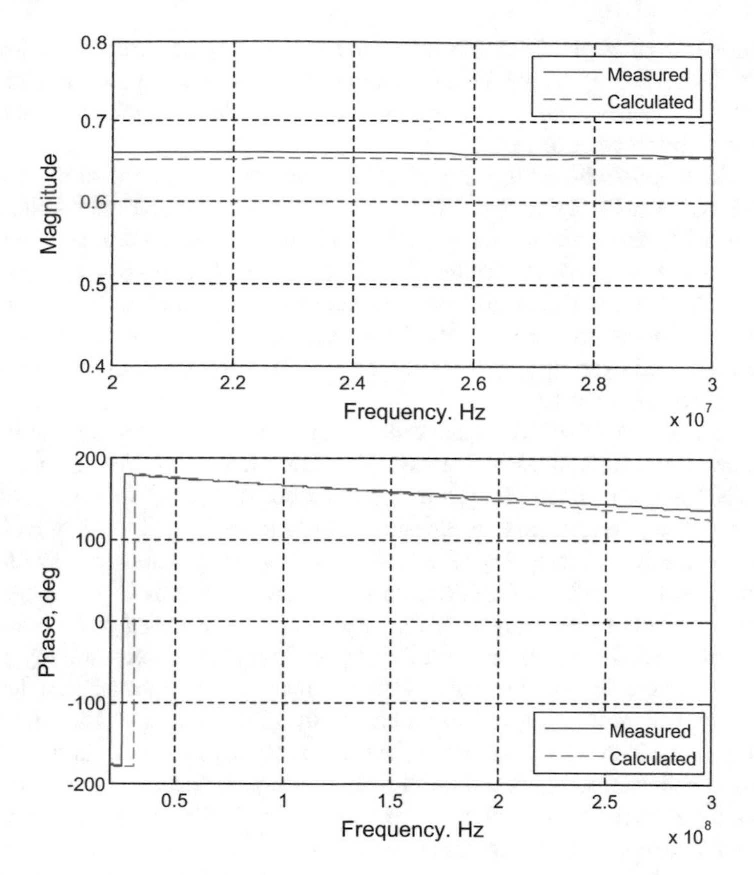

Fig. 4 Frequency response amplitude and phase of the Balun circuit simulated by lumped parameter model (*dashed curve*) compared with the measured ones (*solid curve*)

Table 1 Values of the circuit model parameters of balun as obtained from the CMVN-ELM extraction procedure

L_k (μH)	M (nH)	R_p (kΩ)	R_s (mΩ)	L_s (nH)	C_p (pF)	C_o (pF)
0.869	83.45	1.90	30.0	14.2	3.758	4.585

Relating to (i), seen that testability reaches a maximum, no ambiguity groups appear there. Because of that, all the components should be simultaneously varied in order to "cover" all the possible behaviors of the circuit response. On the other hand, to avoid to make the inversion procedure too huge and seen that this is an identification (and not a fault diagnosis) problem, some parameters are kept to their nominal values, following a pre-elaboration based on sensitivity considerations. Seen that response is much less sensitive to the resistive components of the model, only the reactive parameters are adjusted in the circuit via CMVN-ELM. The calculated results are reported in Table 1.

Figure 4 provides the comparison for the scattering parameter s_{31} (relevant to Magnitude and Phase); the values calculated with the lumped equivalent circuit and the measured values. The corresponding curves are almost perfectly overlapped, with a final mean error under 1%.

In order to give a better interpretation of advantage obtained with the proposed solution, (CMVN-ELM), in the Table 2 it is compared with the ones gotten using classical MLP and standard MLMVNN, with no extended complex inputs, no QR-modified learning rule, neither ELM approach. The results are shown in Table 2. The comparison is given in the number of required learning epochs to reach the minimum error, in the RMSE averaged over all variable components, and RMSE calculated over the frequency response of the corresponding set of values in the given frequency range.

As we see, CMVN-ELM introduced in this paper shows its superiority when compared to the traditional techniques. The error reduction with respect to the MLMVNN is not of huge entity, but the other advantages of this new paradigm should be taken into account to evaluate it. A great advantage of CMVN-ELM is its ability to directly elaborate the complex-valued frequency response. This fact has an impact on the number of hidden neurons necessary to obtain the same result. Moreover, CMVN-ELM being an ELM, requires for the learning just the inversion of a matrix and the necessary learning epoch number is substantially zeroed. Finally, it is interesting to note that has been in many cases demonstrated that ELM systems have a greater generalization capability [25]. This is not a fundamental advantage in an identification problem, but becomes very important in the soft fault diagnosis applications, which can faced with very similar approaches, but where the recognition of anomalous configurations (in fact, the faults), never seen before by the neural network, become an essential feature.

4.2 *Coaxial Cable*

As a further example, the circuit identification of the equivalent circuit of a coaxial cable is shown. It is well known that a coaxial cable is can be described with a lumped model based on the geometrical characteristics of the cable. The section representation of the coaxial cable and its electric scheme drawn in Fig. 5. The corresponding transfer function can be represented in the form of a low pass filter:

Table 2 Comparison of extraction results of balun parameters between the CMVN-ELM and others two networks

	N° of neur	N° of epochs	RMSE over comp. value	RMSE over freq. resp.
CMVN-ELM	20	–	0.00389	0.0440
MLPNN	40	324	0.0570	0.1462
MLMVNN	30	2038	0.00526	0.0545

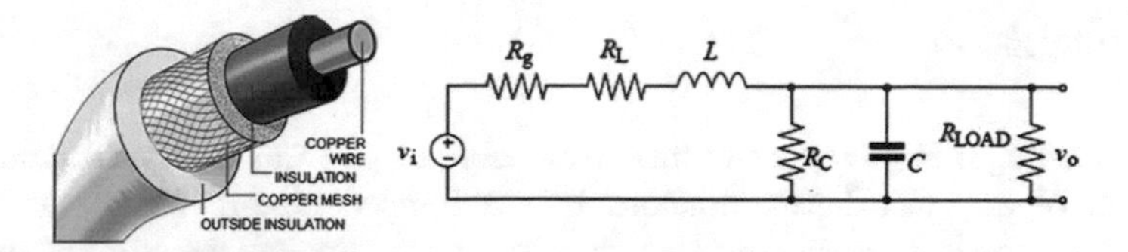

Fig. 5 Section representation of the coaxial cable and electric scheme

Table 3 Comparison of extraction results of coaxial cable parameters between the CMVN-ELM and others two networks

	N° of neur	N° of epochs	RMSE over comp. value	RMSE over freq. resp.
CMVN-ELM	12	–	7.8×10^{-5}	0.0026
MLPNN	40	650	0.00570	0.0354
MLMVNN	30	540	8.5×10^{-4}	0.0155

$$T_V(s) = \frac{R_C R_{LOAD}}{\begin{array}{c} R_{LOAD}(R_g + R_L + R_c) + R_C(R_g + R_L) + \\ + \left[C R_C R_{LOAD}(R_g + R_L) + L(R_{LOAD} + R_C)\right]s \\ + (R_C R_{LOAD} LC)s^2 \end{array}} \tag{16}$$

Also in this case a CMVN-ELM is used to extract and tune the component values, which give a frequency response as close as possible to the measured one. To do that, the following steps are made:

(i) the testability T of the circuit is evaluated, using the Eq. (16); in this case T is not maximum, but it is equal to 3. Anyway, the two reactive parameters L and C, sensitive to the geometric features of the case do not belong to the same ambiguity group (then they belong to a testable group) and they can be independently identified.

(ii) In order to obtain a significant dataset, a number of simulation examples is constituted with SPICE, generated from 36 different (and realistic) combinations of the two geometric parameters internal and external radii of the coaxial cable. For every combination of these parameters the low pass curve is sampled with 200 frequency points in the range 10 MHz–1 GHz.

(iii) a CMVN-ELM is trained over the set of only 200 generated examples, using a part of that (150 samples) for the training, and the other portion of 50 samples for the validation.

(iv) the measurements are finally used for extracting the identified components of the model and to test the quality of the global approximation.

The same comparison of the results done with the previous example is shown in Table 3 for the coaxial cable.

Also in this case, CMVN-ELM shows a superior performance when compared to the other techniques.

5 Conclusions

A new paradigm is presented in this work, able to perform a very accurate identification of any distributed structure into a lumped circuit. Given the simple architecture, the low number of parameters in the network and the no-prop nature of the ELM, which furtherly simplify the learning algorithm the systems is very "low-energy" and presents excellent results when compared with other already tested neurocomputing methods, also by virtue of the modification introduced to directly accept input in complex form associated with the frequency response. The natural extension of this approach will be to the "soft" fault diagnosis of the analog circuit, seen the obvious similitude between the problems and the naturally good behavior of this kind of networks to face the generalization issue.

References

1. Hirose, A.: Complex-Valued Neural Networks, 2nd edn. Springer, Berlin (2012)
2. Nakano, Y., Hirose, A.: Improvement of plastic landmine visualization performance by use of ring-CSOM and frequency-domain local correlation. IEICE Trans. Electron. **E92-C**, 102–108 (2009)
3. Goh, S.L., Chen, M., Popovic, D.H., Aihara, K., Obradovic, D., Mandic, D.P.: Complex valued forecasting of wind profile. Renew. Energy **31**, 1733–1750 (2006)
4. Handayani, A., Suksmono, A.B., Mengko, T.L.R., Hirose, A.: Blood vessel segmentation in complex-valued magnetic resonance images with snake active contour model. Int. J. E-Health Med. Commun. **1**, 41–52 (2010)
5. Avitabile, G., Chellini, B., Fedi, G., Luchetta A., Manetti, S.: A neural architecture for the parameter extraction of high frequency devices. In: 2001 IEEE International Symposium on Circuits and Systems (ISCAS), pp. 577–580. IEEE Press, New York (2001)
6. Rashtchi, V., Rahimpour, E., Rezapour, E.M.: Using a genetic algorithm for parameter identification of transformer R-L-C-M model. Electr. Eng. **88**, 417–422 (2006)
7. Shinterimov, A., Tang, W.J., Tang, W.H., Wu, Q.H.: Improved modelling of power transformer winding using bacterial swarming algorithm and frequency response analysis. Electr. Power Syst. Res. **80**, 1111–1120 (2010)
8. Tang, W.H., He, S., Wu Q.H., Richardson, Z.J.: Winding deformation identification using a particle swarm optimiser with passive congregation of power transformers. Int. J. Innov. Energy Syst. Power **11**, 46–52 (2006)
9. Aizenberg, N.N., Aizenberg, I.N.: CNN based on multi-valued neuron as a model of associative memory for gray-scale images. In: 2nd IEEE International Workshop on Cellular Neural Networks and their Applications, pp. 36–41: IEEE Press, New York (1992)
10. Aizenberg, I., Moraga, C.: Multilayer feedforward neural network based on multi-valued neurons (MLMVN) and a backpropagation learning algorithm. Soft. Comput. **11**, 169–183 (2007)
11. Huang, G.B., Zhu, Q.Y., Siew, C.K.: Extreme learning machine: theory and applications. Neurocomputing **70**, 489–501 (2006)
12. Sen, N., Saeks, R.: Fault diagnosis for linear system via multifrequency measurement. IEEE Trans. Circuits Syst. **26**, 457–456 (1979)
13. Huang, G., Huang, G.B., Song, S., You, K.: Trends in extreme learning machines: a review. Neural Netw. **61**, 32–48 (2015)

14. Aizenberg, I.: Complex-Valued Neural Networks with Multi-valued Neurons. Springer Publishers, Berlin (2011)
15. Aizenberg, I., Paliy, D., Zurada, J., Astola, J.: Blur identification by multilayer neural network based on multivalued neurons. IEEE Trans. Neural Netw. **19**, 883–898 (2008)
16. Aizenberg, I., Luchetta A., Manetti, S.: A modified learning algorithm for the multilayer neural network with multi-valued neurons based on the complex QR decomposition. Soft Comput. **16**, 563–575 (2012)
17. Aizenberg, I., Luchetta, A., Manetti, S., Piccirilli, M.C.: System identification using FRA and a modified MLMVN with arbitrary complex-valued inputs. In: 2016 IEEE Joint Conference on Neural Networks (IJCNN). IEEE Press, New York (2016)
18. Lawson, C.L., Hanson, R.J.: Solving Least Squares Problems. Prentice-Hall, Englewood Cliffs (1974)
19. Householder, A.S.: Unitary triangularization of a nonsymmetric matrix. J. ACM **5**, 339–342 (1958)
20. Golub, G.H., Van Loan, C.F.: Matrix Computations, 3rd edn. Johns Hopkins University Press (1996)
21. Fontana, G., Luchetta, A., Manetti S., Piccirilli. M.C.: An unconditionally sound algorithm for testability analysis in linear time-invariant electrical networks. Int. J. Circuit Theory Appl. **44**, 1308–1340 (2016)
22. Bennett, W.S.: Properly applied antenna factors. IEEE Trans. Electromagn. Compat. **28**, 2–6 (1986)
23. Grasso, F., Luchetta, A., Manetti, S., Piccirilli, M.C., Reatti, A.: SapWin 4.0–a new simulation program for electrical engineering education using symbolic analysis. Comput. Appl. Eng. Educ. **24**, 44–57 (2016)
24. SapWin download page on Department of Information engineering of the University of Florence. http://www.sapwin.info
25. Chen, H., Peng, J., Zhou, Y., Li, L., Pan, Z.: Extreme learning machine for ranking: generalization analysis and applications. Neural Netw. **53**, 119–126 (2014)

Quaternion Extreme Learning Machine

Hui Lv and Huisheng Zhang

Abstract Quaternion signal processing has been an increasing popular research topic for its application in a wide range of fields, and extreme learning machine (ELM) is an emerging training strategy for the generalized single hidden layer feedforward neural networks. However, extreme learning machine could not fully explore its potentials in quaternion signal processing. To this end, this paper propose an quaternion ELM model, which retain the essential characters of the ELM such as the fast learning and universal approximation capability, while enjoying advantages originated from the quaternion algebra. Two simulation examples are provided to support our analysis and to exhibit the enhanced performance of the proposed model over ELM when dealing with the 3D and 4D signal processing problems.

Keywords Quaternion extreme learning machine · Quaternion signal processing · Generalized Moore-Penrose inverse · Chaotic time series

1 Introduction

The popularity of quaternion signal processing has increased in recent years due to its applications in image processing [1–3], computer graphics [4], modeling of wind profile [5], processing of polarized waves [6], etc. As one of the natural models for quaternion signal processing, quaternion multilayer perceptrons has been proposed and their universal approximation capability has been proved [7]. Recently, many new or improved models for quaternion neural networks (QNNs) have also been established, such as the quaternion adaptive neural filters [8], quaternion Kalman filtering [9], quaternion echo state networks [10], and quaternion ICA [11]. Owing

This work is supported by the National Natural Science Foundation of China (Nos. 61101228, 61402071, 61671099), the Liaoning Provincial Natural Science Foundation of China (No. 2015020011), and the Fundamental Research Funds for the Central Universities of China.

H. Lv · H. Zhang (✉)
Department of Mathematics, Dalian Maritime University, Dalian 116026, China
e-mail: zhhuisheng@163.com

J. Cao et al. (eds.), *Proceedings of ELM-2016*, Proceedings in Adaptation, Learning and Optimization 9, DOI 10.1007/978-3-319-57421-9_3

to the power of quaternion algebra, QNNs usually exhibited enhanced performance over vector-based neural models [12, 13].

Training algorithms for the above quaternion neural models are basically gradient-based. However, quaternion gradient method suffers from the drawbacks of slow convergence and easily trapping into local minimum. Moreover, the generalized Cauchy-Riemann condition [14] admits only linear functions and constants as globally analytic quaternion-valued functions, which makes a great restriction on using gradient training method for quaternion neural networks.

Extreme learning machine has become a popular training strategy for single hidden layer neural network [15, 16]. The essence of ELM is that the input layer weights are randomly determined and then the hidden layer weights can be simply calculated by least squares optimization. This approach avoids the iterative computing process and has excellent learning accuracy/speed in various application. In order to process complex-valued signals, complex-valued extreme learning machine (CELM) has been proposed by extending ELM to complex domain [17]. However, ELM model for quaternion signal processing is still lacking. To this end, the aim of this paper is to propose a quaternion extreme learning machine (QELM) model for quaternion signal processing. This model retains the inherent properties of the original ELM such as the fast learning and universal approximation capability, meanwhile gaining advantages from the quaternion algebra. Simulations in the prediction setting on both the 3D and 4D time series support our analysis.

The rest of this paper is organized as follows. A brief introduction of the quaternion algebra is provided in the next section. The quaternion ELM model is derived in the third section. In Sect. 4 two simulation examples are given. Section 5 concludes this paper.

2 Quaternion Algebra

The quaternion was introduced by Hamilton in 1843 to expand the complex numbers from two-dimensional space to four-dimensional space. Though the quaternion domain is a noncommutative extension of the complex domain, it provides a natural framework for the processing of three and four dimensional signals [18].

A quaternion variable $q \in \mathbb{H}$ comprises a real part $\mathfrak{R}(q) = q_r$ and a vector part, also known as a pure quaternion $\mathfrak{I}(q) = q_i i + q_j j + q_k k$. In this way, a quaternion can be expressed as

$$q = q_r + q_i i + q_j j + q_k k,$$

where $q_r, q_i, q_j, q_k \in \mathbb{R}$. When $q_r = 0, q_i = q_j = 0$, or $q_i = q_j = q_k = 0$, a quaternion is reduced to a pure quaternion, complex number or real number, respectively.

The properties of the orthogonal unit vectors i, j, k describing the three vector dimensions of a quaternion are listed as follows

$$i^2 = j^2 = k^2 = ijk = -1,$$
$$ij = -ji = k, jk = -kj = i, ki = -ik = j.$$

Given two quaternions $a, b \in \mathbb{H}$, the quaternion addition and subtraction are computed as

$$\begin{aligned} a \pm b &= (a_r \pm b_r) + (a_i \pm b_i)i + (a_j \pm b_j)j + (a_k \pm b_k)k, \\ ab &= (a_r b_r - a_i b_i - a_j b_j - a_k b_k) + (a_r b_i + a_i b_j + a_j b_k - a_k b_j)i \\ &+ (a_r b_j - a_i b_k + a_j b_r + a_k b_i)j + (a_r b_k + a_i b_j - a_j b_i + a_k b_r)k. \end{aligned}$$

As show in the above rules, the multiplication of quaternions is non-commutative.

The conjugate of a quaternion is defined by $q^* = q_r - q_i - q_j - q_k$, and the modulus $\|q\| = \sqrt{qq^*} = \sqrt{q_r^2 + q_i^2 + q_j^2 + q_k^2}$.

3 Quaternion Extreme Learning Machine

Given a series of quaternion-valued training samples $\{(\mathbf{x}_s, \mathbf{t}_s)\}_{s=1}^{S}$, we train a quaternion single hidden layer feedforward network (QSHLFN) which is mathematically modeled by

$$\sum_{m=1}^{M} \boldsymbol{\beta}_m g_q(\mathbf{w}_m \cdot \mathbf{x}_s + b_m) = \mathbf{o}_s, s = 1, 2, \ldots, S, \tag{1}$$

where $\mathbf{x}_s \in \mathbb{H}^L$ is the input vector, $\mathbf{t}_s \in \mathbb{H}^N$ is the corresponding target output vector, $\mathbf{w}_m \in \mathbb{H}^L$ is the input weight vector connecting the input layer neurons to the mth hidden neuron, $b_m \in \mathbb{H}$ is the quaternion bias of the mth hidden neuron, $\boldsymbol{\beta}_m \in \mathbb{H}^N$ is the quaternion output weight vector connecting the mth hidden neuron and the output neurons, $\mathbf{o}_s$ is the network output for an input vector $\mathbf{x}_s$, $\mathbf{w}_m \cdot \mathbf{x}_s$ denotes the inner product of vectors $\mathbf{w}_m$ and $\mathbf{x}_s$, $g_q(\cdot)$ is a quaternion-valued activation function.

As a QSHLFN can be used to approximate any quaternion-valued continuous function [7], we try to find the appropriate network weights to satisfy

$$\sum_{m=1}^{M} \boldsymbol{\beta}_m g_q(\mathbf{w}_m \cdot \mathbf{x}_s + b_m) = \mathbf{t}_s, s = 1, 2, \ldots, S. \tag{2}$$

The above S equations can be written in a compact form

$$\mathbf{H}\boldsymbol{\beta} = \mathbf{T}, \tag{3}$$

where

$$\mathbf{H}(\mathbf{w}_1, \mathbf{w}_2, \ldots, \mathbf{w}_M, \mathbf{x}_1, \mathbf{x}_2, \ldots, \mathbf{x}_S, b_1, b_2, \ldots, b_S)$$

$$= \begin{bmatrix} g_q(\mathbf{w}_1 \cdot \mathbf{x}_1 + b_1) & \ldots & g_q(\mathbf{w}_M \cdot \mathbf{x}_1 + b_M) \\ \vdots & \ldots & \vdots \\ g_q(\mathbf{w}_1 \cdot \mathbf{x}_S + b_1) & \ldots & g_q(\mathbf{w}_M \cdot \mathbf{x}_S + b_M) \end{bmatrix}_{S\times M},$$

$$\boldsymbol{\beta} = \begin{bmatrix} \boldsymbol{\beta}_1^T \\ \vdots \\ \boldsymbol{\beta}_M^T \end{bmatrix}_{M\times m}, \text{and } \mathbf{T} = \begin{bmatrix} \mathbf{t}_1^T \\ \vdots \\ \mathbf{t}_M^T \end{bmatrix}_{M\times m}.$$

Similar to the theoretical analysis of ELM [16, 17], we can easily prove that the input weights $\mathbf{w}_m$ and hidden layer biases b_m are in fact not necessarily tuned and can be randomly chosen based on some continuous distribution probability. Thus we only need to determine the weight matrix $\boldsymbol{\beta}$. If the hidden layer output matrix $\mathbf{H}$ is invertible, then $\boldsymbol{\beta}$ can be directly obtained as $\mathbf{H}^{-1}\mathbf{T}$. However, in practical applications the number of hidden neurons M is usually less than the number of samples S. In this case one can not expect the exact solution of (3). Instead, we devote to solving the least-squares problem $\min_{\boldsymbol{\beta}} \|\mathbf{H}\boldsymbol{\beta} - \mathbf{T}\|$, and obtain the explicit solution

$$\widehat{\boldsymbol{\beta}} = \mathbf{H}^{\dagger}\mathbf{T}, \tag{4}$$

where the quaternion matrix $\mathbf{H}^{\dagger}$ is the Moore-Penrose generalized inverse of $\mathbf{H}$ [19].

Now, QELM algorithm can be summarized as follows:

Algorithm QELM

Given a training set $X = \{(\mathbf{x}_s, \mathbf{t}_s) | \mathbf{x}_s \in \mathbb{H}^L, \mathbf{t}_s \in \mathbb{H}^M, s = 1, 2, \ldots, S\}$, quaternion activation function $g_q(\cdot)$ and hidden neuron number M:

Step 1 Randomly choose the quaternion input weight $\mathbf{w}_m$ and the quaternion bias b_k;
Step 2 Calculate the quaternion hidden layer output matrix $\mathbf{H}$;
Step 3 Calculate the quaternion output weight $\boldsymbol{\beta}$, where

$$\widehat{\boldsymbol{\beta}} = \mathbf{H}^{\dagger}Y.$$

Remark 1 As the only analytic quaternion function is a linear quaternion function (Sudbery 1979), it is very difficult for the fully quaternion neural networks trained by traditional gradient algorithms to choose an eligible activation function. However, for QELM, there is no need to do the gradient operation, which allows almost all the activation functions used for real-valued neural networks or complex-valued neural networks to be still qualified as activation functions of QELM.

Remark 2 As shown in the appendix, Moore-Penrose generalized inverse of a quaternion matrix is defined and computed in a same way as the real or complex domain.

Remark 3 Similar to the original ELM, by choosing the number of hidden neurons M bo be equal to the number of samples S, we can prove with probability one equation (3) has an accurate solution.

4 Simulation Results and Discussion

In this section two simulation examples are presented. The first example is the Lorenz attractor and the second one is the Saito's chaotic circuit, which are benchmark 3D and 4D signal processing problems respectively. In both examples, the times series are predicted by QELM and ELM separately, and the corresponding performances comparison is conducted. The root mean square error (RMSE) is used to characterize the accuracy of prediction:

$$\mathrm{RMSE} = \sqrt{\frac{1}{S}\sum_{s=1}^{S}(\mathbf{o}_s - \mathbf{t}_s)^H(\mathbf{o}_s - \mathbf{t}_s)},$$

where $\mathbf{o}_s$ indicates the sth sample of actual output, $\mathbf{t}_s$ indicates the sth sample of the forecast output, and S is the number of samples.

4.1 Lorenz Chaotic Time Series

The Lorenz attractor is used originally to model atmospheric turbulence, but also to model lasers, dynamos, and the motion of waterwheel [20]. Mathematically, the Lorenz system is a three-dimensional nonlinear system and can be expressed as a system of coupled differential equations:

$$\begin{cases} \dot{x} = \sigma(y - x) \\ \dot{y} = (\rho - z)x - y \\ \dot{z} = xy - \gamma z \end{cases}$$

where $\sigma, \gamma, \rho > 0$. Taking $\rho = 10, \gamma = 8/3, \rho = 28, x(0) = 1, y(0) = 0$, and $z(0) = 1$, the fourth-order Runge-Kutta method is used to generate the tripartite time series, and 250 samples are obtained. For the convenience of processing using QELM, we represent Lorenz attractor as a pure quaternion: $xi + yj + zk$.

The parameter setting is as follows: the activation functions of QELM and ELM algorithm are chosen as the tan and atan functions respectively, and the number of hidden nodes is 70. In this simulation, we conduct one step ahead prediction using

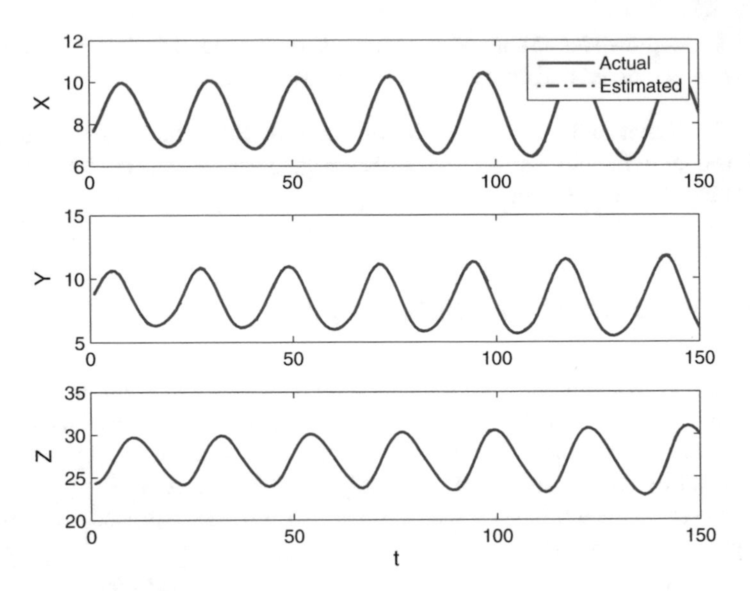

Fig. 1 The estimated and actual time series for QELM(3D)

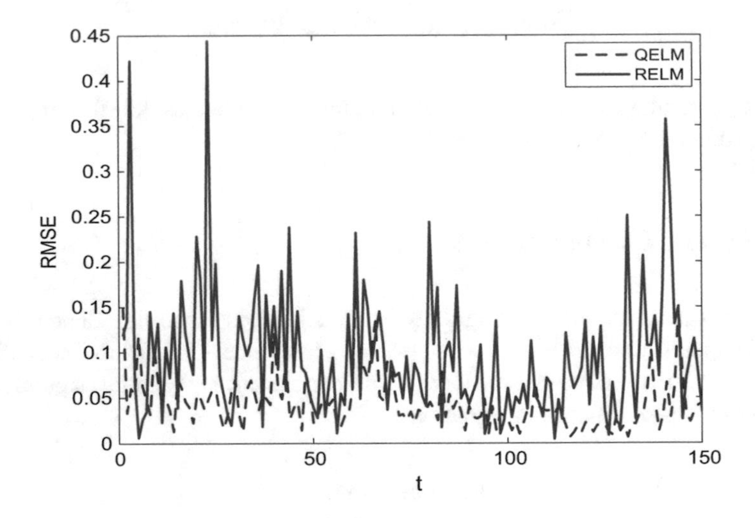

Fig. 2 Prediction error curves for QELM(3D) and ELM

100 training samples, that is to say, $\{x(t), y(t), z(t)\}_{t=n+1}^{n+100}$ series are used together to predict $x(t+\eta)$, where $\eta = 1$, and $n = 0, 1, \ldots, 150$.

Figure 1 shows the prediction curves based on the proposed QELM, and it can be seen from Fig. 1 that the predicted curve and the actual curve are matched well. Figure 2 compares the prediction error curves for QELM and ELM. It can be observed that, the QELM generates smaller prediction errors than the traditional ELM in real domain.

4.2 Q-Improper Four-Dimensional Saitos Circuit

The Saito's chaotic circuit is governed by four state variables and five parameters, these variables comprise full quaternion. The equations are as follows [21, 22]

$$\begin{bmatrix} \partial x_1/\partial t \\ \partial y_1/\partial t \end{bmatrix} = \begin{bmatrix} -1 & 1 \\ -\alpha_1 & -\alpha_1\beta_1 \end{bmatrix} = \begin{bmatrix} x_1 - \eta\rho_1 h(z) \\ y_1 - \eta\frac{\rho_1}{\beta_1}h(z) \end{bmatrix},$$

$$\begin{bmatrix} \partial x_2/\partial t \\ \partial y_2/\partial t \end{bmatrix} = \begin{bmatrix} -1 & 1 \\ -\alpha_2 & -\alpha_2\beta_2 \end{bmatrix} = \begin{bmatrix} x_2 - \eta\rho_2 h(z) \\ y_2 - \eta\frac{\rho_2}{\beta_2}h(z) \end{bmatrix},$$

where t is the time constant of the chaotic circuit and $h(z)$ is the normalized hysteresis value which is given by [22]

$$h(z) = \begin{cases} 1 & z \geq -1, \\ -1 & z \leq 1. \end{cases}$$

The variables z, ρ_1, ρ_2 are given as

$$z = x_1 + x_2,$$
$$\rho_1 = \frac{\beta_1}{1-\beta_1},$$
$$\rho_2 = \frac{\beta_2}{1-\beta_2}.$$

In this example, Saito's chaotic signal is initialized with the following standard parameters:

$$\eta = 1.3, \alpha_1 = 7.5, \alpha_2 = 15, \beta_1 = 0.16, \beta_2 = 0.097.$$

The fourth-order Runge-Kutta method is used to generate the quadrupled time series, and 600 samples are obtained.

The activation functions for both QELM and ELM algorithm are chosen as tan and the number of hidden nodes is 10. Similar to the former example, we conduct one step ahead prediction using 100 training samples, that is to say, $\{x(t), y(t), z(t)\}_{t=n+1}^{n+100}$ series are used together to predict $x(t+\eta)$, where $\eta = 1$,and $n = 0, 1, \ldots, 500$.

Figure 3 shows the prediction curves based on the proposed QELM. It can be seen from Fig. 3 that the predicted curve and the actual curve are so identical that it is difficult to distinguish. Figure 4 plots the prediction error curves for both QELM and ELM, and it can be observed that, the QELM again generates smaller prediction errors than the traditional real-valued ELM.

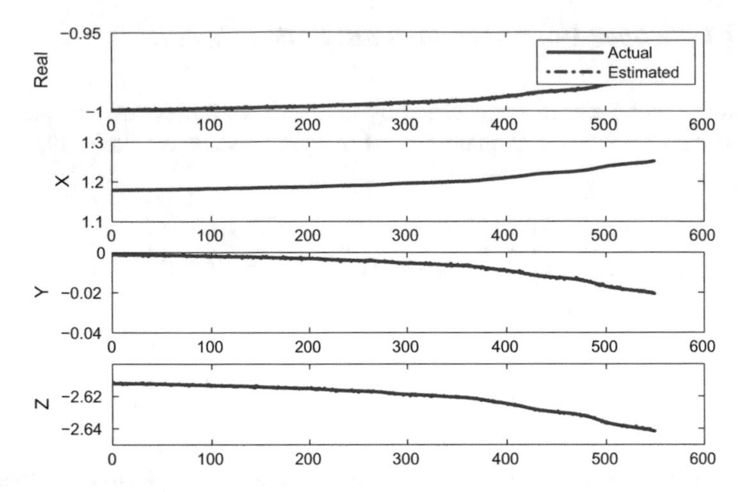

Fig. 3 The estimated and actual series for QELM(4D)

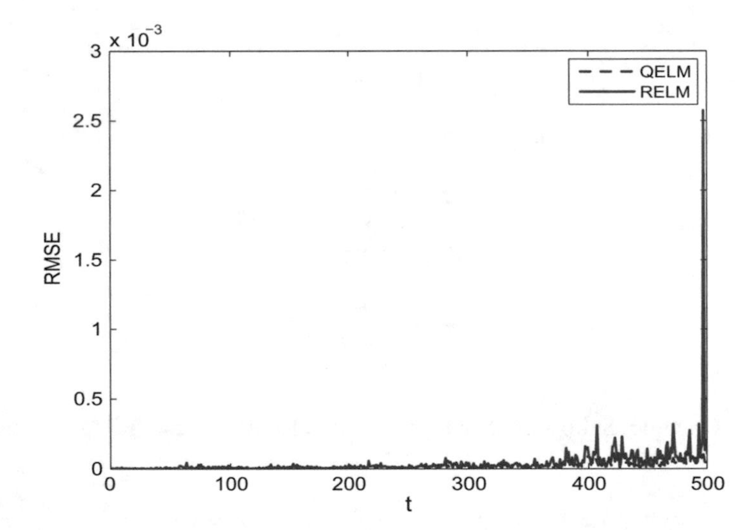

Fig. 4 Prediction error curves for QELM(4D) and ELM

5 Conclusions

Quaternion signal processing has become an increasingly popular research topic in recent years. In this paper we have proposed a quaternion extreme learning machine (QELM) model to cater for the needs of quaternion signal processing. The merits of this model is to retain the welcoming properties of ELM like the fast learning and universal approximation capability, while gaining the new advantages originated for quaternion algebra. Moreover, QELM releases the restrictions on the quaternion activation functions for the traditional gradient-based quaternion learning algorithms.

Two simulation examples have been provided to exhibit the enhanced performance of the proposed model in dealing with the 3D and 4D signal processing problems over the extreme learning machine in real domain.

Appendix: Moore-Penrose Generalized Inverse

Definition [19, 23]: Let quaternion matrix $A \in \mathbb{H}^{m \times n}$ and $G \in \mathbb{H}^{m \times n}$. If G meets the following conditions:

$$\begin{aligned}
&(1)\, AGA = A,\\
&(2)\, GAG = G,\\
&(3)\, (AG)^* = AG,\\
&(4)\, (GA)^* = GA,
\end{aligned}$$

then G is the Moore-Penrose generalized inverse of quaternion matrix A.

References

1. Pei, S.C., Cheng, C.M.: Color image processing by using binary quaternion-moment-preserving thresholding technique. IEEE Trans. Image Process. **8**(5), 614–628 (1999)
2. Moxey, C., Sangwine, S., Ell, T.: Hypercomplex correlation techniques for vector images. IEEE Trans. Signal Process. **51**(7), 1941–1953 (2003)
3. Bulowand, T., Sommer, G.: Hypercomplex signals-a novel extension of the analytic signal to the multidimensional case. IEEE Trans. Signal Process. **49**(11), 2844–2852 (2001)
4. Hanson A.J.: Visualizing Quaternions. Morgan Kaufmann (2005)
5. Took, C.C., Mandic, D.P.: The quaternion LMS algorithm for adaptive filtering of hypercomplex processes. IEEE Trans. Signal Process. **57**(4), 1316–1327 (2009)
6. Miron, S., Bihan, N.L., Mars, J.: Quaternion-MUSIC for vector-sensor array processing. IEEE Trans. Signal Process. **54**(4), 1218–1229 (2006)
7. Arena, P., Fortuna, L., Muscato, G., Xibilia, M.G.: Multilayer perceptrons to approximate quaternion valued functions. Neural Netw. **10**(2), 335–342 (1997)
8. Ujang, B.C., Took, C.C., Danilo, P.M.: Quaternion-valued nonlinear adaptive filtering. IEEE Trans. Neural Netw. **22**(8), 1193–1206 (2011)
9. Dini, D.H., Danilo, P.M.: Class of widely linear complex Kalman filters. IEEE Trans. Neural Netw. Learn. Syst. **23**(5), 775–786 (2012)
10. Xia, Y.L., Cyrus, J., Danilo, P.M.: Quaternion-valued echo state networks. IEEE Trans. Neural Netw. Learn. Syst. **26**(4), 663–673 (2015)
11. Via, J., Palomar, D.P., Vielva, L., Santamaria, I.: Quaternion ICA from second-order statistics. IEEE Trans. Signal Process. **59**(4), 1586–1600 (2011)
12. Nitta, T.: A backpropagation algorithm for neural networks based on 3-D vector product. Proc. Int. Joint Conf. Neural Netw. **1**, 589–592 (1993)
13. Isokawa, T., Haruhiko, N., Nobuyuki, M.: Quaternionic multilayer perceptron with local analyticity. Information **3**(4), 756–770 (2012)

14. Watson, R.E.: The generalized Cauchy-Riemann-Fueter equation and handedness. Compl. Variab. **48**(7), 555–568 (2003)
15. Huang G.B., Zhu Q.Y., Siew C.K.; Extreme learning machine: a new learning scheme of feedforward neural networks. In: Proceedings of International Joint Conference on Neural Networks (IJCNN2004), vol. 2, pp. 985–990 (2004)
16. Li, M.B., Huang, G.B., Zhu, Q.Y., Siew, C.K.: Extreme learning machine: theory and applications. Neurocomputing **70**, 489–501 (2006)
17. Li, M.B., Huang, G.B., Saratchandran, P., Sundararajan, N.: Fully complex extreme learning machine. Neurocomputing **68**, 306–314 (2005)
18. Ward, J.P.: Quaternions and Cayley Numbers: Algebra and Applications. Kluwer Academic, Boston, MA (1997)
19. Liu, B.: The Moore-Penrose generalized inverse of quaternion matrix. J. Guilin Univ. Electron. Techol. **24**(5), 68–71 (2004)
20. Strogatz S.H.: Nonlinear Dynamics and Chaos: With Applications to Physics, Biology, Chemistry and Engineering (Studies in Nonlinearity), 1st edn. Westview Press (2001)
21. Arena, P., Fortuna, L., Muscato, G., Xibilia, M.G.: Neural networks in multidimensional domains. Lect. Notes Control Inf. Sci. **234** (1998)
22. Mitsubori K., Saito T.: Torus Doubling and Hyperchaos in a five dimensional hysteresis circuit. In: Proceedings of 1994 IEEE international symposium on circuit and systems, vol. 6, pp. 113–116 (1994)
23. Serre, D.: Matrices: Theory and Applications. Springer, New York (2002)

Robotic Grasp Stability Analysis Using Extreme Learning Machine

Peng Bai, Huaping Liu, Fuchun Sun and Meng Gao

Abstract Recently, autonomous grasping of unknown objects is a fundamental requirement for robots performing manipulation tasks in real world environments. It is still considered as a challenging problem no matter how process we have made. It is significant that how the robot to judge the stability of grabbing object. In this paper, we analyze the data through process of grabbing 3 objects whether is successful or failed by constructing Global Alignment kernel with Extreme Learning Machine and Support Vector Machine. For comparative analysis, the Barrett hand's finger angles and robot joint angles are also recorded. By processing obtained data in different ways, we have comparative results in various modes. Experiments denote the tactile results achieve better performance than the finger angle's and robot joint angle's.

Keywords Grasp stability · Extreme learning machine · Tactile data

1 Introduction

Recently, autonomous grasping of unknown objects is a fundamental requirement for robots performing manipulation tasks in real world environments. Even though there has been a lot of progress in the area of grasping, it is still considered as an open challenging and even the state-of-the-art grasping methods may result in

P. Bai · M. Gao
Department of Electrical and Electronic Engineering,
Shijiazhuang Tiedao University, Shijiazhuang, China

H. Liu (✉) · F. Sun
Department of Computer Science and Technology, Tsinghua University,
Beijing, China
e-mail: hpliu@tsinghua.edu.cn

P. Bai · H. Liu · F. Sun
State Key Laboratory of Intelligent Technology and Systems,
Tsinghua University TNLIST, Beijing, China

J. Cao et al. (eds.), *Proceedings of ELM-2016*, Proceedings in Adaptation, Learning and Optimization 9, DOI 10.1007/978-3-319-57421-9_4

failures [1]. A reliable prediction of grasp stability helps to avoid such failures and provides an option to re-grasp the object safely. Since the majority of grasping failures happen at the contact points, which are occluded for vision systems, tactile feedback plays a major role for predicting grasp stability.

When an object is grasped, there are two constraints from the task, the object and the hand. Within these constraints, Cutkosky [2] defined various analytical measures used to describe a grasp as stability, compliance, connectivity, isotropy, etc. Besides, he also classified shapes of manufacturing grasps. He believes if the overall stiffness matrix is positively definite that the grasp is stable at low speeds and at higher speeds dynamic stability must be considered. When being disturbed by external forces and moments, the grasp stability shows it maintain balance and without slipping form robot hand. Because the stability is the ability which is able to resist from disturbance, and gives many effects to the grasp relationship between robot and object, many researchers have interest and advanced researches regarding a stability with various methods [3]. Funahashi et al. [4] analyzed to consider the curvatures of both hand and object at contact points by using potential energy, and showed that the grasp using round fingers was more stable than using sharp fingers. Jenmalm et al. [5] verified grasp stability change with different surface curvatures by tests. Howard and Kumar [6] classified the categories of equilibrium grasps and established a general framework for the determination of the stability of grasps by using stiffness matrix. Yamada et al. [7] analyzed the stability of 3D grasps by using potential energy of a three-dimensional spring model by a multi-fingered hand. Yamada et al. [8] analyzed stability of simultaneous grasps of two objects in two dimensions by using potential energy method. Sudsang and Phoka [9] proposed a method of testing whether three contact points form a three-fingered force-closure grasp in two dimensions.

In brief, many studies have analyzed grasp stability by using potential energy and stiffness matrix. However, potential energy and stiffness matrix methods requested experiences about the work and have a weak point of complex calculation because these methods have to know active force and moved displacement after grasp [3]. In our work, we just need to analyze the data which collected in grasping process and then send them into machine learning such as ELM, SVM, and neural networks. In computational intelligence techniques, SVM and neural networks have been dominant roles. However, it is known that both them have some challenging issues such as: (1) slow learning speed (2) trivial human intervene (3) poor computational scalability [10], on the contrary, ELM which has better generalization performance at a much faster learning speed and with least human intervene compared with those traditional computational intelligence techniques has attracted the attention from more and more researchers. This work has not yet found the relevant application.

In this work, we use the tactile data, finger angles and robot joint angles to analyze the stability of grasped object with Global Alignment kernel-ELM. In order to compare, we also use SVM. The main contributions are listed as follows:

(1) We use Global Alignment kernel to process the tactile data, finger angles and robot joint angles which is better than dynamic time warping kernel.
(2) In this work, we prefer to choose ELM to be a classifier.
(3) We have extensive comparative experimental results.

The rest of the paper is organized as follows: In Sect. 2 we give a description of the problem. A brief introduction about ELM classification and tactile modeling is described in Sect. 3. Section 4 presents the experimental results and the conclusions are given in Sect. 5.

2 Problem Formulation

How to judge the stability of an object? Suppose a robot hand approaching and grasping the object using the force grip controller, the robot picks the object up and performs a range of extensive shaking motions in all directions to ensure that the grasp is stable. In this process, we will obtain various data like tactile data, finger angle, and robot joint angle. Through analyzing these data, and sending them into different machine learning to classify. The stability of grasping problem is transformed into a classification problem using ELM and SVM as showed in Fig. 1.

3 ELM Classification

3.1 Tactile Modeling

The method of stable classification mentioned in our paper can be analyzed by Global Alignment kernel. In kernel methods, both large and small similarities matter, since they all contribute to the Gram matrix. Global Alignment (GA) kernels, which are positive definite, seem to do a better job of quantifying all similarities coherently, because they consider all possible alignments. Triangular Global Alignment (TGA) kernels consider a smaller subset of such alignments. They are faster to compute and positive definite, and can be seen as trade-off

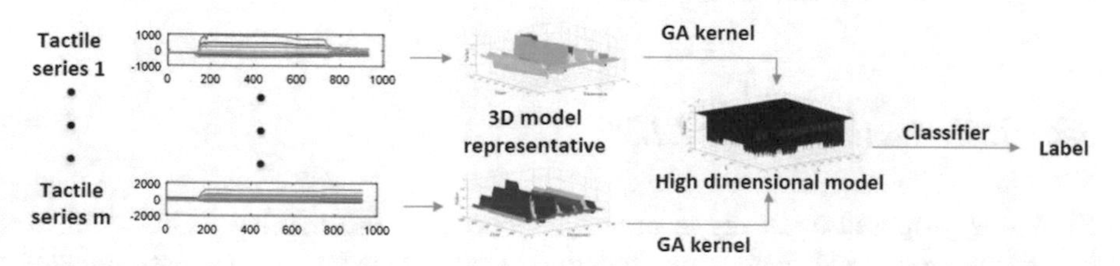

Fig. 1 Work flow

between the full GA kernel (accurate, versatile but slow) and a Gaussian kernel (fast but limited) as discussed below.

Suppose we have two time series S_i and S_j, of the length T_i and T_j, we define S_i and S_j as follows:

$$S_i = [S_{i,1}, S_{i,2}, \ldots, S_{i,T_i}], \tag{1}$$

$$S_j = [S_{j,1}, S_{j,2}, \ldots, S_{j,T_j}], \tag{2}$$

Global Alignment kernels compare two time-series using the Kernel bandwidth and Triangular parameter, When Triangular is set to 0, the routine returns the original GA kernel, defined as follows:

$$k(x, y) = \sum_{\pi \in A(n,m)} \prod_{i=1}^{|\pi|} k(x_{\pi_1(i)}, y_{\pi_2(i)}), \tag{3}$$

where $A(n, m)$ is the set of all possible alignments between two series of length n and m. In this new implementation we do not use the Gaussian kernel for $k(x, y)$ and consider instead as:

$$k(x, y) = e^{-\phi_\sigma(x, y)}, \tag{4}$$

$$\phi_\sigma(x, y) = \frac{1}{2\sigma^2} \|x - y\|^2 + \log(2 - e^{-\frac{\|x-y\|^2}{2\sigma^2}}). \tag{5}$$

When Triangular is bigger than 1 the routine only considers alignments for which $-T < \pi_1(i) - \pi_2(i) < T$ for all indices of the alignment. When this parameter is set to 1, the kernel becomes the kernel as:

$$K_{T=1} = (x, y) = \delta(|x| = |y|) \prod_{i=1}^{|x|} e^{-\phi_\sigma}(x_i, y_i), \tag{6}$$

between time series, which is non-zero for series of the same length only. It is a slightly modified Gaussian kernel between vectors which does not take into account the temporal structure of time series. When $T \to \infty$ the Triangular kernel's values converge to that of the usual Global Alignment kernel. The smaller T the shorter runtime for each iteration of log*GAK*.

3.2 Introduction About ELM

ELM was proposed in Huang et al [10]. Suppose we are training SLFNs with K hidden neurons and activation function g(x) to learn N distinct samples

Fig. 2 Model of basic ELM

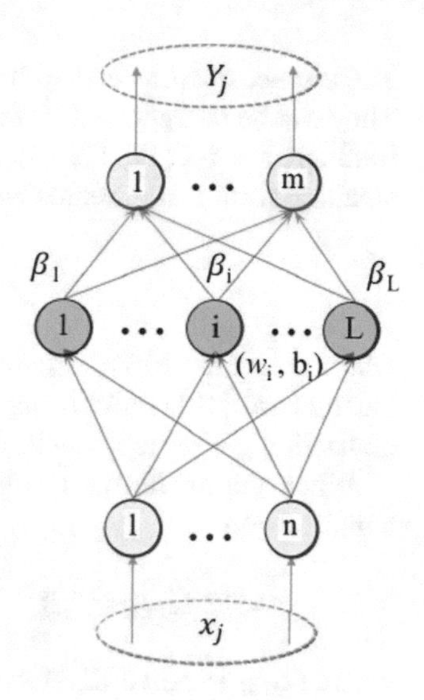

$\{X, T\} = \{X_j, t_j\}_{j=1}^{N}$ where $x_j \in R^n$ and $t_j \in R^n$. In ELM, the input weights and hidden biases are randomly generated instead of tuned. By doing so, the nonlinear system has been converted to a linear system (Fig. 2).

$$Y_j = \sum_{i=1}^{L} \beta_i g_i(x_j) = \sum_{i=1}^{L} \beta_i g(w_i^T x_j + b_i) = t_j, j = 1, 2, \ldots N, \tag{7}$$

where $Y_j \in R^m$ is the output vector of the j-th training sample, $W_i \in R^n$ is the input weight vector connecting the input nodes to the i-th hidden node, b_i denotes the bias of the i-th hidden neuron;$\beta_i = (\beta_{i1}, \beta_{i2}, \ldots \beta_{im})^T$ denotes the weight vector connecting the i-th hidden neuron and output neurons; $g(\cdot)$ denotes hidden nodes nonlinear piecewise continuous activation functions. The above N equations can be written compactly as:

$$H\beta = T, \tag{8}$$

where the matrix T is target matrix,

$$H = \begin{pmatrix} g(w_1^T x_1 + b_1) & \ldots & g(w_L^T x_1 + b_L) \\ \vdots & \ddots & \vdots \\ g(w_1^T x_N + b_1) & \cdots & g(w_L^T x_N + b_L) \end{pmatrix} \tag{9}$$

$$\beta = \begin{bmatrix} \beta_1^T \\ \vdots \\ \beta_L^T \end{bmatrix}, T = \begin{bmatrix} t_1^T \\ \vdots \\ t_N^T \end{bmatrix}. \tag{10}$$

The essence of ELM is that the hidden nodes of SLFNs can be randomly generated. They can be independent of the training data. The output weight β can be obtained in different ways [11, 12, 13]. For example, a simple way is to obtain the following smallest norm least-squares solution [12]

$$\hat{\beta} = H^{\dagger} T, \tag{11}$$

where $H^{\dagger}$ is the Moore-Penrose generalized inverse of matrix H. As analyzed by Huang et al [10], ELM using such MP inverse method tends to obtain good generalization performance with dramatically increased learning speed.

When the hidden nodes are unknown, kernels satisfying Mercer's conditions could be used:

$$\Omega_{ELM} = HH^T : \Omega_{ELM}(x_i, x_j) = h(x_i)h(x_j)T = K(x_i, x_j), \tag{12}$$

where Ω_{ELM} is called ELM kernel matrix. Then the output function of ELM can be written as:

$$f(x) = h(x)H^T(\frac{1}{\lambda} + HH^T)^{-1}T = \begin{bmatrix} K(x, x_1) \\ \vdots \\ K(x, x_N) \end{bmatrix}^T \left(\frac{1}{\lambda} + \Omega_{ELM}\right)^{-1} T. \tag{13}$$

In this specific kernel implementation of ELM, the hidden layer feature mapping $h(x)$ need not be known to users, instead its corresponding kernel $K(u, v)$ is given to users.

4 Experimental Results

4.1 Data Description

In this paper, we analyze grasping stability through obtained data by the sensors installed on the robot hand. The human-inspired Biomimetic Tactile sensor (BioTac) [14] is equipped with a 19-electrode array and a hydro-acoustic sensor surrounded by silicon skin inflated with incompressible and conductive liquid. This design provides rich tactile feedback similar to the slowly-adapting and fast-adapting afferents present in the human skin [15]. Latest developments in classification algorithms [16] allow us to explore the potential of large amounts of data from these sensors. Meanwhile, the BioTac is consisted with three complementary sensory modalities: force, pressure, and temperature. When the skin is in contact with an object, the liquid is displaced, resulting in distributed impedance changes in the electrode array on the surface of the rigid core. The impedance of each electrode tends to be dominated by the thickness of the liquid between the electrode and the immediately overlying skin. Slip-related micro-vibrations in the skin propagate through the fluid and are detected as AC signals by the hydro-acoustic pressure sensor. Temperature and heat flow are transduced by a thermistor near the surface of the rigid core (Fig 3).

For the cylindrical object, there are 1000 grasps, out of which 46% resulted in failures and 54% succeeded. For the box object, there are 500 grasps, out of which we had 31% successes and 69% failures. For the ball, there are 500 grasps—52.6% successes, 47.4% failures. In each experiment, as Fig. 4 showed, the bowl is used to bring the object up right if it falls out of the gripper during the extensive shaking motions that are performed later in the experiment. The Biotac sensor will obtain the force, pressure and temperature, then we use the electrode values which is a 57 dimension matrix and meanwhile the finger angles is a 3 dimension matrix, the robot joint angles is a 7 dimension matrix to analyze the stability of the object. In our work, we define bandwidth as 1, 3, 5, 7, 9, 10, 11, 13, 15 and Triangular as 0.

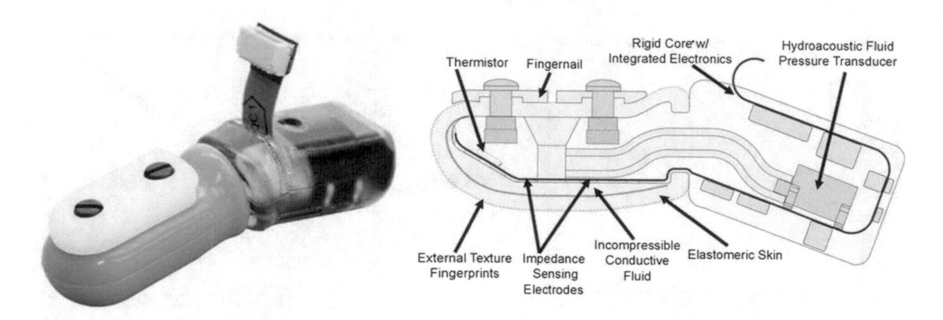

Fig. 3 Biomimetic tactile sensor [17]

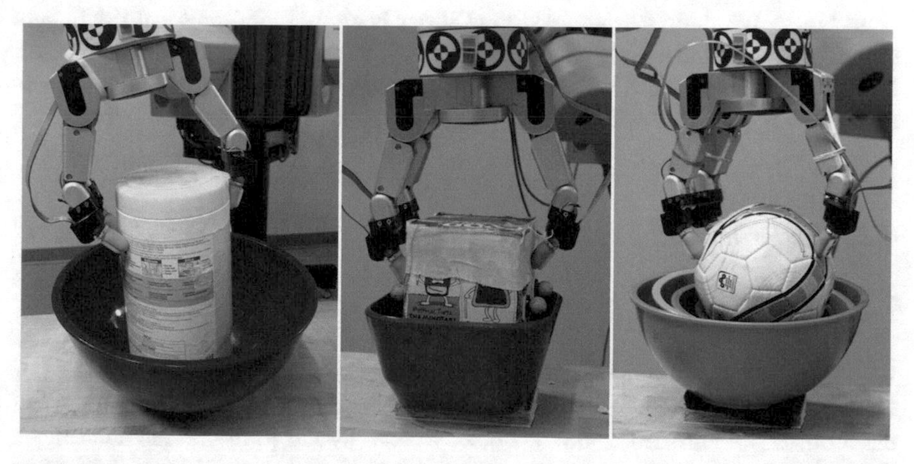

Fig. 4 Grasping process using Barrett hand with Biotac, if the object falls, then label the data as a failure (0), otherwise label it as a success (1) [18]

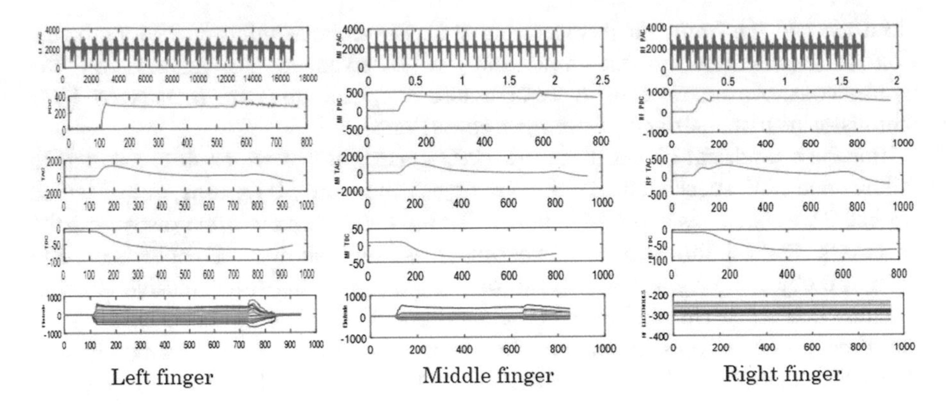

Fig. 5 Obtained data is described as PAC, PDC, TAC, TDC, Electrode (19 * 3 dimension) from *top* to *bottom*. *P* represent pressure while *T* represent temperature. We only use the electrode as tactile in this paper

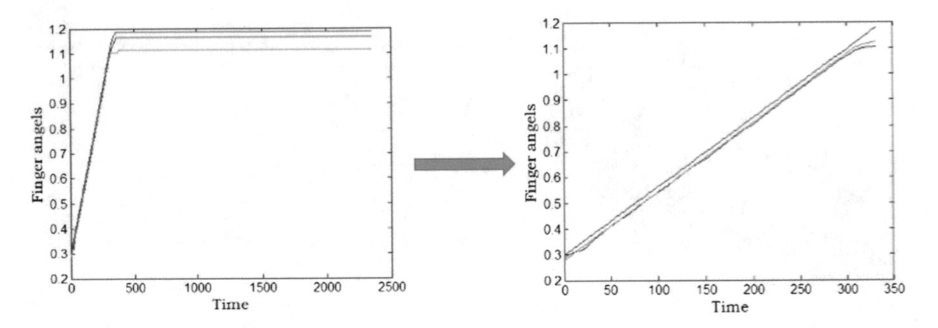

Fig. 6 Processed finger data. We remain data from 0 until the curve is nearly smooth

In Fig. 5, we show the provided time series begin with the moment the fingers start closing around the object and end 2 s after starting picking up the object. Meanwhile, finger angles data is presented in Fig. 6. Robot joint angles data is showed in Fig. 7.

4.2 Experimental Results

Tactile Results.

We use the Global Alignment to process the electrode values to structure the GA kernels. In each object we random divide into train samples and test samples. Then the 3 object are regarded as a new object we call hybrid to divide the dataset. As showed in Fig. 8, ball's accuracy with different *C* in kernel-ELM and SVM in tactile data.

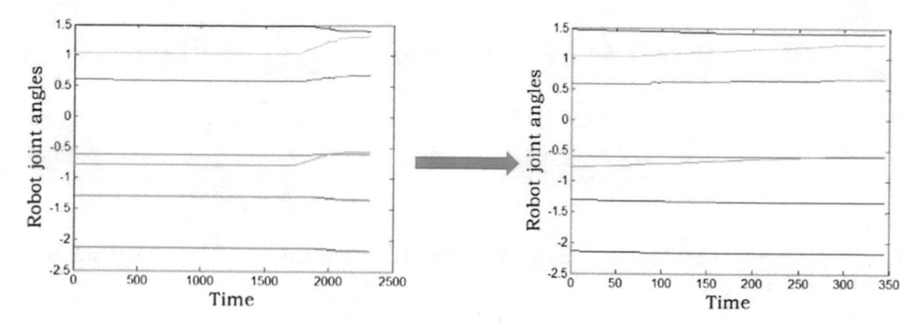

Fig. 7 Processed robot joint angles (shoulder flexion extension, shoulder abduction adduction, humeral roll, elbow flexion extension, wrist roll, wrist flexion extension, wrist abduction adduction). we only remain the data when it begin to change until the last moment

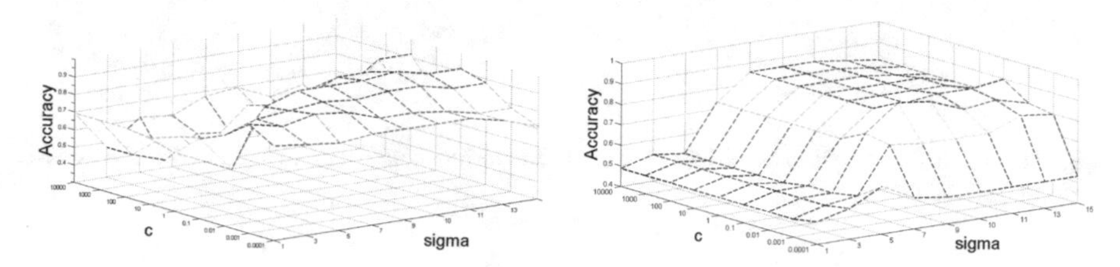

Fig. 8 Accuracy in different sigma (bandwidth) with different *C* in GA kernel-ELM (*left*) and SVM (*right*)

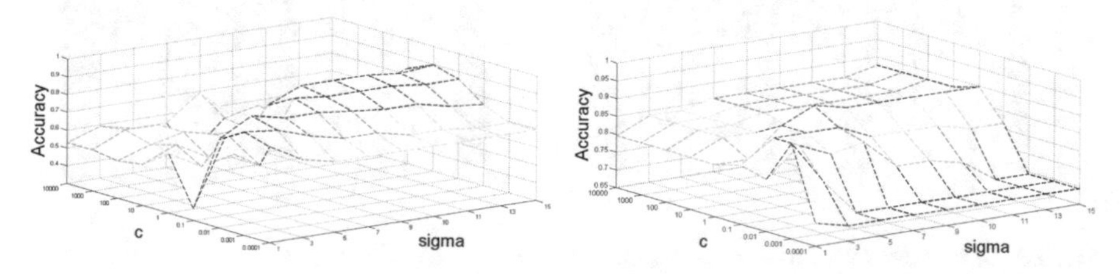

Fig. 9 Accuracy of box in different sigma (bandwidth) with different *C* in GA kernel-ELM (*left*) and SVM (*right*)

Figures 9, 10 and 11 respectively shows the accuracy of box, cylinder and hybrid. With the increased of sigma the trend becomes smooth in ELM.

As presented in Fig. 12, we compare the best accuracy in different sigma with different object. It is showed each object's best results. Ball-95.92%, box-89.8%, cylinder-91%, hybrid-96.48%. We can clearly see that in three object, ball has the best performance no matter how sigma is, but the hybrid's play better than any object. All results are above 90%. Otherwise in SVM the result has great fluctuation, they are not as well as ELM.

In Fig. 13 we can clearly see that the ELM have better performance than SVM, but SVM is generally smooth.

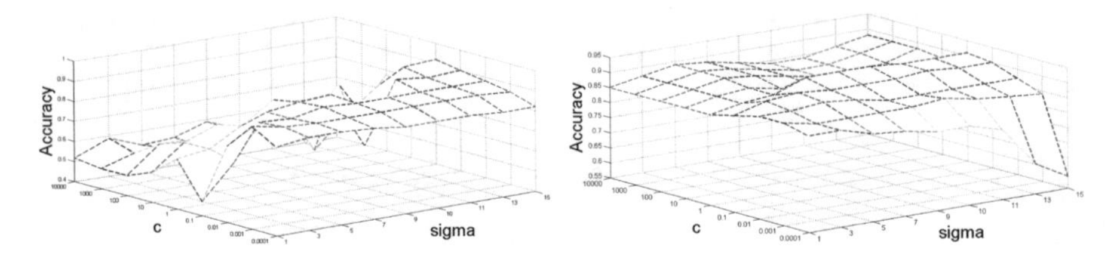

Fig. 10 Accuracy of cylinder in different sigma (bandwidth) with different C in GA kernel-ELM (*left*) and SVM (*right*)

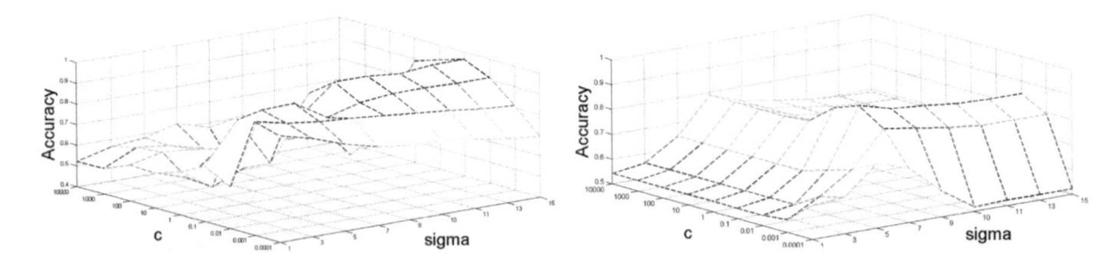

Fig. 11 Accuracy of hybrid in different sigma (bandwidth) with different C in GA kernel-ELM (*left*) and SVM (*right*)

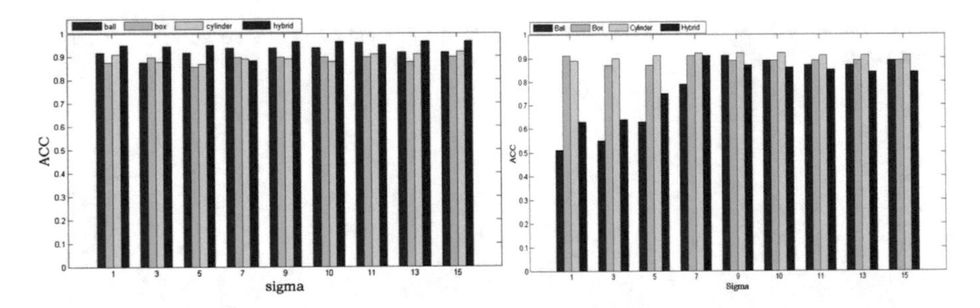

Fig. 12 Best accuracy with different sigma in ELM (*left*) and SVM (*right*)

Fig. 13 Comparison between ELM and SVM

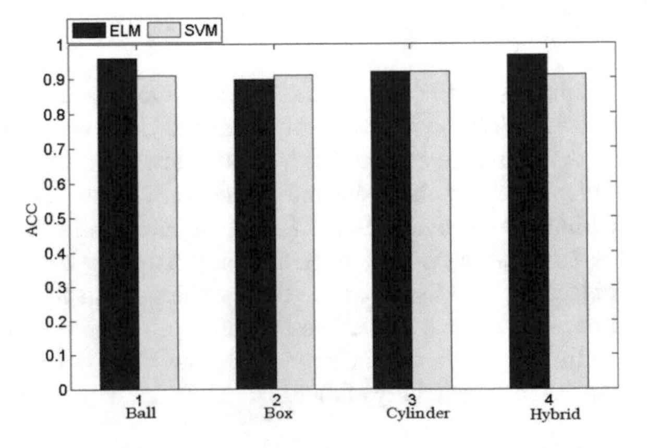

Finger Angle Results

We use 2 methods to process finger angles data. The first one, as mentioned in Fig. 6, is we intercept a part of the data as time series, then combine the 3 finger's data together to form a new data. The second one is we only take the final value.

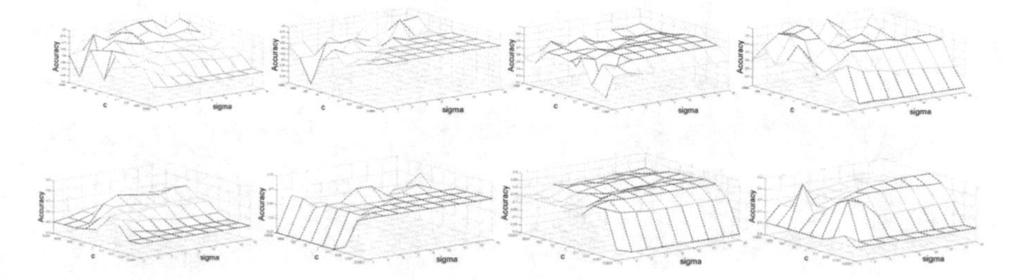

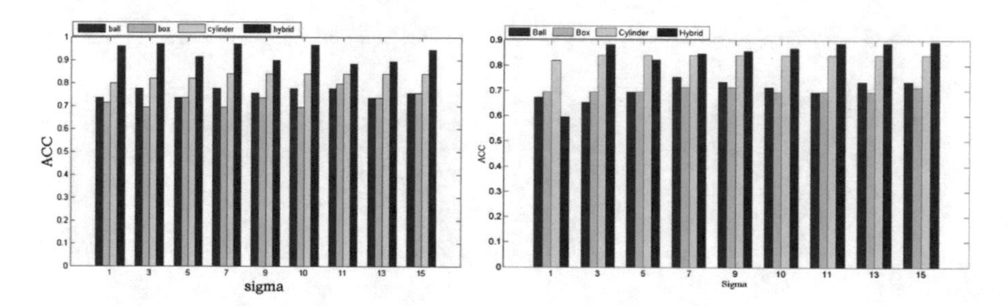

Fig. 14 4 object's accuracy in ELM and SVM (The first line is ELM, the second is SVM. From *left* to *right* is ball, box, cylinder, hybrid)

Fig. 15 Best accuracy with different sigma in ELM (*left*) and SVM (*right*)

Fig. 16 Comparison between ELM and SVM

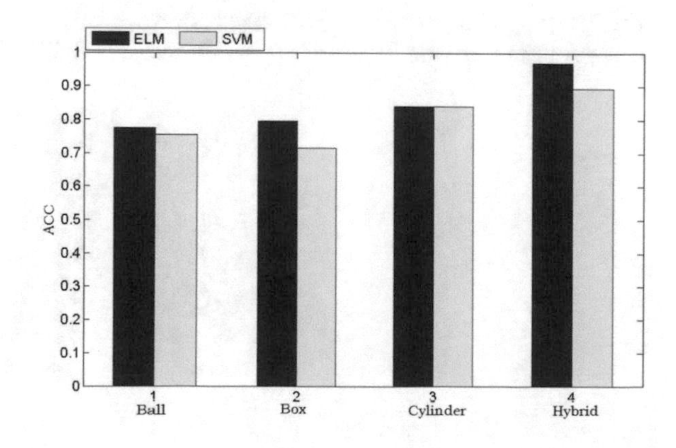

Figures 14 and 15 respectively showed the accuracy of the first processing method in ELM and SVM.

The first method.

The second method.

Compared from Figs. 14, 15, 16, 17, 18 and 19, we can clearly see that the first process is much better than we take final value. In the first method, each object's

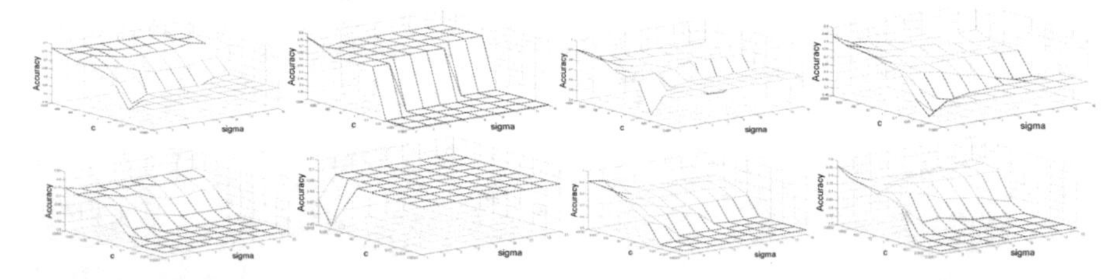

Fig. 17 4 object's accuracy in ELM and SVM (The first line is ELM, the second is SVM. From *left* to *right* is ball, box, cylinder, hybrid)

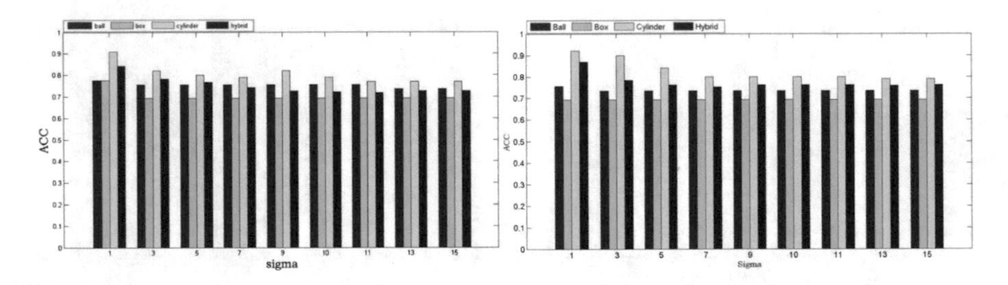

Fig. 18 Best accuracy with different sigma in ELM (*left*) and SVM (*right*)

Fig. 19 Comparison between ELM and SVM

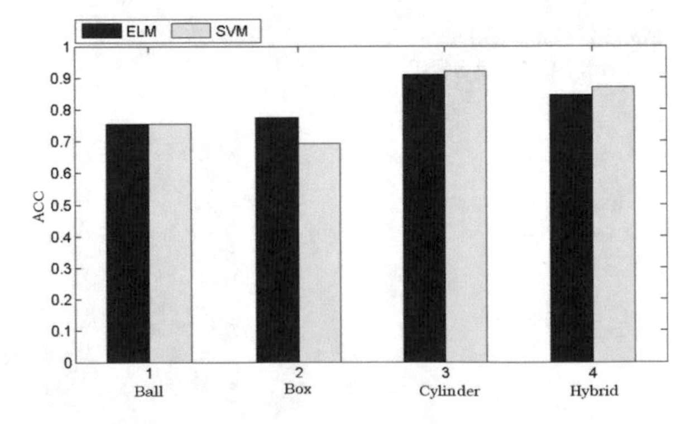

best results: Ball-77.55%, box-79.59%, cylinder-84%, hybrid-96.97%. It is obviously seen that in three object, cylinder has the best performance no matter how sigma is, but the hybrid's play better than any object. All results are above 90%. Besides, ELM has better results than SVM. But in the second method, cylinder has the best results while ELM and SVM results are not much different. It is clearly indicated that the first processing is better.

Compared the tactile results, only the hybrid's performance is better than using tactile information. The overall situation of the tactile experiments have better performance than finger angles.

Robot Joint Angle Results

As mentioned in Fig. 7, we deal with robot joint angles as time series to send into ELM and SVM. The results are as followed.

From Figs. 20, 21 and 22, we can see clearly that it is as much excellent as the result using tactile information in hybrid, but the 3 object's performance is worse than it. On the other hand, the hybrid's result is better than finger angle's, but the 3 object has the almost same results. Besides, ELM has a better result than SVM.

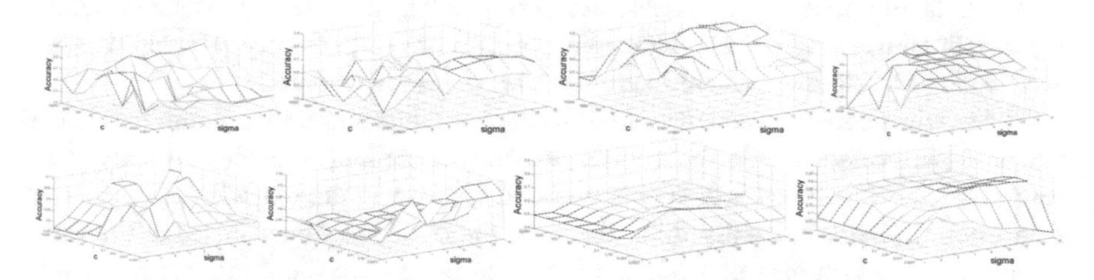

Fig. 20 4 object's accuracy in ELM and SVM (The first line is ELM, the second is SVM. From *left* to *right* is ball, box, cylinder, hybrid)

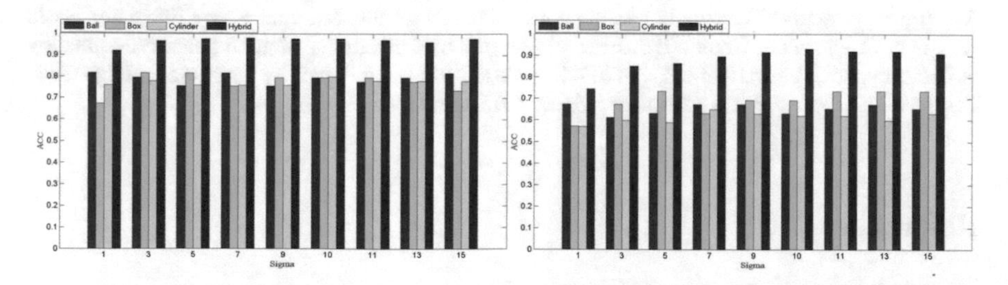

Fig. 21 Best accuracy with different sigma in ELM (*left*) and SVM (*right*)

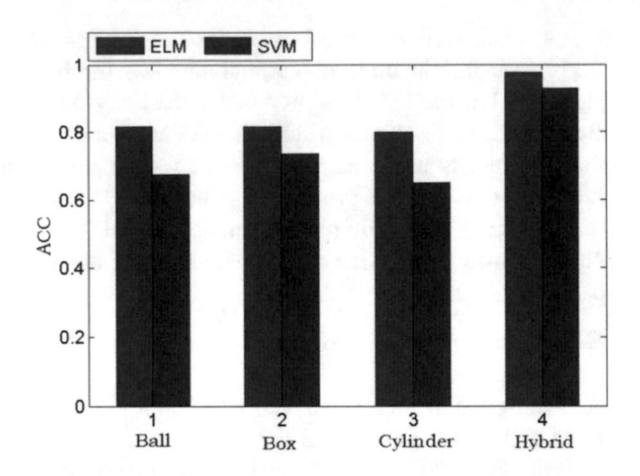

Fig. 22 Comparison between ELM and SVM

5 Conclusions

The database which obtained by BioTac form ball, box and cylinder in this work include force, pressure and temperature. In our work we process the electrode values as tactile time series to construct Global Alignment kernel. Meanwhile, the robot data (finger angles and robot joint angles) are respectively processed. Thus these kinds of data are sent into kernel-ELM and SVM separately.

The results denote the accuracy in different handling ways. In general, tactile has the best performance. All the results are very excellent. Besides, the hybrid's performance is much better than ball, box and cylinder in the majority of cases. It is worth noting that we usually think the final value of the finger angle is the most useful, because in a grasping process it has no effect. On the contrary, the whole process has a better result than only the final moment. In our future work, this algorithm is expected to fuse various data to achieve discrimination of grasping stability.

Acknowledgements This work was supported in part by the National Key Project for Basic Research, China, under Grant 2013CB329403, in part by the National Natural Science Foundation of China under Grants 61673238, 91420302 and 61327809, and in part by the National High-Tech Research and Development Plan under Grant 2015AA042306.

References

1. Bohg, J., et al.: Data-Driven Grasp Synthesis—A Survey. IEEE Trans. Rob. **30**(2), 289–309 (2013)
2. Cutkosky, M.R., et al.: On grasp choice, grasp models, and the design of hands for manufacturing tasks. IEEE Trans. Robot. Autom. **5**(3), 269–279 (1989)

3. Myeong, E., et al.: A new method for grasp stability analysis. In: Wseas International Conference on Signal Processing World Scientific and Engineering Academy and Society, pp. 164–170 (2007)
4. Funahashi, Y., et al.: Grasp stability analysis considering the curvatures at contact points. IEEE Int. Conf. Robot. Autom. **4**, 3040–3046 (1996)
5. Jenmalm, P., et al.: Control of grasp stability when humans lift objects with different surface curvatures. J. Neurophysiol. **79**(4), 1643–1652 (1998)
6. Howard, W.S., et al.: On the stability of grasped objects. IEEE Trans. Robot. Autom. **12**(6), 904–917 (1996)
7. Yamada, T., et al.: Stability analysis of 3D grasps by a multifingered hand. IEEE Int. Conf. Robot. Autom. 2466–2473 (2000)
8. Yamada, T., et al.: Grasp stability analysis of two objects in two dimensions. IEEE Int. Conf. Robot. Autom. 760–765 (2005)
9. Sudsang, A., et al.: Geometric reformulation of 3-fingered force-closure condition. In: Proceedings of the 2005 IEEE International Conference on Robotics and Automation, pp. 2338–2343 (2005)
10. Huang, G.B., et al.: Extreme learning machines: a survey. Int. J. Mach. Learn. Cybernet. **2**(2), 107–122 (2011)
11. Huang, G.B., et al.: Universal approximation using incremental constructive feedforward networks with random hidden nodes. IEEE Trans. Neural Networks **17**(4), 879–892 (2006)
12. Huang, G.B., et al.: Extreme learning machine: theory and applications. Neurocomputing **70** (1-3), 489–501 (2006)
13. Widrow, B., et al.: The no-prop, algorithm: a new learning algorithm for multilayer neural networks. Neural Netw. Off. J. Int. Neural Netw. Soc. **37**(1), 182–188 (2012)
14. Wettels, N., et al.: Biomimetic tactile sensor array. Adv. Robot. **22**(8), 829–849 (2008)
15. Johansson, R.S., et al.: Coding and use of tactile signals from the fingertips in object manipulation tasks. Nat. Rev. Neurosci. **10**(5), 345–359 (2009)
16. Madry, M., et al.: ST-HMP: unsupervised spatio-temporal feature learning for tactile data. IEEE Int. Conf. Robot. Autom. 2262–2269 (2014)
17. Su, Z., et al.: Force estimation and slip detection/classification for grip control using a biomimetic tactile sensor. In: IEEE International Conference on Humanoid Robots, pp. 297–303 (2015)
18. Chebotar, Y., et al.: Bigs: biotac grasp stability dataset. In: ICRA 2016 Workshop on Grasping and Manipulation Datasets (2016)

Extreme Learning Machine for Intent Classification of Web Data

Yogesh Parth and Wang Zhaoxia

Abstract Web search engines return a large amount of results for a user search query. Understanding the intent of these search queries can help us to narrow down the search results based on the type of information needed. In the research reported in this paper, we implemented machine learning algorithms to validate the accuracy of the classification of user search query. Broad categories of web query data are used from two different sources. Feature sets extracted solely from the web query are used to train the machine learning classifier. Classification results reveal that the performance of extreme learning machine (ELM) is much better when classifying user query intent than other machine learning classifiers.

Keywords Extreme learning machine · Web search engines · Web query · Intent classification

1 Introduction

Web search engines are the most widely used tools for access to the Internet. According to [1], more than 70% of the people use search engines for web access, and these search engines process billions of search results to user per week in response to their queries. The highly dynamic, diverse, and abruptness of web queries (usually about two or three terms) make categorization of queries difficult. Understanding intentions behind the queries can potentially improve the relevance and effectiveness of search engines.

Intent classification mainly aims to classify queries based on their intent. Broder's taxonomy method [2] classifies queries according to their intent into three main categories i.e., navigational, informational, and transactional. Informational search

Y. Parth (✉)
Space Applications Centre (SAC), ISRO, Ahmedabad 380015, India
e-mail: yogeshp@sac.isro.gov.in; yogeshparthiist@gmail.com

W. Zhaoxia
Institute of High Performance Computing (IHPC), A*STAR, Singapore 138632, Singapore
e-mail: wangz@ihpc.a-star.edu.sg

J. Cao et al. (eds.), *Proceedings of ELM-2016*, Proceedings in Adaptation, Learning and Optimization 9, DOI 10.1007/978-3-319-57421-9_5

Table 1 Category for web query intent classification

User intent	Purpose	Example
Request	Recipient is asked to perform some activity	*Do you have any data on intent classification?*
Proposal	Recipient and sender are committed to perform some activity	*There is meeting at 6 o'clock on big data analysis. Hope to see you there!*
Navigational	To look for a specific URL to reach a particular website	*Tumblr, Facebook*
Informational	To look to obtain data or information available on the web	*how to do scuba diving?*
Transactional	To look for interaction with site	*Buying air asia ticket to malaysia*

queries have intent of finding information about a particular topic, navigational search queries look for a specific website or webpage, and transactional search queries are intended towards making a purchase or transaction. The findings of Jasen and Booth research [1] revealed that more than 80% of web queries are informational in nature, with about 10% each being navigational and transactional. We have defined intent primarily for categories of request, proposal, navigational, informational, and transactional based on ontology of sentences as per Table 1. The query logs, the anchor text, and the results returned from the search engines have not been used to extract features to represent a query.

Extreme learning machine (ELM) is a feedforward neural network with a single-hidden layer, in which hidden layer nodes parameters are randomly assigned. The output weights between the hidden nodes and the output of network are determined using simple generalized inverse operation of the hidden layer output matrices. ELM tends to learn thousand of times faster with a better generalized performance compared to other gradient-based networks trained using backpropagation [3]. Sentiment classification based on ELM has been successfully implemented in various research such as [4–8]. However, to our knowledge, almost none of the research incorporates the use of ELM for intent classification.

The main contribution of this work is to study the classification accuracy of different machine learning methods such as ELM, Support Vector Machine(SVM), Maximum Entropy, and Naïve Bayes for intent classification of query leveraging on the feature extraction method. The experimental results reveal that the ELM performs much better than the other machine learning classifiers.

The rest of the paper is structured as follows. Section 2 introduces the relevant work done in the field followed by the implementation of such methods for Intent Classifications in Sect. 3. Performance evaluation by using different datasets is depicted in Sect. 4. Section 5 concludes this paper.

2 Relevant Work

Understanding intent behind user's query can help us to improve the relevant search results, which in turn reduce search time and improve users satisfaction. According to Hu et al. [9], intent representation, domain coverage, and semantic interpretation are three major challenges to intent classification. Instead of using machine learning based classifier, they have proposed a graph based methodology to overcome these challenges. There are three main categories of queries classification [10]. Category I augments the queries with query return search results for intent classification. Category II helps to improve the accuracy of supervised learning by leveraging on unlabeled query data. The last category, category III, adaptively trains itself using the training data via a self-training-like approach and then automatically labels queries data. Purohit et al. [11] employ pattern-set creation from a variety of knowledge sources to overcome ambiguity and sparsity challenge. They have done intent classification of tweets using knowledge-guided patterns for top-down processing along with bag-of-tokens model for bottom-up processing. Jansen and Booth [12] took more than 20,000 web query data categorized by topics and proposed an approach to automatically classify other queries based on existing queries. Their approach was to code a set of queries with attributes and then leverage the enriched data set to classify other web queries. Some of the researches include multi-faceted approach to query intent classification. Carlo et al. [13], proposed a hypothesis that the performance of single-faceted classification of queries can be improved by introducing information of multi-faceted training samples into the learning process. Their results show that the combination of correlated facets can improve the quality of classification results.

3 Implementation of ELM and Approach Towards Intent Classification

ELM, as proposed in Huang et al. [14] is a tuning free feedforward neural network with a single layer of hidden nodes in which the hidden nodes parameters are randomly assigned. For a feedforward neural network (single hidden layer) having n input nodes, o output nodes, and $\tilde{m}$ neurons in the hidden layer, the output function can be written as

$$f(x) = \sum_{i=1}^{\tilde{m}} \beta_i g(w_k.x) \quad (1)$$

where w_k are the weights of input (x) nodes which are randomly assigned, $g(.)$ is an activation function, e.g.,multiquadratic, sigmoid, etc. and β_i are the output layer weights which can be determined after finding the least-squares solution to the matrix equation,

$$\hat{\beta} = G^{\dagger}T, where\ G_{ik} = (w_k.x_i) \quad (2)$$

where $G^{\dagger}$ is the moore-penrose pseudoinverse [15] of the output matrix G.

For a dataset of N distinct samples, with x input belonging to c number of distinct classes, we define T as the matrix of output targets such that $T_{ik} = 1$ if and only if the sample belongs to a particular class c, otherwise its value is 0. ELM can be tuned with respect to the activation function and number of nodes to obtain a higher classification accuracy.

In order to train the machine learning methods, pre-processing steps are required for extraction of feature words. The collected corpus is pre-processed through a cleaning process which includes the removal of usernames ("@username"), punctuations, whitespaces, and hashtags. The cleaned data is tokenized [16] and stemmed [17] to convert them into structured text. Structured texts are further tokenized with labels to create word features lists. Using frequency distribution methods such as chi-square, the word features are assigned a score and based on the score, feature words are extracted. The so obtained feature words are used for training the ELM and other machine learning classifiers.

4 Experiment and Results

4.1 Data Collection

For testing our approach, the data were collected from two independent sources. The first source consisted of the email datasets that were taken from the "Enron database[1]", which contains labeled training and test data for email intent machine learning. We used a subset of email query of around 3657 cases for training and 992 cases as test data from the datasets.

In the second case, the data were extracted through the perl[2] wrapper around the Twitter API by using the different keywords such as,"citibank", "business", "buy", etc. over the region of Singapore. The collected tweets were annotated manually into informational, navigational, and transactional query in accordance with Broder's taxonomy to obtain a gold standard data to be used in machine learning based methods. The same dataset was also annotated automatically after training the machine learning classifiers.

[1]https://www.cs.cmu.edu/./enron/.

[2]https://metacpan.org/release/MMIMS/Net-Twitter-4.00003.

4.2 Performance Metric

Machine learning classifiers such as, SVM, ELM, Naïve Bayes, and Maximum Entropy were trained and tested using the training and test datasets respectively. The main performance metric which we use is accuracy.

The accuracy of the classification can be considered as one of the key metric as it provides the degree to which the obtained results is close to the correct value or a standard. However, by looking into the other performance metrics like precision, recall, and F-measure, we make sure that accuracy paradox would never arise in the evaluation.

Accuracy can be calculated using the following formula:

$$Accuracy = Samples\ correctly\ classified/Total\ number\ of\ sample\ cases. \quad (3)$$

4.3 Results

Table 2, presents the results of the accuracy of the classifiers for the email corpus. During the experiments, we applied our pre-processing and feature selection method on training dataset to select the best 130 feature vectors that were good discriminators between categories. Machine learning classifiers were trained on these feature vectors, and the accuracy of the test dataset was obtained.

In Table 3, we present the results of manual classification of twitter datasets. Two annotators were employed to label the corpora. The annotators, after labeling, cross-verified each other's labels. Those queries for which both the annotators had given the same label were taken into account.

In our classification process we used four machine learning algorithms namely ELM, SVM, Maximum Entropy, and Naïve Bayes. We have implemented the multi-quadratic function as activation function in the case of ELM and sequential minimal optimization algorithm for training a support vector classifier with linear kernel in the case of dual classification, and PolyKernel in the case of multi class classifica-

Table 2 Comparison of machine-learning algorithms for classifying the email intent

Datasets	Number of feature	ELM	SVM	Maximum entropy	Naïve bayes
Email	130	**82.50%**	79.03%	70.76%	67.54

Table 3 Manual classification of twitter queries by user intent

Datasets	Informational query	Navigational query	Transactional query
Twitter	56.0%(140)	22.4%(56)	21.60%(54)

Table 4 Accuracy of the automatic classification of twitter query data

Datasets	Naïve Bayes		
	Informational	Transactional	Navigational
Twitter data	87.76%	81.48%	87.71%

Table 5 Accuracy of the automatic classification of twitter query data

Datasets	Maximum entropy		
	Informational	Transcational	Navigational
Twitter data	87.76%	72.22%	92.10%

Table 6 Accuracy of the automatic classification of twitter query data

Datasets	Support vector machine		
	Informational	Transcational	Navigational
Twitter data	89.21%	70.37%	84.21%

Table 7 Accuracy of the automatic classification of twitter query data

Datasets	Extreme learning machine		
	Informational	Transcational	Navigational
Twitter data	**96.4%**	**87.75%**	**94.7%**

tion. We have used accuracy to evaluate the performance of the algorithms for intent classification.

Tables 4, 5, 6 and 7 shows the accuracy of the automatic classification of twitter datasets by user intents. The collected user queries through twitter were used to train and test the machine learning classifier and dataset respectively.

4.4 Discussions

The results in Tables 2 and 7 clearly prove that the overall performance of ELM is much better as compared to traditional machine learning methods. The accuracy of ELM in all the three query categories are much better as compared to that of Naïve Bayes, SVM, and Maximum Entropy. The Naïve Bayes method obtained better results on transactional query as compared to SVM and Maximum Entropy while SVM and Maximum Entropy were better for informational and navigational query data respectively. While manually classifying query data, we noticed that there is a very thin line between categorization of navigational and informational data. While selecting features, just an entity recognition of URL, website, or webpage is not enough to isolate navigational queries from informational queries as most of the informational queries also contain website or webpage references for information. Therefore, we have used bigram to clearly differentiate between these two categorization, thus improving the overall classification results.

5 Conclusions

In this paper, we have presented an approach to classify intent of social media text(twitter), as well as web data (email) with semantic feature based on bigram. Our experiment on two event datasets statistically prove the significant gain in accuracy of machine learning classifier. The results also reveal that in the case of dual as well as multi-class classifications, the performance in terms of accuracy is much better for ELM. We compared the feature words per class obtained from manual classification with the feature words per class extracted by automatic classification of different machine learning classifiers, and we found the similarity index close to 0.93 or 93% for ELM. We conclude that the overall performance or classification accuracy of ELM is significantly better (with less training time) and the feature words obtained just from queries content word can give better classification results when selection of features are done wisely.

In our current experiment, the second dataset, i.e. twitter data is quite a small dataset. We are working on obtaining larger datasets confirmed by more annotators for our future work.

Acknowledgements This work is supported by the Social Technologies+ Programme, which is granted by the Joint Council Office (JCO) at the Agency for Science Technology and Research (A*STAR). We thank Prof. G.B. Huang from Nanyang Technological University (NTU) for his insight and expertise that has greatly helped the research, and Ms. Neha H. Panchal for her assistance with data annotations and data analysis.

References

1. Jansen, B.J., Booth, D.L., Spink, A.: Determining the informational, navigational, and transactional intent of web queries. Inf. Process. Manage. **44**(3), 1251–1266 (2008)
2. Broder, A.: A taxonomy of web search. In: ACM Sigir Forum, vol. 36, no. 2, pp. 3–10. ACM (2002)
3. Huang, G.-B.: Extreme learning machine: theory and applications. Neurocomputing **70**(1), 489–501 (2006)
4. Wang, Z., Parth, Y.: Extreme learning machine for multi-class sentiment classification of tweets. In: Proceedings of ELM-2015 Volume 1, pp. 1–11. Springer (2016)
5. Liang, N.-Y., Saratchandran, P., Huang, G.-B., Sundararajan, N.: Classification of mental tasks from eeg signals using extreme learning machine. Int. J. Neural Syst. **16**(01), 29–38 (2006)
6. Handoko, S.D., Keong, K.C., Soon, O.Y., Zhang, G.L., Brusic, V.: Extreme learning machine for predicting hla-peptide binding. In: Advances in Neural Networks-ISNN 2006, pp. 716–721. Springer (2006)

7. Yeu, C.-W., Lim, M.-H., Huang, G.-B., Agarwal, A., Ong, Y.-S.: A new machine learning paradigm for terrain reconstruction. Geosci. Remote Sens. Lett. IEEE **3**(3), 382–386 (2006)
8. Kim, J., Shin, H., Lee, Y., Lee, M.: Algorithm for classifying arrhythmia using extreme learning machine and principal component analysis. In: 29th Annual International Conference of the IEEE Engineering in Medicine and Biology Society, 2007. EMBS 2007, pp. 3257–3260. IEEE (2007)
9. Hu, J., Wang, G., Lochovsky, F., Sun, J.-T., Chen, Z.: Understanding user's query intent with wikipedia. In: Proceedings of the 18th International Conference on World Wide Web, pp. 471–480. ACM (2009)
10. Cao, H., Hu, D.H., Shen, D., Jiang, D., Sun, J.-T., Chen, E., Yang, Q.: Context-aware query classification. In: Proceedings of the 32nd International ACM SIGIR Conference on Research and Development in Information Retrieval, pp. 3–10. ACM (2009)
11. Purohit, H., Dong, G., Shalin, V., Thirunarayan, K., Sheth, A.: Intent classification of short-text on social media. In: 2015 IEEE International Conference on Smart City/SocialCom/SustainCom (SmartCity), pp. 222–228. IEEE (2015)
12. Jansen, B.J., Booth, D.: Classifying web queries by topic and user intent. In: CHI'10 Extended Abstracts on Human Factors in Computing Systems, pp. 4285–4290. ACM (2010)
13. González-Caro, C., Baeza-Yates, R.: A multi-faceted approach to query intent classification. In: International Symposium on String Processing and Information Retrieval, pp. 368–379. Springer (2011)
14. Huang, G.-B., Zhu, Q.-Y., Siew, C.-K.: Extreme learning machine: a new learning scheme of feedforward neural networks. In: Proceedings of the 2004 IEEE International Joint Conference on Neural Networks, 2004, vol. 2, pp. 985–990. IEEE (2004)
15. Rao, C.R., Mitra, S.K.: Generalized Inverse of Matrices and Its Applications, vol. 7. Wiley, New York (1971)
16. Maršík, J., Bojar, O.: Trtok: a fast and trainable tokenizer for natural languages. Prague Bull. Math. Linguist. **98**, 75–85 (2012)
17. Willett, P.: The porter stemming algorithm: then and now. Program **40**(3), 219–223 (2006)

Reinforcement Extreme Learning Machine for Mobile Robot Navigation

Hongjie Geng, Huaping Liu, Bowen Wang and Fuchun Sun

Abstract Obstacle avoidance is a very important problem for autonomous navigation of mobile robot. However, most of existing work regards the obstacle detection and control as separate problem. In this paper, we solve the joint learning problem of perception and control using the reinforcement learning framework. To address this problem, we propose an effective Reinforcement Extreme Learning Machine architecture, while maintaining ELM's advantages of training efficiency. In this structure, the Extreme Learning Machine (ELM) is used as supervised laserscan classier for specified action. And then, the reward function we designed will give a reward to mobile robot according to the results of navigation. The Reinforcement Extreme Learning Machine is then conducted for updating the expected output weights for the final decision.

Keywords Q-learning · Navigation · Reinforcement extreme learning machine · Obstacle avoidance

1 Introduction

Navigation is an important field in the research of mobile robot [1]. The traditional obstacle avoidance methods, such as the visual map method [2, 3], the grid method [4] and the free space method, can deal with the obstacle avoidance [5] problem when the environmental information is known [6]. Although such methods which are proved to be highly robust have great performance on mobile robot navigation, they all work on the assumption that a map of the scenario is accurate as well as

H. Geng · B. Wang
School of Electrical Engineering, Hebei University of Technology, Tianjin, China

H. Liu (✉) · F. Sun
Department of Computer Science and Technology, State Key Laboratory of Intelligent Technology and Systems, Tsinghua University, TNLIST, Beijing, People's Republic of China
e-mail: hpliu@tsinghua.edu.cn

J. Cao et al. (eds.), *Proceedings of ELM-2016*, Proceedings in Adaptation, Learning and Optimization 9, DOI 10.1007/978-3-319-57421-9_6

available beforehand. Normally, such a map is obtained by labor operation or autonomous search [7]. This can however be very hard to complete in many circumstances: rescue or harmful operations are just few examples where teleoperating a mobile robot could be remarkably time-consuming or even impossible [8, 9]. When the obstacle information is unknown, the traditional navigation methods will not be able to avoid obstacles.

In the vast majority of the actual case, the robot's environment are dynamic [10], changeable and unknown. To solve the above problem, it introduces the artificial intelligence algorithms. Also thanks to the development of processor calculation ability and sensor technology, some complex arithmetic operations on the mobile robot platform has become easier, resulting in a series of intelligent obstacle avoidance method [11], algorithms such as neural network algorithm, genetic algorithm and fuzzy algorithm are very topical.

Though some learning algorithms based on gradient descent method (such as SVM and BP Neural network) have been extensively used in the training of multilayer feed-forward neural network [12], these traditional learning algorithms get stuck in a local minima easily, and the learning process are fairly slow [13, 14]. Furthermore, the activation functions used in these tuning algorithms based on gradient descent method should be differentiable.

In order to improve the inadequacy of these traditional learning algorithms based on gradient descent method, Huang, et al. proposed an efficient training algorithm for single-hidden layer feed forward neural network (SLFN) [15, 16] called Extreme Learning Machine (ELM). ELM randomly generating input weights and hidden biases which exert considerable influence on increasing the learning speed, and the output weights are solved by calculating Moore-Penrose (MP) generalized inverse. Compared with the traditional gradient-based learning algorithms, ELM not only learns much faster but also has higher generalization performance. In addition, ELM get out of troubles brought by learning algorithms based on gradient descent which include stopping criteria, learning rate and local minima [17].

Reinforcement learning [18] is very closely similar to the theory of classical optimal control and is generally a difficult problem and many of its challenges are particularly apparent in the robotics setting. A wide variety of methods of value-function-based reinforcement learning algorithms have been developed and can be divided mainly into three classes: methods with dynamic programming-based optimal control approaches such as policy iteration or value iteration, methods with rollout-based Monte Carlo methods and methods with temporal difference methods such as TD (Temporal Difference learning), Q-learning [19], and SARSA (State-Action-Reward-State-Action). Here, we choose the Q-learning method because it is a model-free reinforcement learning technique and it can be used to find an optimal action-selection policy for any given Markov decision process (MDP).

We address the problem of performing mobile robot navigation tasks relying upon reinforcement learning [20] which is widely researched in many aspects. Reinforcement learning is a continuous process of decision-making, the essence of mathematics is a Markov decision process. The ultimate objective is to get the optimal expectation of return function overall decision-making process. A successful achievement has made in locating active object with deep reinforcement learning proposed by Girshick et al. [21], which combined CNN features and reinforcement learning.

In this paper, Extreme Learning Machines (ELM) [22–24], used as a classifier [15, 16, 25], and Reinforcement Learning will be combined to train the navigation model on mobile robot. This algorithm just need to be set up the number of hidden neuron layer node, and it does not have to adjust the input weights and the bias of hidden neuron in executive process. The output of ELM has only one optimal solution, ELM has a fast learning speed and good generalization performance. What's more, more and more deep ELM learning algorithms have been proposed [26, 27]. Recently, a new biologically inspired ELM framework was proposed in [28] called ELM-LRF, which was implemented by introducing the local receptive field concept in neuroscience [29, 30]. Thus, in this mobile robot navigation work, we adopt the ELM-RL method. The contributions of this work are summarized as follows:

1. We propose to use reinforcement learning to solve mobile robot active obstacle avoidance. Mobile robot strengthen the navigation learning strategies through trial and error interactions.
2. We use an architecture ELM-RL framework, to learn representations from the input laser data for the navigation. The important merit of such a method is that the label prediction accuracy is improved and the training time is greatly shortened.
3. We evaluate our proposed navigation system on the ROS simulate platform. The obtained results show that the proposed method obtains rather promising results on mobile robot navigation.

The remainder of this paper is organized as follows: Sect. 2 introduces the Reinforcement Extreme Learning Machine system for Mobile Robot Navigation; Sect. 3 describes the process of the algorithm including the fundamental concepts and theories of ELM-RL structure and Q-ELM process; Sect. 4 present some experimental results of the proposed structure; while Sect. 5 concludes this paper.

2 Architecture

For navigation problems, there is such a solution. We design a reward function, if agent learning (Mobile Robot) move forward one step and get closer to the goal, we give the agent a positive reward. On the contrary, if agent move back and get farther away from goal position, the agent will receive a negative reward. Now we can evaluation for each step and get the sum of corresponding rewards, it would be easy to find a return value maximum path which is the best path (Fig. 1).

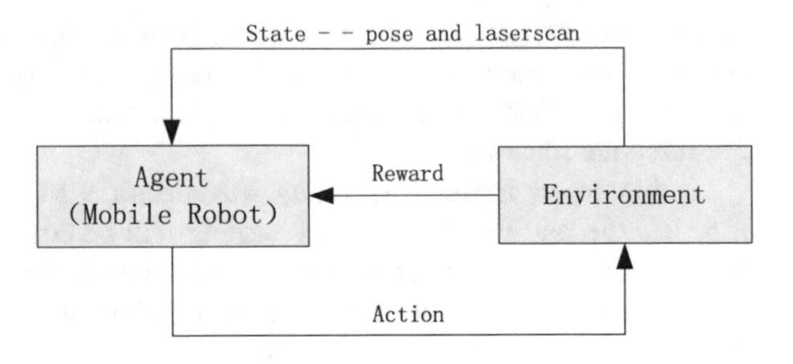

Fig. 1 Reinforcement learning architecture

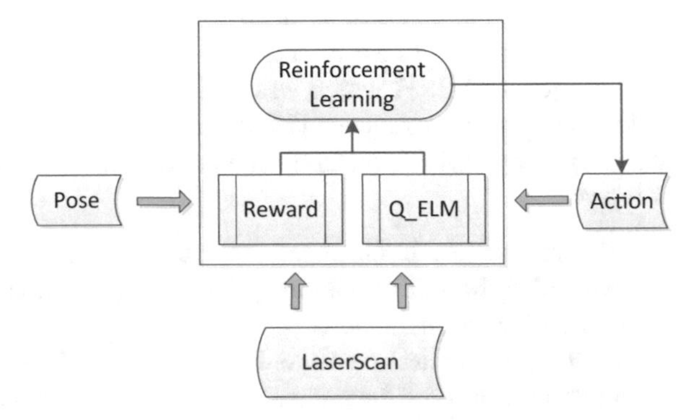

Fig. 2 The proposed reinforcement extreme learning machine architecture

Our architecture, which is depicted in Fig. 2, employs the ELM and Reinforcement Learning as the learning unit to learn navigation. The Reinforcement Extreme Learning Machine for Mobile Robot Navigation is structurally divided into three phases: generating action, retrieving reward by collecting laser data and updating output weights of ELM.

Three simple kinds of action beforehand, including going straight, turning left and turning right, should be provided to mobile robot to chose. Now the mobile robot should determine which action to take next step according to its state in the environment. Here we use ELM as a classifier to classify the laser data, input is the laserscan, the output is the action, so mobile robot is able to take action next step in terms of laserscan.

We expect a class-specific active navigation model that is able to avoid obstacles detected by the laser data. Mobile robot navigation not only travel from one position to the goal, but also go round obstacles, just as water flows round an obstruction. Obstacles avoidance is an important part of robot's path planning. Figure 3 illustrates some steps of the dynamic decision process to bypass obstacle ideally. The sequence of obstacle avoidance is decided by the agent that detect the locations of obstacle by distance sensor. Agent should keep both obstacles and goal in mind when considering how to move.

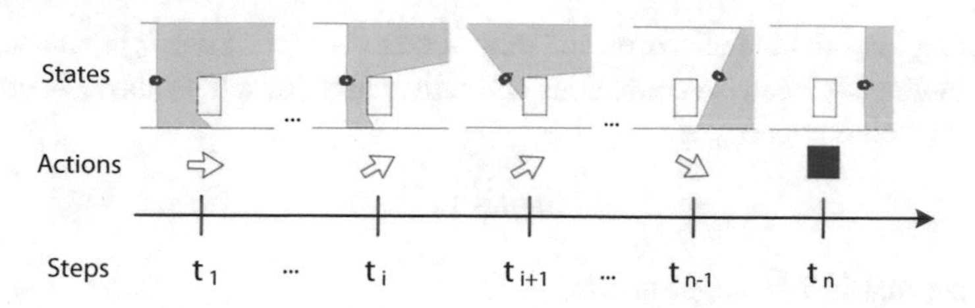

Fig. 3 A sequence of actions taken by the proposed algorithm to avoid obstacle ideally

Fig. 4 The model of basic ELM

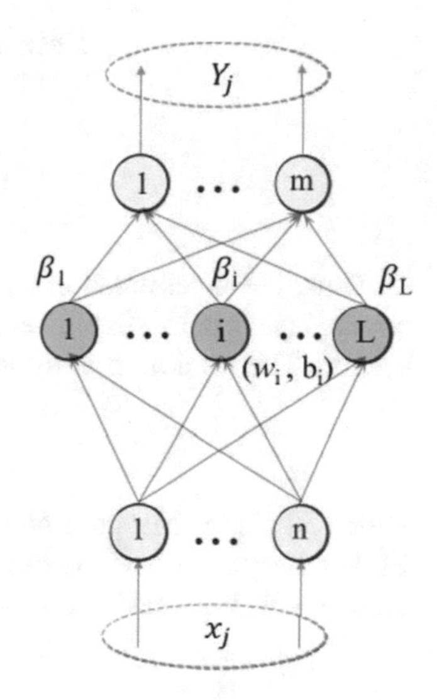

3 ELM Q-learning

3.1 Breif Review for ELM

Extreme learning machine which was a single-hidden layer feed forward neural network (SLFN) learning algorithm was proposed in Huang, et al. Suppose we are training SLFN with K hidden neurons and activation function $g(x)$ to learn N distinct samples $\{X,T\} = \{X_j, t_j\}_{j=1}^{N}$, where $x_j \in R^m$ and $t_j \in R^m$. In ELM, the input weights and hidden biases are randomly generated instead of tuned. By doing so, then onlinear system has been converted to a linear system (Fig. 4).

$$Y_j = \sum_{i=1}^{L} \beta_i g_i(x_j) = \sum_{i=1}^{L} \beta_i g_i\left(w_i^T x_j + b_i\right) = t_j, j = 1, 2, \ldots, N, \tag{1}$$

where $Y_j \in R^m$ is the output vector of the j-th training sample, $W_i \in R^n$ is the input weight vector connecting the input nodes to the i-th hidden node, b_i denotes the bias of the i-th hidden neuron; $\beta_j = (\beta_{i1}, \beta_{i2}, \ldots, \beta_{im})^T$ denotes the weight vector connecting the i-th hidden neuron and output neurons; $g_i(w_i^T x_j + b_i)$ denotes hidden nodes nonlinear piecewise continuous activation functions. The above N equations can be written compactly as:

$$H\beta = T, \tag{2}$$

where the matrix T is target matrix,

$$H = \begin{bmatrix} g(w_1^T x_1 + b_1) & \cdots & g(w_L^T x_1 + b_L) \\ \vdots & \ddots & \cdots \\ g(w_1^T x_N + b_1) & \cdots & g(w_L^T x_N + b_L) \end{bmatrix}, \tag{3}$$

$$\beta = \begin{bmatrix} \beta_1^T \\ \vdots \\ \beta_L^T \end{bmatrix}, T = \begin{bmatrix} t_1^T \\ \vdots \\ t_N^T \end{bmatrix}. \tag{4}$$

Thus, the determination of the output weights (linking the hidden layer to the output layer) is as simple as finding the least-square solution to the given linear system. The minimum norm least-square (LS) solution to the linear system (1) is

$$\widehat{\beta} = H^+ T, \tag{5}$$

where H^+ is the MP generalized inverse of matrix H. As analyzed by Huang et al., ELM using such MP inverse method tends to have good generalization performance with learn fast enormously.

3.2 Q-Elm

We cast the problem of mobile robot navigation as a Markov decision process (MDP) since this setting provides a formal framework to model an agent that makes a sequence of decisions. The agent has a state transition with information of the currently laserscan and past actions, and receives positive and negative rewards for every decision in the training.

Formally, the MDP has actions A, states S, and a reward function R beforehand. This section presents details of these three components and Reinforcement Extreme Learning Machine for Mobile Robot navigation algorithm which combines Reinforcement Learning and ELM.

Three actions including go straight, turn left and turn right are prepared previously to be chosen by agent. The velocity of the mobile robot represented by angular velocity and the linear velocity: α_a and α_l. And the actions could be represented by α_a and α_l, as shown in Table 1, α is a parameter which could be adjusted as we want. When mobile robot is going straight line, the angular velocity should be set zero. Angular velocity and linear velocity must be coordinated when

Table 1 Angular velocity and linear velocity of mobile robot

Velocity	Go straight	Turn right	Turn left
α_l	$\alpha_l \times \alpha$	$\alpha_l \times \alpha$	$\alpha_l \times \alpha$
α_a	0	$-\alpha_a \times \alpha$	$\alpha_a \times \alpha$

turning. Perhaps these kinds of action could not meet the requirements of obstacle avoidance perfectly in some environment, in the future work we will increase the types of action.

S is the laser data provided by distance sensor on mobile robot. If agent bump into obstacles, we give the agent a negative reward. On the contrary, we give mobile robot a positive reward after arriving at the target according to agent's present state.

We use the ELM to train the orientation of mobile robot in previous work. The input is laser data, and output is the orientation of mobile robot, which can be attributed as a regression problem. The results are not very good, robot encounter obstacles frequently after making heading angle regression. The reason is that the accuracy obtained by ELM regression is not good enough, floating deviation reach about 10°, while the orientation's maximum is 180°. So the robot hardly avoid obstacle in terms of ELM regression.

In order to solve this problem, we apply reinforcement learning to mobile robot's navigation on the basis of ELM, and we use the reward to correct the output of ELM, the correct process is as follows:

$$Y_t = r_t + \gamma max\widehat{Q}\left(S_{t+1}, a_{t+1}; \widehat{\beta}\right), \tag{6}$$

where the matrix $\widehat{\beta}$ is output weights last moment, $\widehat{Q}(.)$ is ELM's classification process corresponding to $\widehat{\beta}$, and γ is a constant parameter.

Here is the algorithm process. At first, mobile robot will make a choice in the preset actions randomly. After finishing this action, we record the reward, the action and the laser data which represent agent's state, we take three things mentioned above and the next time state as a set of data, one set of data will be put into buffer D after one step. Wait until the data set is full, now ELM's input and Label are there, training could be carried out. At the beginning of training, mobile robot often encounter obstacles, so agent will get a negative reward (e.g. −1), after that, robot go back to the starting position and initialize state. A new epoch will begin, thus learning the ability of obstacle avoidance. In each epoch, agent will update the buffer D including the current state, reward, action, next moment state. In addition, agent also update the output weights applied in ELM, the output weights is updated by

$$\beta = H^{+} Y_t. \tag{7}$$

Combining Eqs. 6 and 7, we get

$$\beta = H^{+}(r_t + \gamma \max \widehat{Q}(S_{t+1}, a_{t+1}; \widehat{\beta})). \tag{8}$$

In the algorithm, the input weights of ELM and the hidden biases which are unchanged throughout are generated randomly. This algorithm inherits the advantages of fast calculating speed. Moreover, we use the reward obtained from reinforcement learning to correct and update the output weights of ELM. This improves the accuracy of the original ELM algorithm. Algorithm can be concluded as follow:

Algorithm: reinforcement extreme learning machine for robot navigation.
Generate the input weights w, hidden bias b and output weights β randomly
For epoch = 1 ~ M do
    Initialize position of mobile robot
    For t = 1 ~ T do
        Get original Laser Data
        With probability ε select a random positive action a_t
        otherwise select $a_t = max_a Q(S_t; \beta)$
        Execute action a_t and observe reward r_t and LaserScan S_{t+1}
        Store transition ($S_{t+1}, a_t, r_t, S_{t+1}$) in Buffer D
        Sample random minibatch of transition (a_j, r_j, S_{j+1}) from D
        Set the target $y_j = r_j + \gamma max_a \hat{Q}(S_{j+1}, a_{j+1}; \hat{\beta})$
        Perform a generalized inverse on $H^{+}(r_j + \gamma max_a \hat{Q}(S_{t+1}, a_{t+1}; \hat{\beta}))$ to update β
        Every C steps reset $\hat{Q} = Q$
        If action a_t is trigger, break the loop
    End For
End For

4 Experimental Results

4.1 Data Set

Simulation experiments base on ROS simulation platform. In ROS, a node is a process that performs an operation task. The nodes communicate by sending messages. The robot can be released from the node to make the robot move. Laser, as a separate node, will publish a topic whose name is laserscan. This algorithm is used as a separate node training robot's navigation by subscribing laserscan topic. In simulation, the distance sensor is installed on the mobile robot, and sensor and

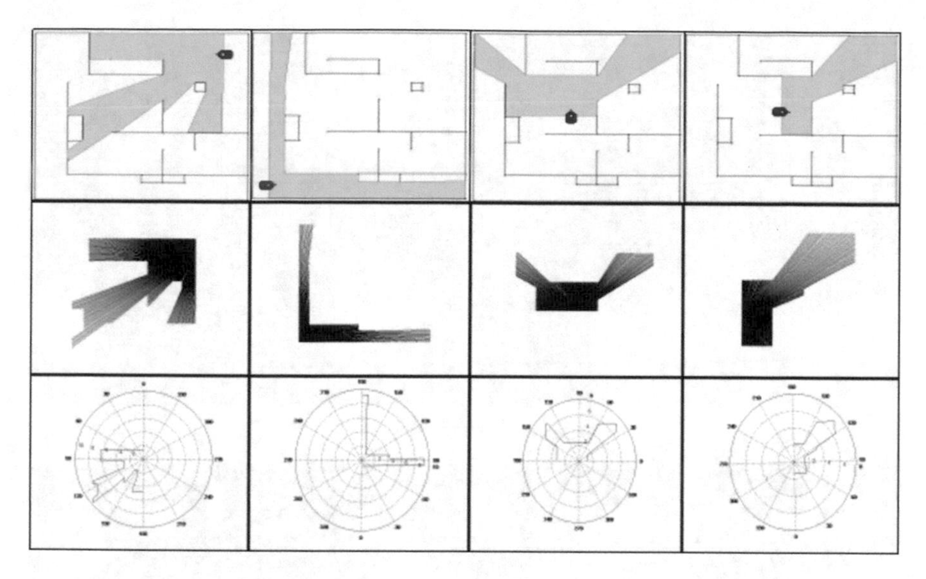

Fig. 5 Laser data visualization

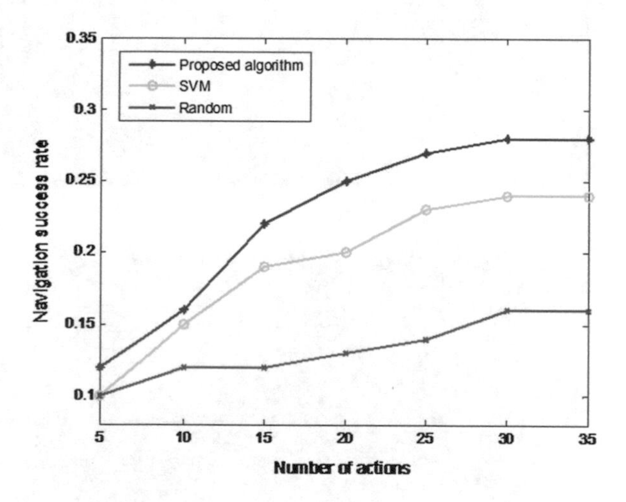

Fig. 6 Performance comparison between our method, SVM and random policies

robot are almost coincident, so the distance information collected by distance sensor is able to represent the state of the robot. The distance data of the robot is divided into two kinds: one is the pointcloud from Kinect, the other is laserscan from the laser. We convert pointcloud data to Laserscan data which could be used as the input data of ELM-RL. The parameters of the laser data are as follows, Angular rang and resolution: 180° and 1°. So every group of laser data is 180 dimensional.

Among the distance information obtained from the laser, if some data are approximately zero, we believe that the robot encounters obstacles. Each experiment includes 100 epochs. In one epoch, when the data group number reach twenty

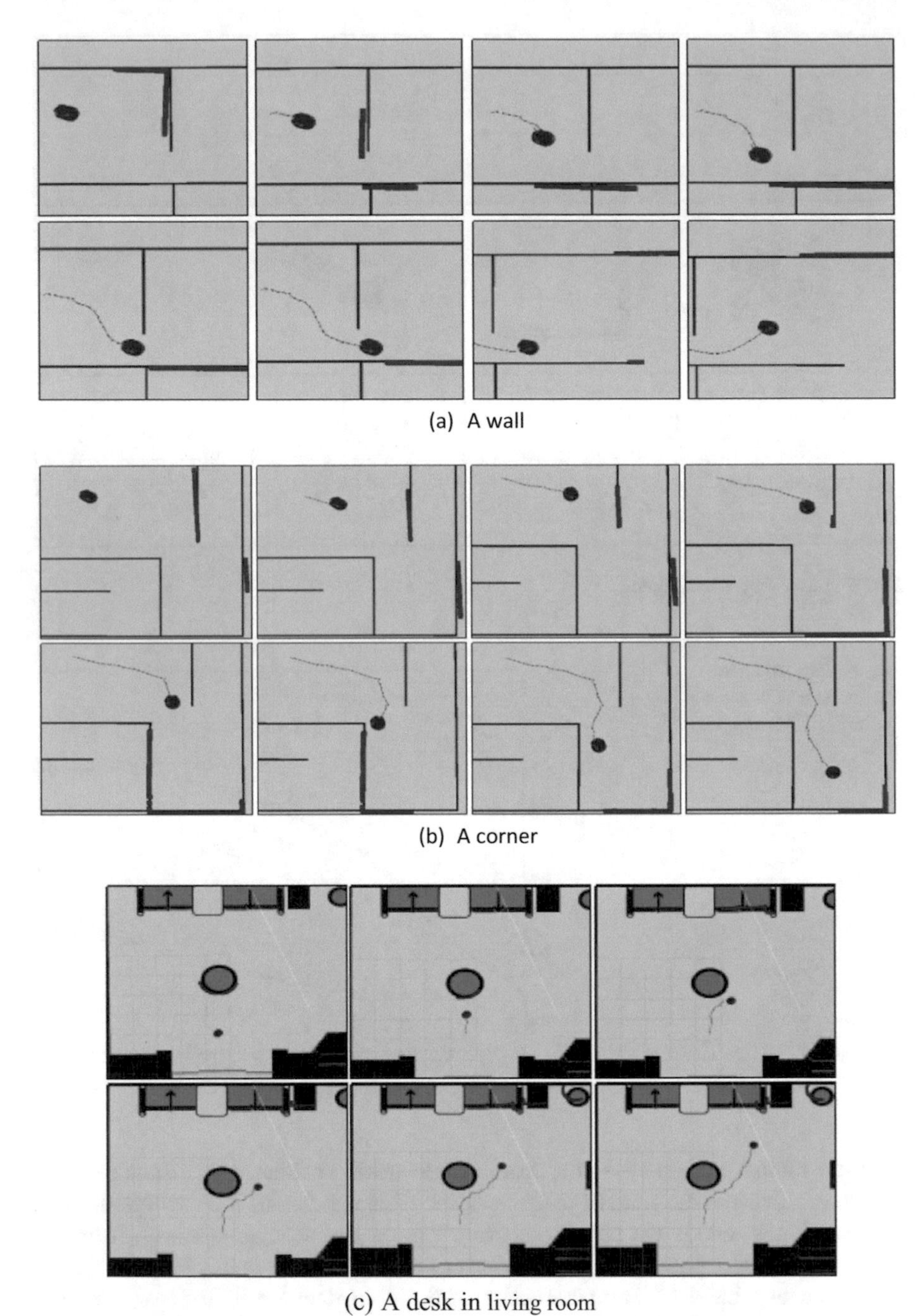
(a) A wall

(b) A corner

(c) A desk in living room

Fig. 7 The sequence of actions chosen by the proposed method on different scenes in buffer D, ELM-RF algorithm to start updating output weights. The data of the laser is shown in Fig. 5. The upper level of the graph is the original data, the middle layer show the distance of the obstacles at each angle, the lower layer represents the ideal feasible pathways.

4.2 Navigation Results

The results contain two main tasks, the SVM and the ELM-RL task. And the navigation success rate of the proposed method was evaluated by comparing it with the two alternative policies, that is, SVM and random, which shows in Fig. 6. This performance is averaged over the entire set of laser data in all epochs. The random policy selects a random action with uniform probability, while the SVM is used to classify the laser radar data by regarding the action as the category label. Figure 6 shows that compared with sequential and random policies, our method is able to avoid obstacle efficiently and improve the navigation success rate greatly. What's more, the navigation success rate of our proposed system is higher than that of the SVM.

In order to better understand our proposed architecture, we visualize the consecutive actions in navigation. Figure 7 shows the sequence of actions chosen by the proposed method on different scenes. The red line is laser data, robot's odometry will record the walking route of the robot and display it. For example, when obstructed by a wall or obstacle, mobile robot find the right way after many tries. Robot keep moving on after avoiding obstacles. It means that the training model we obtained could be used to do similar navigation.

5 Conclusions

In this paper, we proposed a algorithm for mobile robot navigation based on Reinforcement Extreme Learning Machine, which takes full advantage of the local receptive field to learn distance information of the input laser. Maybe three kinds of action could not avoid some obstacles which are of odd shape in actual environment. To address this problem, we will increase more kinds of action in next work. By introducing this architecture at the early stage in the system and employing ELM to the Q-learning process, the proposed approach outperforms sequential and random action selection policies.

Acknowledgements This work was supported in part by the National Key Project for Basic Research, China, under Grant 2013CB329403, in part by the National Natural Science Foundation of China under Grants 61673238, 91420302 and 61327809, and in part by the National High-Tech Research and Development Plan under Grant 2015AA042306.

References

1. Khatib, O.: Real-time obstacle avoidance for manipulators and mobile robots. Int. J. Robot. Res. 90–98 (1986)
2. Setalaphruk, V., Ueno, A., Kume, I., Kono, Y., Kidode, M.: Robot navigation in corridor environments using a sketch floor map. In: Proceedings of the IEEE International Symposium on Computational Intelligence in Robotics and Automation (CIRA), pp. 552–557 (2003)
3. Shah, D.C., Campbell, M.E.: A robust qualitative planner for mobile robot navigation using human-provided maps. In: Proceedings of the IEEE International Conference on Robotics and Automation (ICRA), pp. 2580–2585 (2011)
4. Konolige, K., Marder, E., Marthi, B.: Navigation in hybrid metric-topological maps. In: Proceedings of the IEEE International Conference on Robotics and Automation (ICRA), pp. 3041–3047. IEEE (2011)
5. Simmons, R.: The curvature-velocity method for local obstacle avoidance. In: Proceedings of IEEE International Conference on Robotics and Automation, pp. 3375–3382 (1996)
6. Feiten, W., Bauer, R., Lawitzky, G.: Robust obstacle avoidance in unknown and cramped environments. In: Proceedings of IEEE International Conference on Robotics and Automation, pp. 2412–2417 (1994)
7. Koenig, S., Simmons, R.G.: Passive distance learning for robot navigation. In: ICML (1996)
8. Malmir, M., Forster, D., Youngstrom, K., Morrison, L., Movellan, J.R.: Home alone: social robots for digital ethnography of toddler behavior. In: Proceedings of the Computer Vision Workshops, pp. 762–768 (2013)
9. Torralba, A., Murphy, K., Freeman, W., Rubin, M.: Context-based vision system for place and object recognition. In: Proceedings of the International Conference on Computer Vision (ICCV) (2003)
10. Fox, D., Burgard, W., Thrun, S.: Markov localization for mobile robots in dynamic environments. J. Artif. Intell. Res. 11 (1999)
11. Mnih, V., Kavukcuoglu, K., Silver, D., Rusu, A., Veness, J.: Human-level control through deep reinforcement learning. Nature 529–33 (2015)
12. Yang, Y., Wu, Q.M.J.: Mutilayer extreme learning machine with subnetwork nodes for representation learning. IEEE Trans. Cybern. 1–14 (2015)
13. Ulrich, I., Borenstein, J.: VFH+: reliable obstacle avoidance for fast mobile robots. In: Proceedings of IEEE International Conference on Robotics and Automation, pp. 1572–1577 (1998)
14. Ulrich, I., Borenstein, J.: VFH*: local obstacle avoidance with look-ahead verification. In: Proceedings of IEEE International Conference on Robotics and Automation, pp. 2505–2511 (2000)
15. Huang, G., Zhou, H., Ding, X., Zhang, R.: Extreme learning machine for regression and multi-class classification. In: Proceedings of the IEEE Systems Man and Cybernetics Society, pp. 513–529 (2012)
16. Li, M.B., Huang, G., Saratchandran, P., Sundararajan, N.: Fully complex extreme learning machine. Neurocomputing 306–314 (2005)
17. Cao, J., Zhao, Y., Lai, X., Yin, C., Koh, Z., Liu, N.: Landmark recognition with sparse representation classification and extreme learning machine. J. Franklin Inst. 4528–4545 (2015)
18. Sutton, R.S., Barto, A.G.: Introduction to Reinforcement Learning, pp. 383–385. MIT Press (1998)
19. Levine, S., Finn, C., Darrell, T., Abbeel, P.: End-to-End training of deep visuomotor policies. Comput. Sci. (2015)
20. Altuntas, N., Imal, E., Emanet, N., Ozturk, C.N.: Reinforcement learning-based mobile robot navigation. J. Electr. Eng. Comput. Sci. 1747–1767 (2016)

21. Hariharan, B., Arbelez, P., Girshick, R., Malik, J.: Simultaneous detection and segmentation. In: Proceedings of the European Conference on Computer Vision (ECCV), pp. 297–312 (2014)
22. Wang, X., Han, M.: Multivariate time series prediction based on multiple kernel extreme learning machine. In: Proceedings of the International Joint Conference Neural Networks (IJCNN), pp. 198–201 (2015)
23. Huang, G., Zhu, Q., Siew, C.: Extreme learning machine: theory and applications. Neurocomputing 489–501 (2006)
24. Huang, G., Zhu, Q., Siew, C.: Extreme learning machine: a new learning scheme of feedforward neural networks. In: Proceedings of the International Joint Conference Neural Networks (IJCNN), pp. 985–990 (2004)
25. Huang, G., Zhu, Q., Siew, C.: Extreme learning machine: theory and applications. Neurocomputing 489–501 (2006)
26. Malmir, M., Sikka, K., Forster, D., Movellan, J., Cottrell, G.W.: Deep q-learning for active recognition of GERMS: baseline performance on a standardized dataset for active learning. In: Proceedings of the British Machine Vision Conference, pp. 161.1–161.11 (2015)
27. Rumelhart, D.E., Mcclelland, J.L.: Parallel distributed processing. Encycl. Database Syst. 45–76 (1986)
28. Huang, G., Bai, Z., Kasun, L.L.C., Chi, M.V.: Local receptive fields based extreme learning machine. In: Proceedings of the IEEE Computational Intelligence Magazine, pp. 18–29 (2015)
29. Huang, G., Babri, H.A.: Upper bounds on the number of hidden neurons in feedforward networks with arbitrary bounded nonlinear activation functions. IEEE Trans. Neural Netw. 224–229 (1998)
30. Leshno, M., Lin, V.Y., Pinkus, A., Schocken, S.: Multilayer feed forward networks with a nonpolynomial activation function can approximate any function. Neural Netw. 861–867 (1993)

Detection of Cellular Spikes and Classification of Cells from Raw Nanoscale Biosensor Data

Muhammad Rizwan, Abdul Hafeez, Ali R. Butt and Samir M. Iqbal

Abstract Nanoscale devices have provided promising endeavors for detecting crucial biomarkers such as DNA, proteins, and human cells at a finer scale. These biomarkers can improve prognosis by detecting dreadful disease such as cancer at an early stage than the current approaches. Analyzing raw data from these nanoscale devices for disease detection is tedious as the raw data suffers from noise. Furthermore, disease detection decisions are made based on manual or semi-automated analysis—which are time-consuming, monotonous and error-prone process. Recent trends show an unprecedented growth in the advancement of nanotechnology for medical diagnosis. These devices generate huge amount of raw data and analyzing raw data in order to classify biomarkers in a fully automated and robust way is a challenge. In this paper, we present an algorithm for identifying cellular spikes, we have adapted extreme learning machines and dynamic time warping for the classification of cancer in raw data collected from nanoscale biosensors, such as solid-state micropores. Our approach can classify cancer cells with an accuracy of 95.6%, and with a precision and recall of 85.7% and 80.0%, respectively.

M. Rizwan (✉)
School of Electrical and Computer Engineering, Georgia Institute of Technology,
Atlanta, GA 30332, USA
e-mail: mrizwan@gatech.edu; muhammadriz@gmail.com

A. Hafeez
Computer Systems Engineering, University of Engineering & Tech, Peshawar, Pakistan

A.R. Butt
Department of Computer Science, Virginia Polytechnic Institute
and State University, Blacksburg, VA 24060, USA

S.M. Iqbal
Department of Electrical Engineering, Nano-Bio Lab,
Nanotechnology Research Center, Department of Bioengineering,
University of Texas at Arlington, Arlington, TX 76019, USA

S.M. Iqbal
Department of Urology, University of Texas Southwestern Medical Center at Dallas,
Dallas, TX 75390, USA

J. Cao et al. (eds.), *Proceedings of ELM-2016*, Proceedings in Adaptation,
Learning and Optimization 9, DOI 10.1007/978-3-319-57421-9_7

Keywords Cancer cells · Spike detection · Extreme learning machines · Dynamic time warping · Biomarkers · Nanoscale biosensing

1 Introduction

Diseases such as cancer can be completely cured, if detected and diagnosed at early stages. Conventional techniques like magnetic resonance imaging (MRI) and cytology are intrusive and are done as part of screening of cancer. However, such techniques suffer from the limitation of decoding the type of cancer whether it's a live or a brain tumor. The advent of nanoscale biosensing devices such as solid-state micropores and nanopores can detect cancer at an early stage by enabling the translocation of biological targets such as human cells, proteins and DNA in a biological assay at a finer granularity. Nanopore exhibits patterns in the output current upon single molecule arrivals and enables segregation of individual polymers. Nonetheless, these devices suffer from the inherit noise and baseline wanders in the output current, short transit time of the molecule, pore clogging and poor biomarkers selectivity.

Computational techniques have been utilized for the analysis of data collected from nanoscale devices [1–3]. Recent work shows the applications of supervised machine-learning algorithms [4–8] in the classification of important patterns in gene expression data [9–12]. Furthermore, a simple threshold based on peak-detection algorithms detect useful patterns in raw data emerging from ECG and mass spectroscopy [13–15]. Parameter for such a threshold can be local minimum/maximum, mean, standard deviation, energy or entropy [16–18]. These strategies inspire the design and development of machine learning methods for the effective detection and classification of biomarkers in the said data collected from nanoscale biosensors.

The sensors in this paper are minuscule channels made in thin silicon membranes and their output is an electrical current signal that is measured in micro and nanoamperes. Research shows that the cancer cells are softer and deform more readily than their healthy counterparts because of their elastic nature [19]. Such behavior of diseased and healthy cells is recorded as distinguishing patterns, i.e., pulse spikes in the output signal stemming from the extent to which they block the pore [20]. The pulses occur at different scales and magnitudes, which stem from the varying size and biophysical properties of cells, i.e., stiffness and viscosity. Nevertheless, the data collected from such sensors suffer from a large amount of raw data coupled with sensor noise and baseline wanders. Moreover, the translocation of a characteristic biological assay—0.5 milliliter of a blood sample through a micropore, results in 10 GB of raw data. The commercial software tools used to analyze the raw data are limited to trivial datasets and even a well-trained technician has to consume a lot of time to process and analyze the data from a typical biological assay.

In order to unlock the potential use of nanosale biosensors in a clinical setup, innovative solutions based on machine learning approach are required, which will leverage new computing models for the disease detection and high-quality decision making. In-situ, we propose a novel cellular spike detection algorithm to identify cellular spike locations in the raw data and use extreme learning machines and dynamic time warping based methods to classify cancer cells. Our method is very robust and can be readily used to infer useful information for disease diagnosis.

2 Related Work

2.1 Pattern Detection Techniques

Recent trends in medical research and modern clinical setups show an increase in recording important physiological signals, which substantially involves the detection of patterns (peaks/troughs) in the target signals. Numerous pattern-detection algorithms have been developed [13, 21–24], however, these are domain specific. Efforts have been made to build a generalized mathematical model [25] for peak detection algorithms. Nonetheless, such model suffers from a large number of false alarms and cannot be used in a particular domain, unless properly tailored and tuned. Our approach has the ability to adapt to the changing characteristics of the noisy micropore data and can efficiently search and identify pulses, followed by an accurate classification into benign and malignant types. Furthermore, our approach is amenable for online monitoring in a clinical setup because of its speed and accuracy.

2.2 Solid-State Micropores

These are tiny orifices in 200 nm thin silicon-based membranes used to measure the passage of human cells through them in the form of electrical pulses [26]. Cells passing through block the micropore and result in translocation events in the output current. The strength of the event is determined from the degree to which the pore is blocked by a target cell. These events are registered as pulses and their features depict patterns specific to the human cell type [1, 27]. Biomechanical properties of the diseased cells such as tumor cells, are known to be different than the normal cells [27–30]. Furthermore, the malignant cells are more elastic than benign and healthy cells. [19, 31–33] and the recorded pulses remarkably differ in the case of tumor cells [27, 34, 35].

The downside of these devices is that the detection and analysis is subjective. Therefore, it is indispensable to automate the identification and classification of different biological targets in high-throughput raw data generated by micropores. Such automation can replace manual analysis with automated analysis in which case the detection in 10 GB of data collected from a biological assay can be accomplished in few minutes rather than innumerable hours spent by a well-trained technician [34]. Nonetheless, the time taken by a micropore for the translocation of a biological assay is critical. For instance, a micropore calibrated at 2 MB/s can translocate an entire biological assay in about 83 minutes, resulting in 10 billion samples of current values. Our work can further benefit from improvements in the translocation speed and reduction in the noise in the output current of a micropore and nanopore technology, as delineated in [1, 35].

3 Extreme Learning Machines

Extreme learning machines (ELMs) use supervised learning approach for training single hidden layer feed-forward neural network (SLFNs) [36–39]. Recent research have shown the efficiency and effectiveness of ELMs for multi-class classification problems and its versatility in a gamut of applications such as classification of medical signals, prediction of protein-protein interactions, accurate forecasting of photovoltaic power for grid management, hyperspectral image classification, for the prediction of atrial fibrillation in electrocardiogram (ECG) and intracardiac electrogram (IEGM) [40–44]. The intrinsic nature of ELMs is that the learning parameters including input weights and biases of the hidden nodes are randomly assigned and are tuned until and unless output weights are analytically determined by simple generalized inverse operation. The weights between hidden nodes and output are learned in a single step. Such random assignment of learning parameters makes ELMs extremely faster at learning, achieve better generalized performance and with lesser subjectivity compared to traditional SLFNs [37–39]. Given an observation dataset with N nodes in the hidden layer and the excitation function G, extreme learning machines are given by Eq. 1:

$$f(x) = \sum_{i=1}^{N} \beta_i G(a_i, b_i, x_i) = \beta.h(x) \tag{1}$$

where β_i is the output weight of ith hidden node and the corresponding output neuron, a_i is the input weight of the input neuron and ith hidden layer node. b_i is the offset of ith hidden layer node.

4 Dynamic Time Warping

Dynamic time warping (DTW) is a technique for measuring dissimilarity between temporal sequences. Temporal sequences are widely used in various applications such as health monitoring, climatology, geology, astronomy, etc.

Dissimilarity comparison is a common step in most of the time series data analysis and despite of numerous alternatives available, there is an increasing evidence that DTW has proven itself as the best candidate for similarity mesaure in majority of domains such as medicine, music/speech processing, climatology, aviation, mining of historical manuscripts, geology, astronomy, space exploration, wildlife monitoring, robotics, and cryptanalysis [45, 46].

It computes an optimal match between two given sequences, e.g., time series with certain restrictions. The sequences under observation are warped non-linearly in time dimension to determine their similarity independent of certain non-linear variations in the time dimension. Such technique is frequently used in time series classification.

The DTW distance between any two time series sequences, A and B with samples $s_1, s_2, \ldots, s_m$ and $t_1, t_2, \ldots, t_n$ is $D(m, n)$—typically computed using a dynamic programming approach as shown in Eq. 2:

$$D(i,j) = min\,(D(i,j-1), D(i-1,j), D(i-1,j-1)) + d(x_i, j_i) \tag{2}$$

The two sequences are placed along x-axis and y-axis of a DTW grid. In each step of $D(i,j)$, the minimum among the three neighboring distances is computed, this ensures smooth warping, e.g., no samples left without warping, called local continuity constraint. Once all the possible paths within a warping window are computed, then the final step is to backtrack the best path through the grid, starting from (m, n), yielding the DTW warping path.

In order to constraint the number of paths between two sequences, warping window constraint is employed. This makes sure that the number of possible paths should be within a window width of r, i.e., $j - r <= i <= j + r$. Sakoe-Chiba band constraint [47] ensures that the DTW path is in proximity to the diagonal of DTW grid which contains the $D(i,j)$. This eliminates pathological warping which aligns a short span of one sequence to a large span of another sequence. For further details on the types of continuity constraints, interested readers are referred to [34].

5 Method

In our method, we pre-process the raw data using moving average filter, followed by cellular spike detection algorithm, and use ELM and DTW based classification method to make decision regarding cancer cells. The overall block diagram is shown in Fig. 1.

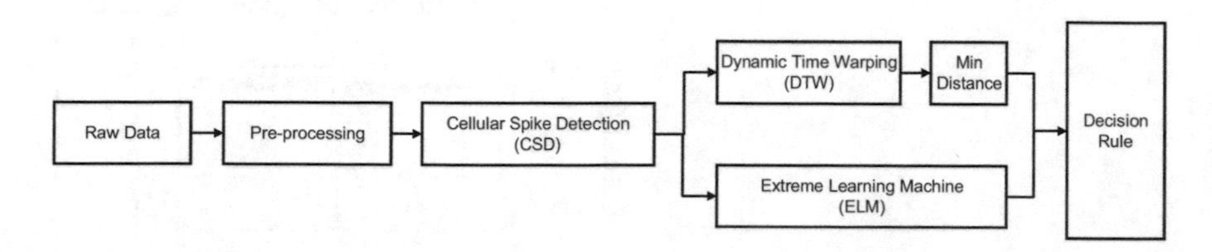

Fig. 1 Block diagram

5.1 Pre-processing

The raw data collected from a micropore is pre-processed by normalizing it to zero mean. Normalized data is passed through 3-point moving average filter to reduce random noise present in the raw data. The information that distinguishes different cell types lies in cellular spikes. These spikes result from the cell passage through micropore, which results in successive current values to fall abruptly in the output current level and reverts quickly back to the original baseline current forming a valley. Typical spikes of cancer cells, white blood cells (WBCs), and red blood cells (RBCs) after pre-processing are shown in Figs. 2, 3, and 4 respectively. The distinguishing information we get is very sparse as there are only a small number of cellular spike samples out of millions samples that we get in pre-processed raw data. In general, classification algorithms are computationally expensive. In order to mitigate the bottleneck arising from the sparsity in the data, identifying cellular spike samples prior to classification in the pre-processed data make the overall process efficient. This helps classification algorithm to process only the useful spike samples (smaller in number) instead of processing all samples in raw data (million in number).

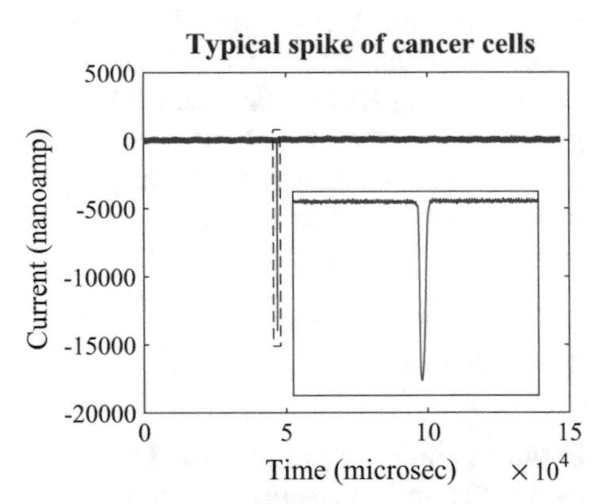

Fig. 2 A typical cancer cell spike in a pre-processed data

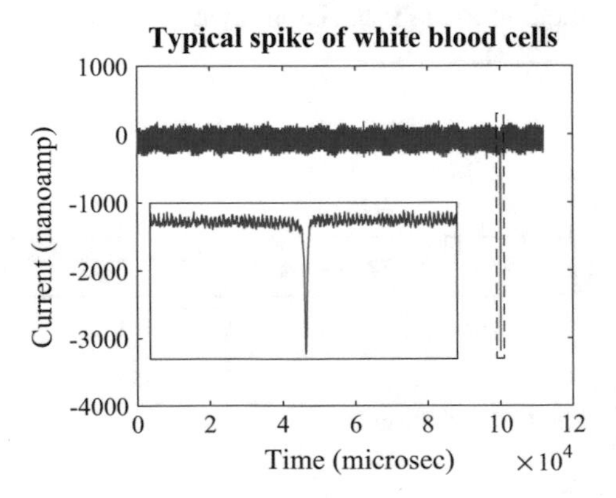

Fig. 3 A typical WBC spike in a pre-processed data

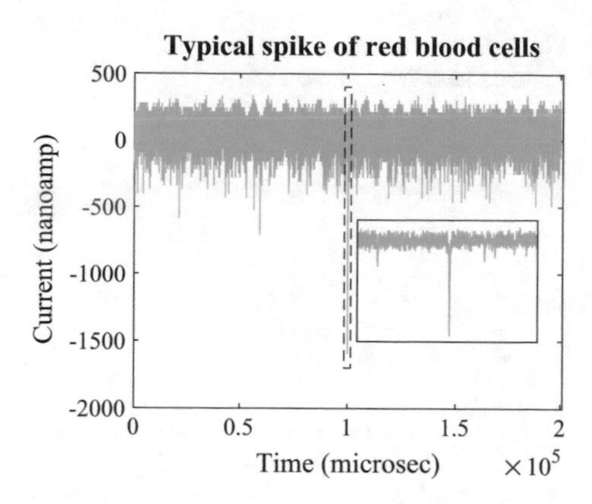

Fig. 4 A typical RBC spike in a pre-processed data

5.2 *Cellular Spike Detection (CSD) Algorithm*

The cellular spike detection algorithm identifies the location of cellular spike samples in the raw pre-processed data (see Algorihm 1). The main steps are as follows:

1. Calculate threshold from pre-process raw data based on standard deviation (σ).
2. Identify samples in pre-process raw data whose amplitude is below the threshold.
3. Compute the longest consecutive sequence of data samples whose amplitude value is below the threshold.

The longest sequence of data samples whose amplitude value is below the threshold identify a cellular spike. All other data samples with amplitude value below the threshold are mainly due to the short burst of noise. The value of threshold used in our experiment is -1.5σ and is learned by trial and error based approach. We picked 50 samples from the center of identified cellular spike samples. The reason for picking 50 samples from center is that it contains more distinguishing information as compared with the samples at the boundaries. Figures 5, 6, and 7 shows the ensemble cellular spikes of cancer cells, white blood cells, and red blood cells respectively.

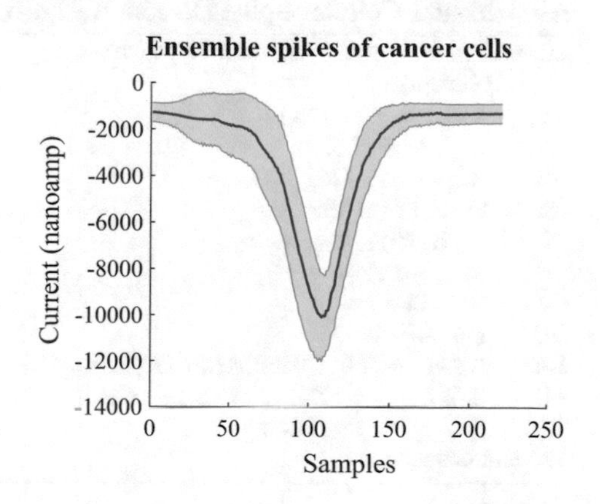

Fig. 5 An ensemble (average) of typical cancer pulses

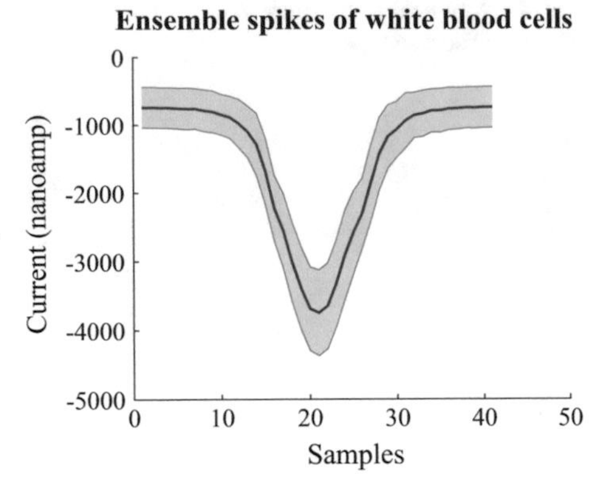

Fig. 6 An ensemble of characteristic WBC pulses

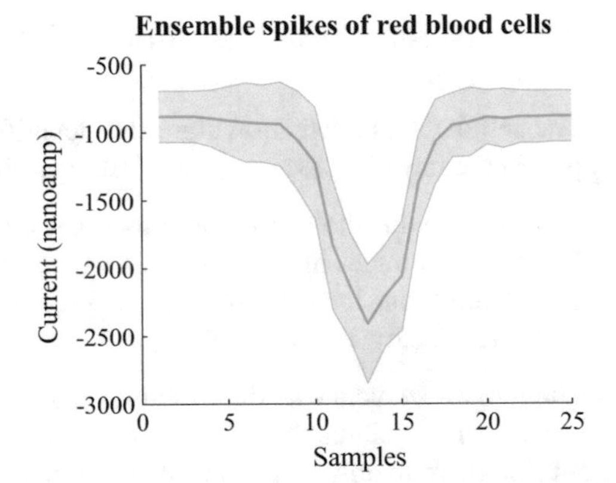

Fig. 7 An ensemble of characteristic RBC pulses

From these figures, it is clearly evident that the most idiosyncratic information is in the central part of the spike waveforms. In case, when the length of identified cellular spike samples is less than 50, we zero pad these in a symmetric way at both ends.

Algorithm 1 Cellular Spike Detection Algorithm

1: **procedure** CELLULAR SPIKE DETECTION(X) ▷ X: Data samples in cellular spike
2: $X_{mean} \leftarrow$ mean(X) ▷ Calculating statistics of data
3: $X_{dev} \leftarrow$ dev(X)
4: $X_{TH} \leftarrow 1.5 \cdot X_{dev}$
5: **for** i $1 \rightarrow$ n **do**
6: **if** $X(i) \leq X_{TH}$ **then**
7: $IDX_{TH} \leftarrow i$
8: **end if**
9: **end for**
10: TEMP = [0 CUMSUM(DIFF(IDX_{TH}) $\neq$ 1)]
11: IDX_{CELLS} = IDX_{TH} (TEMP = MODE(TEMP))
12: X_{CELLS} = X(IDX_{CELLS}) ▷ Detected cellular spike waveform
13: **end procedure**

Table 1 Decision rule

ELM based classification	DTW based classification	Decision
Cancer cell spike	Cancer cell spike	Cancer
Cancer cell spike	RBC/WBC spike	Non-cancer
RBC/WBC spike	Cancer cell spike	Non-cancer
RBC/WBC spike	RBC/WBC spike	Non-cancer

5.3 Cellular Spike Classification

For cellular spike classification, we used ELMs and DTW. In case of ELMs, we used central 50 samples of cellular spike samples of cancer cells, WBCs, and RBCs from the training data. We varied the number of hidden neurons from 10 to 20 and used sigmoid as an activation function. The best ELM model comprises of x neurons and is learned on the validation data. As of DTW based approach, we used cellular spikes in our training and validation data as a reference data for distance comparison. The reference data comprises all the samples of cancer cells, WBCs, and RBCs. For an unknown cellular spike from the test data, it computes Euclidean distance with all the cellular spikes (cancer cells, WBCs, and RBCs) in our reference data using dynamic programing. The DTW based approach computes output class of the unknown cellular spike based on the minimum distance.

5.4 Decision Rule

We combined classification results from ELM and DTW based classifier to make a decision about cancer and non-cancer. The decision rule is based on logical "AND" that decides for cancer only when output of the ELM and DTW classifies cellular spike as cancer. For all other cases, it will decide as non-cancer as shown in Table 1. This results in a greater confidence regarding our decision for cancer.

6 Experiment

6.1 Dataset

The biological raw datasets are collected from a characteristic tumor sample by translocating through a micropore. The assay comprises of cancer cells, WBCs, and RBCs. The collected dataset contain characteristic profiles of the translocated cells. Typical profile of cells contains 4 million samples recorded over a sampling interval of 2.2 μs. Each sample has a resolution of 2 bytes and can measure from 0 to 65, 535 nanoamperes. Overall, the data collected from a typical assay consists of 90 profiles, resulting in 360 million samples, which is equivalent to 10 GB of raw data.

6.2 Micropore Assembly

Micropore with a radius of 12 μm made in 200 nm thin membrane is used to translocate the biological assay and thus, generate the raw dataset. The calibration of sampling frequency needed to operate the micropore is an important factor in achieving the maximum throughput of the pore. Decreasing sampling frequency results in a stable baseline with less noise, but lacks the ability to capture the useful translocation events at a finer granularity. Contrariwise, higher sampling frequency results in noisy data which can suppress some of the useful translocation events. The optimal sampling frequency used was 0.4 MHz.

6.3 Cellular Spike Detection and Classification

The raw data obtained after passing through micropore is normalized as discussed in Sect. 5.1. The cellular spike detection algorithm identifies the location of cellular spikes in the pre-processed data. The central 50 samples from the identified cellular spike location is fed to ELMs and DTW based classification methods. The results of the ELMs and DTW based classifier is combined and decision is made regarding the cancer or non-cancer based on the decision rule as mentioned in Sect. 5.4. Figure 8 summarizes the result for cancer cell classification based on our method.

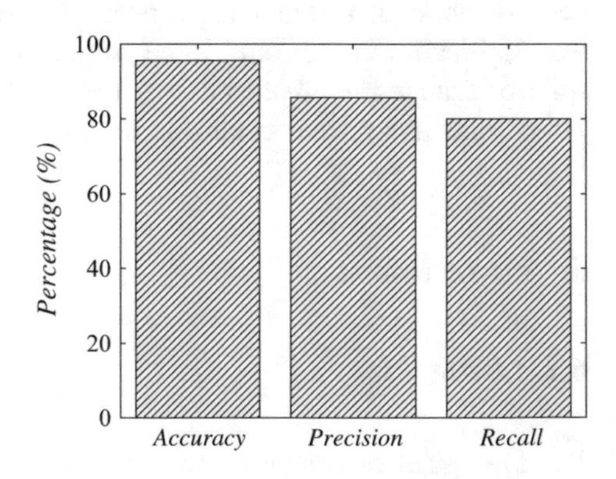

Fig. 8 Extreme learning machines and dynamic time warping based cancer cell classification

6.4 Comparison of Results

In [34], they used threshold based framework for cancer cell classification. The raw data was pre-processed using moving average filter and threshold was calculated using statistics of the pre-processed data. They used width and amplitude of the cellular spikes as features. They obtained a classification accuracy of 63%. In [35], they used average of mean and standard deviation for each cellular spike in addition to the spike width and amplitude. They used k-nearest neighbor as a classifier and obtained a classification accuracy of 70%.

7 Conclusion

We have developed a robust and automated approach for identification of cancer and non-cancer patients from the raw data collected from the translocation of a biological assay through a solid-state micropore. Our novel cellular spike detection algorithm identifies the location of cellular spike samples in the raw data which contains key information about cancer. These cellular spike samples are later passed through extreme learning machines and dynamic time warping based classification methods for cancer and non-cancer decisions. Using above approach, we got a classification accuracy of 95.6% with a precision and recall of 85.7% and 80.0 respectively. Our approach can be used in clinical settings where data can be collected from multiple micropores and can be processed quickly for instantaneous responses to help physicians make diagnostic decisions. This can help in early disease detection and prognosis for patients suffering from cancer. In the future work we are investigating on signal enhancement and noise reduction techniques to overcome high variability in cellular spike data.

References

1. Hafeez, A., et al.: GPU-based real-time detection and analysis of biological targets using solid-state nanopores. Med. Biol. Eng. Comput. **50**(6), 605–615 (2012)
2. Huang, Y., et al.: A high-speed real-time nanopore signal detector. In: 2015 IEEE Conference on Computational Intelligence in Bioinformatics and Computational Biology (CIBCB). IEEE (2015)
3. Eren, A.M., et al.: Pattern recognition-informed feedback for nanopore detector cheminformatics. Advances in Computational Biology, pp. 99–108. Springer, New York (2010)
4. Zhu, X., Ghahramani, Z., Lafferty, J.: Semi-supervised learning using gaussian fields and harmonic functions. In: ICML, vol. 3 (2003)
5. Stormo, G.D., et al.: Use of the Perceptronalgorithm to distinguish translational initiation sites in E.coli. Nucleic Acids Res. **10**(9), 2997–3011 (1982)
6. Weston, J., et al.: Semi-supervised protein classification using cluster kernels. Bioinformatics **21**(15), 3241–3247 (2005)
7. Tarca, A.L., et al.: Machine learning and its applications to biology. PLoS Comput. Biol. **3**(6), e116 (2007)

8. Ben-Hur, A., et al.: Support vector machines and kernels for computational biology. PLoS Comput. Biol. **4**(10), e1000173 (2008)
9. Ye, Q.-H., et al.: Predicting hepatitis B viruspositive metastatic hepatocellular carcinomas using gene expression profiling and supervised machine learning. Nat. Med. **9**(4), 416–423 (2003)
10. Tan, A.C., Gilbert, D.: Ensemble machine learning on gene expression data for cancer classification (2003)
11. Alizadeh, A.A., et al.: Distinct types of diffuse large B-cell lymphoma identified by gene expression profiling. Nature **403**(6769), 503–511 (2000)
12. Shipp, M.A., et al.: Diffuse large B-cell lymphoma outcome prediction by gene-expression profiling and supervised machine learning. Nat. Med. **8**(1), 68–74 (2002)
13. Du, P., Kibbe, W.A., Lin, S.M.: Improved peak detection in mass spectrum by incorporating continuous wavelet transform-based pattern matching. Bioinformatics **22**(17), 2059–2065 (2006)
14. Zong, W., et al.: An open-source algorithm to detect onset of arterial blood pressure pulses. In: Computers in Cardiology. IEEE (2003)
15. Kohler, B.-U., Hennig, C., Orglmeister, R.: The principles of software QRS detection. IEEE Eng. Med. Biol. Mag. **21**(1), 42–57 (2002)
16. Harrell, C.C., et al.: Resistive-pulse DNA detection with a conical nanopore sensor. Langmuir **22**(25), 10837–10843 (2006)
17. Palshikar, G.: Simple algorithms for peak detection in time-series. In: Proceedings of 1st International Conference Advanced Data Analysis, Business Analytics and Intelligence (2009)
18. Lamel, L., et al.: An improved endpoint detector for isolated word recognition. IEEE Trans. Acoust. Speech Signal Proces. **29**(4), 777–785 (1981)
19. Lekka, M., et al.: Elasticity of normal and cancerous human bladder cells studied by scanning force microscopy. Eur. Biophys. J. **28**(4), 312–316 (1999)
20. Ling, X.S.: Addressable nanopores and micropores including methods for making and using same. US Patent 7,678,562, 16 Mar 2010
21. Yang, C., He, Z., Yu, W.: Comparison of public peak detection algorithms for MALDI mass spectrometry data analysis. BMC Bioinform. **10**(1), 4 (2009)
22. Azzini, I., et al.: Simple methods for peak detection in time series microarray data. In: Proceedings of CAMDA04 (Critical Assessment of Microarray Data) (2004)
23. Kleinberg, J.: Bursty and hierarchical structure in streams. Data Min. Knowl. Disc. **7**(4), 373–397 (2003)
24. Pan, J., Tompkins, W.J.: A real-time QRS detection algorithm. IEEE Trans. Biomed. Eng. **3**, 230–236 (1985)
25. Todd, B.S., Andrews, D.C.: The identification of peaks in physiological signals. Comput. Biomed. Res. **32**(4), 322–335 (1999)
26. Iqbal, S.M., Akin, D., Bashir, R.: Solid-state nanopore channels with DNA selectivity. Nature nanotechnol. **2**(4), 243–248 (2007)
27. Asghar, W., et al.: Electrical fingerprinting, 3D profiling and detection of tumor cells with solid-state micropores. Lab Chip **12**(13), 2345–2352 (2012)
28. Lee, G.Y.H., Lim, C.T.: Biomechanics approaches to studying human diseases. Trends Biotechnol. **25**(3), 111–118 (2007)
29. Brandao, M.M., et al.: Optical tweezers for measuring red blood cell elasticity: application to the study of drug response in sickle cell disease. Eur. J. Haematol. **70**(4), 207–211 (2003)
30. Evans, E.A., La Celle, P.L.: Intrinsic material properties of the erythrocyte membrane indicated by mechanical analysis of deformation. Blood **45**(1), 29–43 (1975)
31. Vona, G., et al.: Isolation by size of epithelial tumor cells: a new method for the immunomorphological and molecular characterization of circulating tumor cells. Am. J. Pathol. **156**(1), 57–63 (2000)
32. Zabaglo, L., et al.: Cell filtrationlaser scanning cytometry for the characterisation of circulating breast cancer cells. Cytometry Part A **55**(2), 102–108 (2003)

33. Cross, S.E., et al.: Nanomechanical analysis of cells from cancer patients. Nat. Nanotechnol. **2**(12), 780–783 (2007)
34. Hafeez, A., Rafique, M.M., Butt, A.R.: Distributed detection of cancer cells in high-throughput cellular spike streams. In: 14th IEEE/ACM International Symposium on Cluster, Cloud and Grid Computing, pp. 774–783 (2014)
35. Hanif, M., Hafeez, A., Suleman, Y., Rafique, M.M., Butt, A.R., Iqbal, S.M.: An accelerated framework for the classification of biological targets from solid-state micropore data. Comput. Methods Programs Biomed. (2016)
36. Ding, S., Xu, X., Nie, R.: Extreme learning machine and its applications. Neural Comput. Appl. **25**(3–4), 549–556 (2014)
37. Huang, G., et al.: Semi-supervised and unsupervised extreme learning machines. IEEE Trans. Cybern. **44**(12), 2405–2417 (2014)
38. Huang, G.-B.: What are extreme learning machines? Filling the gap between Frank Rosenblatts dream and John von Neumanns puzzle. Cogn. Comput. **7**(3), 263–278 (2015)
39. Huang, G., et al.: Trends in extreme learning machines: a review. Neural Netw. **61**, 32–48 (2015)
40. Mohapatra, P., Chakravarty, S., Dash, P.K.: An improved cuckoo search based extreme learning machine for medical data classification. Swarm Evol. Comput. **24**, 25–49 (2015)
41. You, Z.-H., et al.: Prediction of protein-protein interactions from amino acid sequences with ensemble extreme learning machines and principal component analysis. BMC Bioinform. **14**(8), 1 (2013)
42. Teo, T.T., Logenthiran, T., Woo, W.L.: Forecasting of photovoltaic power using extreme learning machine. In: 2015 IEEE Innovative Smart Grid Technologies-Asia (ISGT ASIA). IEEE (2015)
43. Samat, A., et al.: Ensemble extreme learning machines for hyperspectral image classification. IEEE J. Sel. Top. Appl. Earth Obs. Remote Sens. **7**(4), 1060–1069 (2014)
44. Vizza, P., et al.: A framework for the atrial fibrillation prediction in electrophysiological studies. Comput. Methods Programs Biomed. **120**(2), 65–76 (2015)
45. Berndt, D.J., Clifford, J.: Using dynamic time warping to find patterns in time series. KDD workshop **10**(16), 359–370 (1994)
46. Salvador, S., Chan, P.: Toward accurate dynamic time warping in linear time and space. Intell. Data Anal. **11**(5), 561–580 (2007)
47. National center for biotechnology information. http://www.ncbi.nlm.nih.gov

Hot News Click Rate Prediction Based on Extreme Learning Machine and Grey Verhulst Model

Xu Jingting, Feng Jun, Sun Xia, Zhang Lei and Liu Xiaoning

Abstract Click rate prediction of hot topics contributes to get event tendency, especially for sensitive news. However, click rate prediction is challenge due to inherent features of short-time series such as randomness, uncertainty, volatility and insufficiency of training samples. In this paper, a new hybrid click rate prediction method called Grey Verhulst—Extreme Learning Machine (GVELM) is proposed. Specifically, the raw short-time series data are filled into GV models to acquire stably initial prediction which have incorporated regular pattern of the historic data without noise. Then ELM is employed for prediction refinement for nonlinear space mapping. The experimental results show that the proposed method achieves better prediction accuracy compared with other five state-of-art algorithms.

Keywords Short-term time series prediction · Extreme learning machine (ELM) · Grey verhulst model (GV) · GVELM model

---

X. Jingting · F. Jun (✉) · S. Xia · Z. Lei · L. Xiaoning
School of Information Science and Technology,
Northwest University, Xi'an 710069, China
e-mail: fengjun@nwu.edu.cn

X. Jingting
e-mail: jingtingxu03@gmail.com

S. Xia
e-mail: raindy@nwu.edu.cn

Z. Lei
e-mail: zhlei@nwu.edu.cn

L. Xiaoning
e-mail: xnliu@nwu.edu.cn

J. Cao et al. (eds.), *Proceedings of ELM-2016*, Proceedings in Adaptation, Learning and Optimization 9, DOI 10.1007/978-3-319-57421-9_8

1 Introduction

The Internet provides more free communication platforms for users to discuss all kinds of news with each other. Nowadays, hot news on the Internet can easily attracts the public attentions, which may spreads very fast. However, propagation of some kinds of news such as purported and sensitive topics are detrimental to the society. Media study shows that click rates reflect the hot news topics tendency and guide public opinion in some way. In the other hand, click rates can also help news service providers put the hot news into the most conspicuous place. Therefore, accurate and efficient click rates prediction has vital practical significance for both media servers and governments.

Click rate prediction can be considered as short-term time series analysis due to the characteristics of a short-lived cycle of the hot news from beginning, evolution to termination. Hot news click rate data are generally with the features of randomness, uncertainty and volatility, which are typical non-linear problem that makes traditional short-term time series prediction algorithms failed. In this paper, we propose a novel click rate prediction algorithm for hot news based on the combination of Extreme Learning Machine (ELM) [1] and Grey Verhulst model (GV) [2].

Recently, the research on short-term prediction model mainly cast to statistical or machine learning framework. Autoregressive Integrated Moving Average (ARIMA) [3] is one of the typical methods of statistical model, which has been proven especially useful within time series analysis. It provides an effects of dependency from the data series and allows valid statistical testing, but its computational time is very high. Grey Verhulst model (GV [2]) can present short-term time series samples more simply. Wang et al. [4] applies GV model to predict news click rates, however, the prediction rates are very poor. Actually, GV is intrinsically a linear model, which is hard to extend to the nonlinear mapping underlying the dynamic process.

Meanwhile, the framework of machine learning has drawn more attention due to the ability to recognize nonlinear series. Some learning algorithms such as Back Propagation neural network (BP), Elman Recurrent neural network (ER) and Mixture Density neural network (MD) have been applied to short-term time series prediction learning [5–7]. In order to obtain the optimum structure of neural network, many different adjustable parameters should be examined which is a time consuming and boring task. More importantly, major criticism lies in the fact that it requires a great deal of training data and relatively long training period for robust generalization. Besides, Support Vector Machine (SVM) can be mentioned as outstanding learning models for click rates prediction [8]. However, it trends to suboptimal values [9].

In 2006, Huang [1] presented Extreme Learning Machines (ELM) which has broad capability on approximation of non-linear function. Since ELM requires only a single-pass training stage without any iteration for weights adjustment, learning process is very fast. The number of ELM hidden neurons can be easily determined based on dimensions of input and output vectors. Later, Liang et al. [10] proposed an online sequential ELM (OS-ELM) algorithm for time series prediction. However, this is liable to be trapped in matrix singularity and illposedness, which is not suitable for short-time series analysis.

This paper proposes a new hybrid click rate prediction method called Grey Verhulst—Extreme learning machine (GVELM). In our method, GV is employed to create the initial prediction outputs with n historic data. Then, both the historical data and initial prediction outputs from GV are combined for ELM training. The experimental results show that the proposed method achieves better prediction accuracy compared with other five state-of-art algorithms. Furthermore, the execution time of the proposed method is very low because ELM as a powerful regression tool has extremely fast learning speed without local minimal issues. To the best of our knowledge, it is the first attempt to combine Grey Verhulst with Extreme Learning Machine, and furthermore, for solving short-term time series prediction problem.

2 Click Rate Prediction Based on GVELM

In this section, we describe our click rate prediction solution based on the proposed GVELM model in detail. Since the short-time series of click rates are equal-length, we suppose each click rate time series sample contains m historic data $\{\mathbf{x}_i\}_{i=1}^{m}$ and $n-m$ prediction data $\{\mathbf{x}_i\}_{i=m+1}^{n}$.

From GV model, the next time point $\widehat{x_{k+1}}$ can be roughly predicted by [11].

$$\widehat{x_{k+1}} = \frac{ax_1}{bx_1 + (a - bx_1)e^{ak}}, k = 2, 3, \ldots, n \tag{1}$$

where x_1 is the first time point, a is the awaiting identification parameter and b is grey actuating quantity. **In GV model, all of the prior knowledge of historic data has been incorporated into parameter** a **and** b. Suppose variable $\widehat{a}$ is $\widehat{a} = [a, b]^T$, which can be calculated by

$$\widehat{a} = (B^T B)^{-1} B^T Y \tag{2}$$

$$B = \begin{pmatrix} -z(2) & z^2(2) \\ -z(3) & z^2(3) \\ \vdots & \vdots \\ -z(m) & z^2(m) \end{pmatrix} \tag{3}$$

$$z(k) = 0.5x(k) + 0.5x(k-1) \tag{4}$$

$$Y = \begin{bmatrix} x(2) \\ x(3) \\ \vdots \\ x(4) \end{bmatrix} \tag{5}$$

According to Eq. (1), we can get all of prediction values of $\{\widehat{\mathbf{x}}_{\mathbf{i}}\}_{i=1}^{n}$. Although the prediction value of $\widehat{\mathbf{x}}$ may has deviation from the historical data $\mathbf{x}$, i.e. the true click rate data, $\widehat{\mathbf{x}}$ **does inherit the inherent laws of the historical data based on GV model, and successfully gets rid of the random noise which is very frequently occured in short-time series.**

After got the prediction values of $\mathbf{x}$, ELM is employed as the regression core for prediction. In essence, we tactfully utilize ELM to improve the prediction accuracy by its robust nonlinear mapping capability. Specifically, ELM regression training phase, k **time-point interval prediction values as input** and $k + 1$ historic data as output are filled into ELM for model training. The structure of GVELM model is depicted in Fig. 1.

Furthermore, we describe the algorithm of GVELM based click rate prediction.

Step1: Historic click rate data $\mathbf{x}$ as GV model input.
Step2: For GV model prediction phase, $i = 1 : m$; return a and b which are calculated by Eqs. (2)–(5).
Step3: Get GV predicted values $\{\widehat{\mathbf{x}}_{\mathbf{i}}\}_{i=1}^{n}$ using Eq. (1).
Step4: In ELM training phase, GV's predicted value $\{\widehat{x_i}, \widehat{x_{i+1}}, \widehat{x_{i+2}}\}$, $i = 1, 2, \ldots, m - 3$ are inputs and historic click rate data $\{x_{i+3}\}$ is output.
Step5: In ELM test phase, $\{\widehat{x_i}, \widehat{x_{i+1}}, \widehat{x_{i+2}}\}, i = m-2, m-1, \ldots, m-3+n$ are inputs and $\{\overset{ELM}{x_{i+3}}\}$ is predicted output.

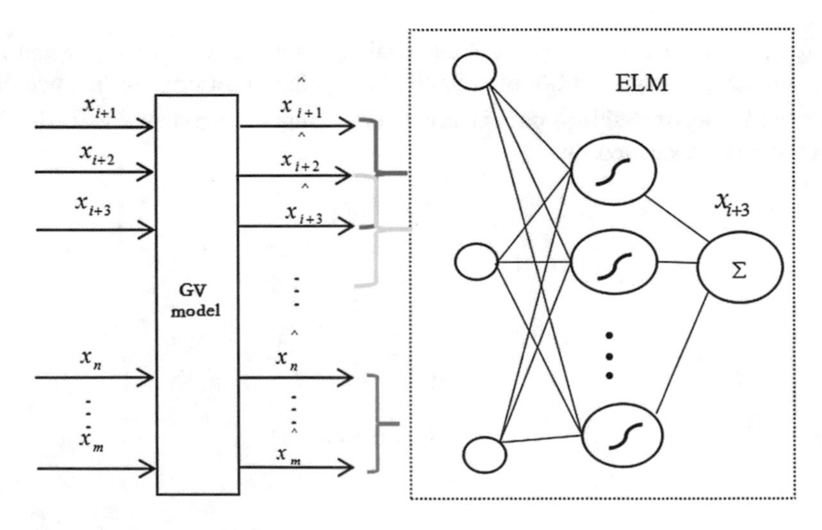

Fig. 1 The pipeline of GVELM

3 Experimental Results

In order to verify the accuracy and robust of the proposed algorithm, several experiments of hot news click rate prediction are conducted compared with GV [4], ES, ARIMA, Verhulst-BP [12] and Verhulst-SVM [13].

We use state-of-art open datasets from [8]. Eight topics are tested, and their titles are listed as following:

(1) "North Korea launches a satellite" (NKLS),
(2) "US. Fiscal Cliff Solutions" (USFCS),
(3) "Messi gets the Golden Globe" (MGG),
(4) "Zimbabwe's President Blasting the US and the UK" (ZPB),
(5) "US government shutdown" (USGS),
(6) "Cultural Relics stolen by US army to be Return to South Korea" (CRRSK),
(7) "A Robber Wounded by his Gun off accidentally in Peru (RWGP)",
(8) "University Dormitory Attacked in Nigerian (UDAN)".

In our experiments, the click rates during 6 h are recorded. Table 1 shows a segmentation of time points of these eight topics. In the following experiments, $n = 18$, $m = 14$. After trail-and-error, k is set to 3.

In this paper, three indicators are applied to evaluate prediction Accuracy, i.e. mean squared error (MSE), mean absolute percentage error (MAPE) and mean absolute error (MAE). These measures are defined as follows. Thus $\varepsilon^{(1)}$ is said to be the residual error series, satisfied that $\varepsilon(k) = \widehat{x(k)} - x(k)$, $k = 1, 2, \ldots, n$. The relative residual error or Relative Error (RE) satisfies that $\Delta_k = \left|\frac{\varepsilon^{(1)}(k)}{x^{(1)}(k)}\right|$, $k = 1, 2, \ldots, n$. So,

- $MAPE = \frac{1}{n}\sum_{k=1}^{n}\Delta_k$, $MAE = \frac{1}{n}\sum_{k=1}^{n}\varepsilon^{(1)}(k)$, $MSE = \frac{1}{n}\sum_{k=1}^{n}(\varepsilon^{(1)}(k))^2$.

We compare our GVELM with five state-of-art algorithms: Grey Verhulst (GV) [4], Exponential Smoothing (ES), ARIMA, Verhulst-BP [12], and Verhulst-SVM [13]. Tables 2, 3, 4, 5, 6, 7, 8 and 9 list the comparison results of 8 news topics. It should be noted that optimal value of ES and ARIMA is showed in expert model.

From Tables 2, 3, 4, 5, 6, 7, 8 and 9, we can find that the proposed model achieves better prediction accuracy than other models. It is also worth to notice that the range of the prediction error by Verhulst—BP is worse due to BP always need a large number of training data, however hot topic is the nature of the short life cycle with sparse samples. Verhulst—SVM is not so good because it is likely to fall into suboptimal values. At the same time, as the linear processing Verhulst cannot represent the time-varying nonlinear dynamics underlying hotness prediction precisely. From the above experiments, we can conclude that proposed algorithm can better represent dynamics short-time series characterized by nonlinearity and short-term time series samples.

Table 1 Original data of 8 topics

Time series	NKLS	USFCS	MGG	ZPB	USGS	CRRSK	RWGP	UDAN
1	310,296	6669	9481	564	1652	208	725	17,861
2	1001,373	13,733	26,621	854	3419	319	1254	36,438
3	1665,051	18,385	63,793	2157	6416	654	1997	58,546
4	1744,956	25,813	116,006	5649	10,263	1256	2854	85,424
5	1875,550	30,509	157,578	10,503	15,880	2072	3795	125,713
6	1920,620	33,736	178,157	22,952	23,535	3219	4574	182,665
7	1965,291	34,579	185,985	45,247	33,271	4518	5659	250,164
8	1971,032	35,913	188,658	70,245	41,069	5524	6420	316,510
9	2016,361	36,289	189,535	96,789	46,867	6256	6905	367,556
10	2084,417	36,683	189,820	105,614	49,597	6785	7344	403,874
11	2126,428	36,819	189,912	115,423	51,761	7121	7675	448,563
12	2132,204	37,072	189,942	122,548	53,792	7302	7835	468,257
13	2159,912	37,127	189,951	128,245	54,514	7569	8031	476,552
14	2195,116	37,289	189,954	131,598	55,246	7643	8198	481,276
15	2218,435	37,359	189,955	134,672	55,895	7701	8235	487,764
16	2222,316	37,397	189,955	134,964	56,386	7756	8307	496,725
17	2224,009	37,412	189,956	135,447	56,774	7798	8384	499,141
18	2225,109	37,498	189,956	135,910	60,183	7825	8421	512,766

Table 2 MAPE, MAE and MSE in NKLS news compared with 5 state-of-art algorithms (Superscript represents the ranking)

	Topic 1—NKLS	MAPE (%)	MAE (10^4%)	MSE (10^6%)
Model 1	Grey Verhulst	$6.67^{(5)}$	$14.81^{(5)}$	$21952.08^{(5)}$
Model 2 and Model 3's optimal value	Expert model (ES\ARIMA)	$2.72^{(2)}$	$6.04^{(2)}$	$5003.58^{(2)}$
Model 3	Verhulst—BP	$4.60^{(4)}$	$10.21^{(4)}$	$10539.53^{(4)}$
Model 4	Verhulst—SVM	$3.99^{(3)}$	$8.86^{(3)}$	$7965.75^{(3)}$
Model 5	GVELM	$2.71^{(1)}$	$5.99^{(1)}$	$3733.85^{(1)}$

Table 3 MAPE, MAE and MSE in USFCS news compared with 5 state-of-art algorithms

	Topic 2—USFCS	MAPE	MAE	MSE
Model 1	Grey Verhulst	$1.26^{(4)}$	$0.05^{(4)}$	$0.23^{(4)}$
Model 2 and Model 3's optimal value	Expert model (ES\ARIMA)	$0.70^{(2)}$	$0.02^{(2)}$	$0.08^{(2)}$
Model 3	Verhulst—BP	$1.57^{(5)}$	$0.06^{(5)}$	$0.35^{(5)}$
Model 4	Verhulst—SVM	$0.89^{(3)}$	$0.03^{(3)}$	$0.1^{(3)}$
Model 5	GVELM	$0.27^{(1)}$	$0.01^{(1)}$	$0.02^{(1)}$

Table 4 MAPE, MAE and MSE in MGG news compared with 5 state-of-art algorithms

	Topic 3—MGG	MAPE	MAE	MSE
Model 1	Grey Verhulst	1.66(5)	0.32(5)	10.29(5)
Model 2 and Model 3's optimal value	Expert model (ES\ARIMA)	0.09(2)	0.02(2)	0.032(2)
Model 3	Verhulst—BP	0.12(3)	0.02(3)	0.59(3)
Model 4	Verhulst—SVM	0.71(4)	0.14(4)	1.88(4)
Model 5	GVELM	0.06(1)	0.01(1)	0.02(1)

Table 5 MAPE, MAE and MSE in ZPB news compared with 5 state-of-art algorithms

	Topic 4—ZPB	MAPE	MAE	MSE
Model 1	Grey Verhulst	4.15(5)	0.56(4)	31.57(4)
Model 2 and Model 3's optimal value	Expert model (ES\ARIMA)	3.49(4)	0.47(3)	33.15(5)
Model 3	Verhulst—BP	2.34(2)	0.32(2)	13.77(2)
Model 4	Verhulst—SVM	2.78(3)	0.38(3)	14.35(3)
Model 5	GVELM	0.72(1)	0.09(1)	0.99(1)

Table 6 MAPE, MAE and MSE in USGS news compared with 5 state-of-art algorithms

	Topic 5—USGS	MAPE	MAE	MSE
Model 1	Grey Verhulst	1.74(4)	0.05(4)	0.23(4)
Model 2 and Model 3's optimal value	Expert model (ES\ARIMA)	0.87(3)	0.02(3)	0.08(3)
Model 3	Verhulst—BP	3.559(5)	0.0976(5)	0.96(5)
Model 4	Verhulst—SVM	0.36(1)	0.01(1)	0.01(1)
Model 5	GVELM	**0.37(2)**	**0.01(1)**	**0.02(2)**

Table 7 MAPE, MAE and MSE in CRRSK news compared with 5 state-of-art algorithms

	Topic 6—CRRSK	MAPE	MAE	MSE
Model 1	Grey Verhulst	2.15(3)	0.017(3)	0.029(3)
Model 2 and Model 3's optimal value	Expert model (ES\ARIMA)	5.41(5)	0.042(5)	0.238(5)
Model 3	Verhulst—BP	4.63(4)	0.035(4)	0.125(4)
Model 4	Verhulst—SVM	1.81(2)	0.014(2)	0.021(2)
Model 5	GVELM	**1.54(1)**	**0.011(1)**	**0.017(1)**

Table 8 MAPE, MAE and MSE in RWGP news compared with 5 state-of-art algorithms

	Topic 7—RNGP	MAPE	MAE	MSE
Model 1	Grey Verhulst	1.69$^{(5)}$	0.0146$^{(2)}$	0.022$^{(2)}$
Model 2 and Model 3's optimal value	Expert model (ES\ARIMA)	3.34$^{(3)}$	0.0279$^{(4)}$	0.092$^{(4)}$
Model 3	Verhulst—BP	4.22$^{(4)}$	0.0352$^{(5)}$	0.129$^{(5)}$
Model 4	Verhulst—SVM	1.83$^{(2)}$	0.0153$^{(3)}$	0.026$^{(3)}$
Model 5	GVELM	**1.37$^{(1)}$**	**0.0115$^{(1)}$**	**0.019$^{(1)}$**

Table 9 MAPE, MAE and MSE in UDAN news compared with 5 state-of-art algorithms

	Topic 8—UDAN	MAPE	MAE	MSE
Model 1	Grey Verhulst	1.70$^{(3)}$	0.86$^{(3)}$	118.76$^{(3)}$
Model 2 and Model 3's optimal value	Expert model (ES\ARIMA)	1.19$^{(2)}$	0.60$_{(2)}$	52.84$^{(2)}$
Model 3	Verhulst—BP	5.68$^{(5)}$	2.52$_{(5)}$	710.63$^{(5)}$
Model 4	Verhulst—SVM	3.74$^{(4)}$	1.88$^{(4)}$	435.36$^{(4)}$
Model 5	GVELM	**1.07$^{(1)}$**	**0.56$^{(1)}$**	**35.66$^{(1)}$**

4 Conclusion

Specific to hot news click rate short-term time series prediction, this paper proposes GVELM model. In the first step, GV model is employed to create the initial prediction outputs with n historic data. The predicted values are considered as prior knowledge that inherit the inherent laws of the historical data based on GV model, and successfully gets rid of the random noise which is very frequently occured in short-time series. The next step is tactfully utilizes ELM to further minimize the prediction error. Besides, a novel combined strategy is also a research highlight that it is the first attempt to combine GV model with ELM.

Acknowledgements This work was supported by NSFc 61202184, 61305032 and Scientific research plan projects 2015JQ6240.

References

1. Huang, G.B., Zhu, Q.Y., Siew, C.K.: Extreme learning machine: theory and applications. Neurocomputing **70**, 489–501 (2006)
2. Deng, J.L.: Control problems of grey systems [J]. Syst. Control Lett. **1**(5), 288–294 (1982)
3. Lee, Y.S., Tong, L.I.: Forecasting time series using a methodology based on autoregressive integrated moving average and genetic programming [J]. Knowl.-Based Syst. **24**(1), 66–72 (2011)

4. Wang, X., Qi, L., Chen, C., et al.: Grey system theory based prediction for topic trend on internet [J]. Eng. Appl. Artif. Intell. **29**(3), 191–200 (2014)
5. Liu, Y.K., Xie, F., Xie, C.L., et al.: Prediction of time series of NPP operating parameters using dynamic model based on BP neural network [J]. Ann. Nucl. Energ. **85**, 566–575 (2015)
6. Chandra, R.: Competition and collaboration in cooperative coevolution of elman recurrent neural networks for time-series prediction [J]. IEEE Trans. Neural Netw. Learn. Syst. **26**(12), 1 (2015)
7. Men, Z., Yee, E., Lien, F.S., Wen, D., Chen, Y.: Short-term wind speed and power forecasting using an ensemble of mixture density neural networks. Renew. Energ. **87**, 203–211 (2016)
8. Rubio, G., Pomares, H., Rojas, I., et al.: A heuristic method for parameter selection in LS-SVM: application to time series prediction [J]. Int. J. Forecast. **27**(3), 725–739 (2011)
9. Huang, G.B.: An insight into extreme learning machines: random neurons, random features and kernels [J]. Cogn. Comput. **6**(3), 376–390 (2014)
10. Liang, N.Y., Huang, G.B., Saratchandran, P., et al.: A fast and accurate online sequential learning algorithm for feedforward networks [J]. IEEE Trans. Neural Netw. **17**(6), 1411–1423 (2006)
11. Gardner, E.S.: Exponential smoothing: the state of the art [J]. J. Forecast. **4**(1), 1–28 (1985)
12. Zhou, D.: A new hybrid grey neural network based on grey verhulst model and BP neural network for time series forecasting [J]. Int. J. Inf. Technol. Comput. Sci. **5**(10), 114–120 (2013)
13. Zhao, W., Wang, F., Niu, D.: The application of support vector machine in load forecasting [J]. J. Comput. **7**(7) (2012)

Multiple Shadows Layered Cooperative Velocity Updating Particle Swarm Optimization

Hongbo Wang, Kezhen Wang and Xuyan Tu

Abstract In real-time high dimensions optimization problem, how to quickly find the optimal solution and give timely response or decisive adjustment is very important. Inspired by space projection behavior, this paper suggests a new PSO variant, Multiple-shadows Layered Cooperative Velocity Updating Particle Swarm Optimization (ML-CVUPSO) that involves visual instructive projections among multiple shadows. According to several different views, the original problem can be divided into different relevant characteristic sub-problems after feature extraction. The ML-CVUPSO provides a flexible and feasible decomposed mechanism to simplify the high dimensions problem into a series of tractable sub-problems. The proposed variant is examined on several widely used benchmark functions, and the experimental results show that the proposed ML-CVUPSO algorithm improves the existing performance of other algorithms when dealing with the high dimension and multimodal problems.

Keywords PSO · Multiple-shadows layered · Decomposable mechanism

1 Introduction

Particle swarm optimization (PSO) is a relatively new heuristic algorithm which was originally proposed by Kennedy and Eberhart in the mid-1990s [1]. It is one kind of evolutionary algorithm and inspired by the concerted actions of flocks among birds for food in a cooperative way. As an important branch of swarm intelligence, PSO has attracted public attention from many research areas or

H. Wang (✉) · K. Wang · X. Tu
Department of Computer Science and Technology, School of Computer and Communication Engineering, University of Science and Technology Beijing, Beijing, People's Republic of China
e-mail: foreverwhb@ustb.edu.cn; foreverwhb@126.com

J. Cao et al. (eds.), *Proceedings of ELM-2016*, Proceedings in Adaptation, Learning and Optimization 9, DOI 10.1007/978-3-319-57421-9_9

communities, and it has been implemented in various scientific and engineering applications, such as web marketing content [2], traveling salesman problem [3], evaluating the collective user feedback [4], designing large-scale passive harmonic filters [5]. In recent years many variants of PSO have been proposed and can be attributed to two directions: (1) expanding the searching scope [6, 7] and (2) reducing the computational complexity [8, 14–23].

The remainder of this paper is organized in the following. Section 2 systematically sets forth the novel ML-CVUPSO variant from multiple perspectives. Section 3 makes comparative experiments on some well-known benchmark functions in CEC2015 and analyzes the related experimental results. Conclusions are made in Sect. 4.

2 Multiple Shadows Layered CVUPSO

Multiple Shadows Layered Cooperative Velocity Updating Particle Swarm Optimization algorithm (ML-CVUPSO), which provides a layered mechanism to simplify the high dimensions and divided the problem into several low dimensions sub-problems to decrease the number of operation iterations and save the operating time. It uses Cooperative Velocity Updating Particle Swarm Optimization [9], which is an improved PSO variant and good at solving complex problems.

2.1 *The Shadows Layered Mechanism*

Assuming a target is a D dimensions problem, the optimization function is $f(x)$, where $x = [x_1, x_2, x_3 \ldots x_D]$ indicates it's all the possible solutions. The progress of finding the optimal feasible solution x^* can be also described as choosing one position which makes the value of the function f smallest or biggest in the space R^D. As to all the PSO variants, it is hard to find the best solution x^*, what we get is always an acceptable result, especially to the high dimensions problems. For the actual optimal problem, different variables affect the final results with different weight and the all variables can be shadowed into different view-planes. Usually the number of the planes should be in a range and the detail value of G is shown in Formula (1). If the G is too big, the complexity may be increased and vice versa, if the G is too small, the effect will be not evident.

$$3 \leq G \leq \lfloor log_2^D \rfloor \tag{1}$$

When G equals to 3, the original optimal result x will be divided to three shadows in Formula (2), namely.

$$
\begin{aligned}
x_A &= [x_1, x_2, \ldots, x_A], \\
x_B &= [x_{A+1}, x_{A+2}, \ldots, x_B], \\
x_C &= [x_{B+1}, x_{B+2}, \ldots, x_D], \\
x &= x_A + x_B + x_c
\end{aligned}
\tag{2}
$$

To explore the optimum equals to pursue of respectively. In the process of iterations, whether the results are better or not should be judged by the optimal function, which involves some inherent questions with a D dimensions solution space. x_A, x_B, x_C are assumed as three projection vectors, when the function f is observed from plane x_A, f_A is a reflection of f and $x_A^{sub} = [x_1, x_2, \ldots, x_B, 0, 0, \ldots 0]$ becomes the related subset of x. Also if f is observed from plane x_B or x_C, $x_B^{sub} = [0, 0, \ldots, x_{A+1}, x_{A+2}, \ldots, x_C, 0, 0, \ldots 0]$ or $x_C^{sub} = [0, 0, \ldots, x_{B+1}, x_{B+2}, \ldots, x_D]$ will be related subsets of x, its fitness function becomes f_B and f_C, respectively. The process is shown as Fig. 1.

The original problem has three projections (shadows) plane x_A, x_B, x_C, one original is divided into three characteristic sub-problems after feature extraction, the relationship between the original problem and its sub-problems is shown in

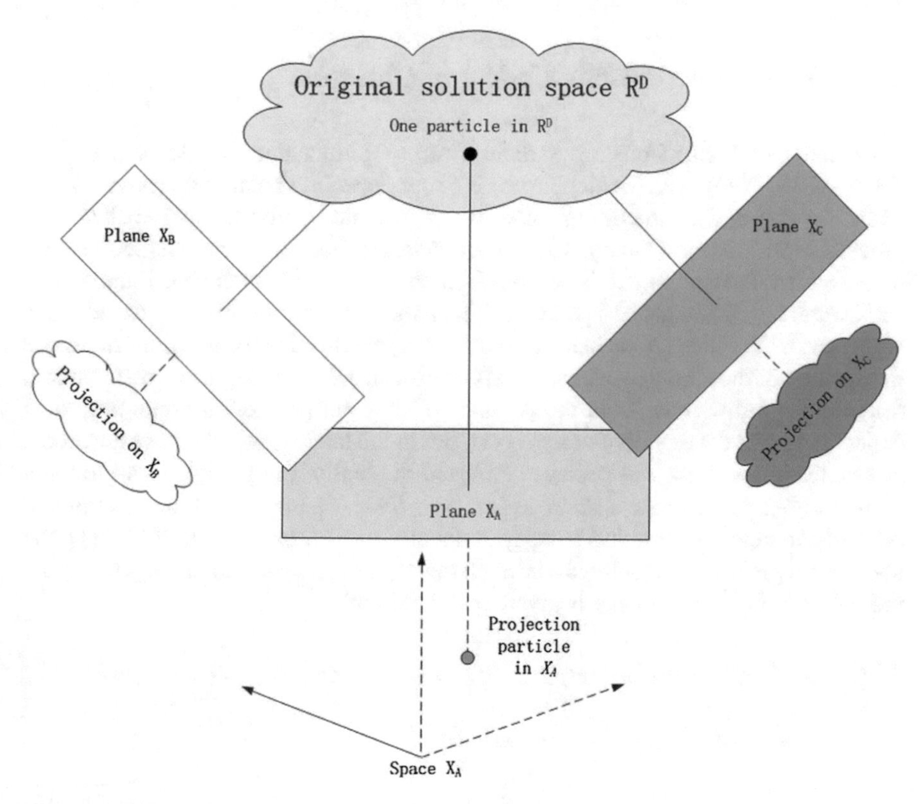

Fig. 1 High dimensional function projection diagram

Formula (3), α, β, γ are numerical (or vector) constant coefficients. There are two kinds of situations:

Situation 1: if these sub-problems are independent, the original problem is equal to the sum of sub-problems, the value of the α, β, γ would be set to 1.0. The layered mechanism makes it easily to handle the problem of high dimension and multi-attribute complex data.

Situation 2: if the sub-problems are not independent, the original problem is not equal to the sum of sub-problems; however, if the optimum of sub-problem for example f_A is appropriated between the dimensions 1 and A of problem f, so the combination of sub-problems is also a reflection of the original problem, especially the subgroup is divided by the weight of every dimension. The value of the α, β, γ will be generated by the s-shaped functiorespectively.

$$\begin{aligned} f(x) &\approx \alpha f_A + \beta f_B + \gamma f_C \\ f_A(x_A) &= f(x_A^{sub}) \\ f_B(x_B) &= f(x_B^{sub}) \\ f_C(x_C) &= f(x_C^{sub}) \end{aligned} \tag{3}$$

2.2 Cooperative Velocity Updating Algorithm

Cooperative Velocity Updating Particle Swarm Optimization is a developed algorithm (CVUPSO) [9], which records four special positions ($pbest$, $lbest$, $gbest$, $lworst$) to describe or remember where it is along with its own evolutionary process. Considering a swarm with m sub-swarms, personal best position ($pbest$) represents the current particle's best position, the best position which is found in this sub-swarm is called local best position ($lbest$), the best position which is found in the whole swarm is called global best position ($gbest$) and local worst position ($lworst$) is the position of the particle with the worst performance. In the early stage of evolution, the particles know little about the population, they get progress through their own experiences and the direction of local best, but in the later stage, all sub-swarm come closer, the global best will become stabilized gradually, the particles tend to learn more information from the $gbest$ rather than $lbest$. Using such four positions, a cooperative velocity updating strategy is developed. Compared with SPSO [1], the updating approach of velocity has been changed, for each particle in the population; the velocity updating strategy is given by formula (4):

$$\begin{aligned} V_i^{k+1} = {} & w \times V_i^k + c_1 \times random_1 \times (pbest_i - X_i^k) + c_2 \times random_2 \times \left(1 - \frac{k}{iter_{\max}}\right)(lbest - X_i^k) \\ & + c_2 \times random_3 \times \frac{k}{iter_{\max}} \times (gbest - X_i^k) \end{aligned} \tag{4}$$

where V_i^k is the current velocity of particle i at iteration k, w is the iteration weight, c_1 and c_2 are the acceleration coefficients, $random_1$ and $random_2$ are between 0 and 1 at random. X_i^k denotes the current position of particle i at iteration k,$pbest_i$ is the best value of particle i, $lbest_i$ is the best evaluation among $j-th$ sub-swarm, $gbest$ representatives the best value among the whole swarm, $lbest$ is the worst. The updating strategy of inertia weight is between 0 and 1 at random.

2.3 The Procedures of ML-CVUPSO

In ML-CVUPSO, the main steps are the decomposing and combining operation, how to divide lot of attributes of the problem into suitable groups is a heart of the matter, the basic process of the ML-CVUPSO is shown below Table 1.

Table 1 Pseudo code of the ML-CVUPSO

```
1   Begin
2     Set Parameters D, size of the swarm S and w, r1, r2, c1, c2
3     Initialize the position P(S,D), the velocity V(S,D)
4     Decide the value of the group number G
5     Divide the original problem into G shadow sub-problems
6     Define the function of the sub-problems FA,FB,...FG
7     For problem Fi from FA to FG
8        Redefine the position Pi and velocity Vi
9        Set the sub-particle swarms nn
10       Set the best particle pbest, gbest, lbest equals Infinity
11       For i=1:max integrations
12         Calculate the Fitness Fiti
13            For each sub-particle swarm 1:nn
14               For each particle 1:S/nn
15                  If fitness is btter than pbest
16                  Update the pbest
17               End
18               Find the best particle PP in sub-particle swarm
19               IF PP is better than the lbest
20               Update lbest
21            End
22         Find the min Fit and the corresponding particle PPP
23            If PPP is good than gbest
24            Update gbest
25         Find the max Fit and the corresponding worst QQQ
26            Throw QQQ
27            Random generate a new particle
28         Velocity_update()
29         Position_update()
30       End
31       Record the results of the sub-problem
32    End
33    Combine all the sub-problems results:
34    Get the final position p = p⃗A + p⃗B + ... + p⃗G
35    Calculate the final fitness F(P)
36  End
```

In Table 1, N dimension is divided into three shadows. The Variable X is the position of particle, Variable V denotes its velocity, Variable G representatives the amount of integration, Variable n is the amount of particles used in sub-problems and Variable nn is the number of sub-swarms. In entire operation process, orthogonal projection resolution method is used by three shadows. In addition, some concise results are obtained by the three injection vectors ($plane_x$ $plane_y$ and $plane_z$) on the subspace, the final optimal solution comes out.

3 Experimentation

A set of 10 well-known benchmark functions has been selected for performance verification of the ML-CVUPSO. The layered mechanism is based on the weight of the different dimension, there are some adjustments to the test functions, giving top 10 dimensions a same weight of 100, intermediate 10 dimensions a same weight of 0.01, and final 10 dimensions a same weight of 10e-6. Then the test function can be divided into three stages: 1–10 dimensions (weighted $plane_x$), 11–20 dimensions (normal weighted $plane_y$) and 21–30 dimensions (less weighted $plane_z$). The CPU time consumed by the variant execution is used to measure its complexity. The following relevant experiments are described briefly, in which the proposed ML-CVUPSO variant is compared with the UPSO [10], SPSO, CLPSO [11], MCPSO [12] and AFPSO [13].

3.1 Experimental Setting and Parameterization

Experimental environment configuration: (1) Operating system: Windows 7; (2) Minimum memory: 1G; (3) Processor Type: Intel Core; (4) Development toolkits: Matlab 7.1. There are some parameters for each particle swarm variant, (1) Population size $\boldsymbol{S}$: the number of particles of all particle swarm algorithm is set to 80; (2) The number and scale of sub-swarm: For multiple-swarm variants (MCPSO, CVUPSO-E, CVUPSO-R), the number of sub-swarm is unified set to 10, which contains 8 particles in each subgroup; (3) Accelerator coefficient: $c_1 = c_2 = 2.05$, $c_3 = 10$; (4) Maximum speed: Set to half of the search range; (5) Maximum iterations are determined by the complexity of the problem; (6) experiments: 30 times.

3.2 Computational Results and Discussion

In each test function, the difference between the theoretical and the actual optimal value is recorded, at the same time the CPU running also is taken down.

Table 2 Mean CPU time of 10 functions in six PSO variants

F#	SPSO	ML-CVUPSO	UPSO	AFPSO	CLPSO	MCPSO
1	**1.485136E + 00**	2.854772E + 00	2.702219E + 00	**1.420709E + 00**	8.614554E + 01	4.515493E + 00
2	**5.163616E + 01**	**5.666467E + 01**	6.044886E + 01	**5.132145E + 01**	5.935384E + 02	7.108243E + 01
3	**3.598875E + 01**	**4.771628E + 01**	4.115785E + 01	**3.634827E + 01**	5.024335E + 02	5.237031E + 01
4	2.170677E + 01	**1.365104E + 01**	2.542917E + 01	2.046511E + 01	2.623511E + 02	2.931979E + 01
5	2.909687E + 02	**1.543750E + 02**	2.956655E + 02	2.923801E + 02	7.227362E + 02	3.076157E + 02
6	1.213021E + 00	**1.169792E + 00**	1.616146E + 00	9.151042E + 01	2.542188E + 01	2.113542E + 00
7	2.726407E + 01	**1.449584E + 01**	3.129688E + 01	2.544896E + 01	2.767250E + 02	3.558177E + 01
8	3.944532E + 01	**2.935365E + 01**	3.406771E + 01	4.652761E + 01	2.765766E + 02	4.958750E + 01
9	3.297709E + 01	**2.906875E + 01**	3.814532E + 01	3.182657E + 01	2.807099E + 02	4.108073E + 01
10	**1.580781E + 01**	**1.534532E + 01**	2.014427E + 01	**1.416719E + 01**	2.611818E + 02	2.389532E + 01

Table 3 Mean CPU time of 10 functions in six PSO variants

F#	SPSO	ML-CVUPSO	UPSO	AFPSO	CLPSO	MCPSO
1	**1.485136E + 00**	2.854772E + 00	2.702219E + 00	**1.420709E + 00**	8.614554E + 01	4.515493E + 00
2	**5.163616E + 01**	**5.666467E + 01**	6.044886E + 01	**5.132145E + 01**	5.935384E + 02	7.108243E + 01
3	**3.598875E + 01**	**4.771628E + 01**	4.115785E + 01	**3.634827E + 01**	5.024335E + 02	5.237031E + 01
4	2.170677E + 01	**1.365104E + 01**	2.542917E + 01	2.046511E + 01	2.623511E + 02	2.931979E + 01
5	2.909687E + 02	**1.543750E + 02**	2.956655E + 02	2.923801E + 02	7.227362E + 02	3.076157E + 02
6	1.213021E + 00	**1.169792E + 00**	1.616146E + 00	9.151042E + 01	2.542188E + 01	2.113542E + 00
7	2.726407E + 01	**1.449584E + 01**	3.129688E + 01	2.544896E + 01	2.767250E + 02	3.558177E + 01
8	3.944532E + 01	**2.935365E + 01**	3.406771E + 01	4.652761E + 01	2.765766E + 02	4.958750E + 01
9	3.297709E + 01	**2.906875E + 01**	3.814532E + 01	3.182657E + 01	2.807099E + 02	4.108073E + 01
10	**1.580781E + 01**	**1.534532E + 01**	2.014427E + 01	**1.416719E + 01**	2.611818E + 02	2.389532E + 01

Table 2 describes the mean difference in ten test functions by using SPSO, ML-CVUPSO, UPSO, AFPSO, CLPSO and MCPSO variants, in order to distinguish the quality of various methods, there are a summary of the data and highlight the better performance with bold and underline, which shows that ML-CVUPSO performances well in 80% of the test functions, the layered mechanism is available to get the same accuracy. It is also helpful to get good scores in some test functions, such as F1, F7, F8 and F9.

Table 3 describes the mean CPU time in ten test functions by using SPSO, ML-CVUPSO, UPSO, AFPSO, CLPSO and MCPSO variants, in order to distinguish the quality of various methods, and it makes a summary of the data and highlight the better performance with bold and underline. Figure 2 shows that ML-CVUPSO can save the CPU time in most of the test functions, especially in F4–F9. From the peak characteristics of the test functions, if the test problem is relative puzzle, the improvement is more obvious for F10.

To determine whether the ML-CVUPSO is more effective than the others, a statistical method need detect the results of CPU time, namely paired *T-test*, paired *F-test* and *Wilcoxon matched-pairs signed-ranks test*. *T-test* is used to compare group means, *F-test* is used to determine whether the 30 independent experiments' results are typical, *Wilcoxon matched-pairs signed-ranks test* is a non-parametric test employed in hypothetical testing situation involving two samples, it is a pair-wise test that can be used to detect significantly differences between the behavior of ML-CVUPSO and UPSO, CLPSO, AFPSO, CVUPSO, usually there is a level of significance $\alpha = 0.05$, if the data in *T-test* and *F-test* is less than α, a cell will mark it with '+', on the contrary, the cell will mark it with '–'. The percent of the '+' during all the results is the effectiveness of the ML-CVUPSO in handing the test problems, for example, as to the test problem 1, only one values (numbers of minus '–') greater than 0.05, 9 (numbers of plus '+') are less than 0.05, which shows the effectiveness of the ML-CVUPSO method in handling Sphere problem is

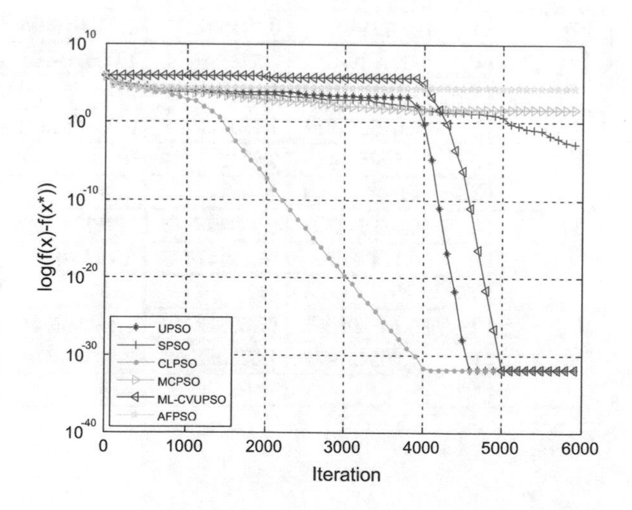

Fig. 2 The convergence curve of **F10** test functions

9/10, equivalent to about 90%. The result of *T-test*, *F-test* and *Wilcoxon matched-pairs signed ranks test* on ML-CVUPSO is shown in Table 4. All the percent is over 50%, which means the experiments are reliable and universal, the **ML-CVUPSO** really has good performance in reducing the computational complexity and running time.

Table 4 T-test, F-test and Wilcoxon-test

#/Rate (%)	Test	SPSO	UPSO	AFPSO	CLPSO	MCPSO
F1/ 90	T	1.34866E-06	0.080224783	2.87622E-16	1.2332E-05	0.00023887
	F	1.76837E-28	5.42422E-06	3.9174E-125	2.34217E-08	7.4543E-166
	T/F	+/+	−/+	+/+	+/+	+/+
F2/ 80	T	0.000745423	2.12505E-16	1.97108E-17	0.297768395	7.76308E-07
	F	0.008705642	4.47317E-06	3.97266E-08	0.225238214	0.038799576
	T/F	+/+	+/+	+/+	−/−	+/+
F3/ 70	T	0.057563588	1.01642E-11	1.15304E-22	0.05743728	2.20472E-19
	F	5.2298E-100	0.109692599	4.13469E-07	0	0.004298327
	T/F	−/+	+/−	+/+	−/+	+/+
F4/ 80	T	0.012597979	3.34647E-07	3.22774E-21	6.91582E-20	0.023828248
	F	0.108272489	0.648845764	0.209814828	2.39754E-11	0.01528636
	T/F	−/−	+/−	+/−	−/+	+/+
F5/ 50	T	0.012597979	3.34647E-07	3.22774E-21	6.91582E-20	0.023828248
	F	0.108272489	0.648845764	0.209814828	2.39754E-11	0.01528636
	T/F	−/−	+/−	+/−	−/+	+/+
F6/ 80	T	1.69648E-09	0.161656977	6.91346E-13	0.440754994	3.8839E-08
	F	2.77967E-09	9.11746E-09	8.279E-113	6.34169E-25	1.6258E-128
	T/F	+/+	−/+	+/+	−/+	+/+
F7/ 70	T	0.4731499	0.107302255	1.04091E-06	0.296344179	1.56013E-06
	F	1.9493E-08	2.59279E-14	1.0067E-105	7.33827E-12	1.53852E-79
	T/F	−/+	−/+	+/+	−/+	+/+
F8/ 60	T	0.016657919	1.11031E-05	0.001568662	1.46045E-23	0.080557893
	F	1.5045E-43	0.434801791	0	0.414002901	6.9597E-272
	T/F	+/−	+/−	+/+	+/−	−/+
F9/ 70	T	0.051543305	3.96349E-32	2.86293E-31	2.49576E-21	1.94659E-17
	F	0.11428693	0.000441478	2.20224E-05	4.69588E-08	0.468893773
	T/F	−/−	+/+	+/+	+/+	+/−
F10/ 60	T	0.163498263	0.162329238	2.18093E-19	0.162790994	6.01318E-08
	F	1.80545E-75	1.67005E-66	2.58358E-94	1	5.98595E-41
	T/F	−/+	−/+	+/+	−/−	+/+

4 Application

Car or Vehicle Routing Problem (CVRP) is a kind of flow optimization in the process of logistics distribution. There are 8 customers, one distribution centers, and two vehicles (8 ton-loaded). The distance between customers and the demand of each customer is shown in Table 5. The total iteration number is 30, and the final result is 67.5. The corresponding transport model is shown in Fig. 3. This is consistent with the known results, proving that the real encoding *ML-CVUPSO* is effective in solving the CVRP problem.

In order to demonstrate the superiority of *ML-CVUPSO* in solving CRVP problem, we also apply other variants with the same encoding mechanism into CRVP. The specific algorithms are SPSO, UPSO, CLPSO, AFPSO and CVUPSO. All variants will be used to solve the benchmark instances of Augerat. According to the different dimensions of benchmark instances, the population size and the

Table 5 Distance between customers and the demand

Customer#	0	1	2	3	4	5	6	7	8
0	0	4	6	7.5	9	20	10	16	8
1	4	0	6.5	4	10	5	7.5	11	10
2	6	6.5	0	7.5	10	10	7.5	7.5	7.5
3	7.5	4	7.5	0	10	5	9	9	15
4	9	10	10	10	0	10	7.5	7.5	10
5	20	5	10	5	10	0	7	9	7.5
6	10	7.5	7.5	9	7.5	7	0	7	10
7	16	11	7.5	9	7.5	9	7	0	10
8	8	10	7.5	15	10	7.5	10	10	0
Demand(tons)	0	1	2	1	2	1	4	2	2

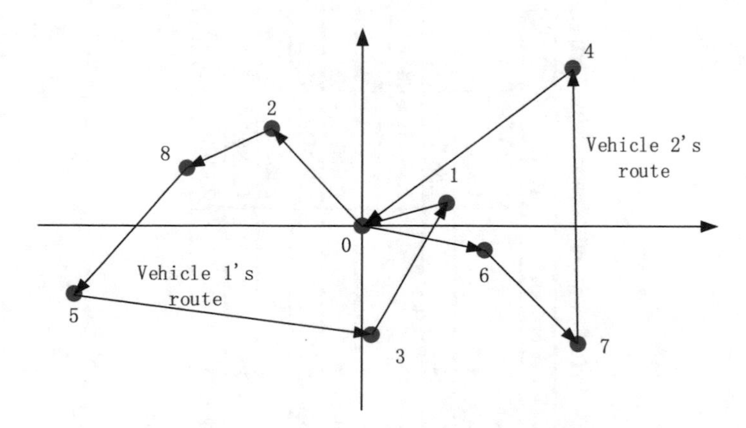

Fig. 3 Small scale CVRP of eight customers transport route model

Table 6 Test results of different Augerat test sets

Benchmark instance	SPSO	UPSO	CLPSO	AFPSO	CVUPSO	ML-CVUPSO
A-n32-k5	1.1213E + 03	1.1411E + 03	1.2286E + 03	1.1975E + 03	1.1862E + 03	**1.0711E + 03**
A-n36-k5	1.2738E + 03	1.2170E + 03	1.2777E + 03	1.2881E + 03	1.2416E + 03	**1.1697E + 03**
A-n44-k6	1.5673E + 03	1.5658E + 03	1.5935E + 03	1.5465E + 03	1.5563E + 03	**1.4080E + 03**
A-n48-k7	1.7371E + 03	1.6991E + 03	1.7929E + 03	1.7983E + 03	1.7970E + 03	**1.6351E + 03**
A-n54-k7	1.7377E + 03	1.7531E + 03	1.8064E + 03	1.7274E + 03	1.7337E + 03	**1.6965E + 03**
A-n60-k9	2.4088E + 03	2.3642E + 03	2.3886E + 03	2.2427E + 03	2.2851E + 03	**2.1777E + 03**
B-n31-k5	6.5571E + 02	6.6932E + 02	6.6357E + 02	6.7280E + 02	7.1756E + 02	**6.1926E + 02**
B-n38-k6	1.2320E + 03	1.1906E + 03	1.2547E + 03	1.2449E + 03	1.3069E + 03	**1.1707E + 03**
B-n50-k7	1.6530E + 03	1.5538E + 03	1.6628E + 03	1.5904E + 03	1.6849E + 03	**1.5438E + 03**
B-n68-k9	2.1215E + 03	2.1883E + 03	2.2270E + 03	2.1961E + 03	2.2656E + 03	**1.9872E + 03**

number of iterations will be adjusted accordingly. All experiments are conducted under the unified test environment: windows7 32 bit system, Matlab7.1.

In the Augerat benchmark instances (6/class A, 4/class B, 30 independent experiments), the best vehicle path and the distance needed for each experiment are recorded. Table 6 shows the average value of the thirty tests, which shows that the ML-CVUPSO is better than the other existing PSO variants in solving the CVRP application problem.

5 Conclusion

Particle swarm optimization is a global meta-heuristic that has shown its good performance in optimizing a wide range of problems. However, in real-time high dimensions environment, how to quickly find the optimal solution and give timely response or decisive adjustment is very important. Inspired by space projection behavior, in order to accelerate the speed of convergence and get timely satisfying solution, a Multiple-shadows Layered Cooperative Velocity Updating PSO (ML-CVUPSO) is proposed in this paper. The ML-CVUPSO provides a shadows layered projection mechanism to evaluate the weight of each dimension and divide them into some independent groups according to the different views (plane $x_A, x_B, x_c \ldots$), then the original high dimension problem turns to serval relevant characteristic sub-problems by feature extraction. The results solved by CVUPSO compose the final result. The proposed ML-CVUPSO is examined on ten widely used benchmark functions and it displays a good performance. The proposed algorithm appears to be especially efficient in decreasing the operation time for the high dimension and multimodal problems.

Acknowledgements Financial supports from the National Natural Science Foundation of China (No. 61572074) and 2012 Ladder Plan Project of Beijing Key Laboratory of Knowledge Engineering for Materials Science (No. Z121101002812005) are highly appreciated.

References

1. Fakhrabadi, M.M.S., Rastgoo, A., Samadzadeh, M.: Multi-objective design optimization of composite laminates using discrete shuffled frog leaping algorithm. J. Mech. Sci. Technol. **27**(6), 1791–1800 (2013)
2. Milani, A., Santucci, V.: Optimizing web content presentation: a online PSO approach. In: Proceedings IEEE/WIC/ACM International Conference on Web Intelligence and Intelligent Agent Technology, pp. 26–29 (2009)
3. Sordo, M., Ochoa, G., Murphy, S.N.: A PSO/ACO approach to knowledge discovery in a pharmacovigilance context. In: Proceedings of 11th Annual Conference Companion on Genetic and Evolutionary Computation Conference, Montreal, Canada, pp. 2679–2684 (2009)

4. Niasar, N.S., Perdam, M.M., Shanbezade, J., Mohajeri, M.: Discrete fuzzy particle swarm optimization for solving traveling salesman problem. In: Proceedings of ICIFE, pp. 162–165 (2009)
5. Trigueros, D.E.G., Módenes, A.N., Ravagnani, M.A.S.S., Espinoza-Qui˜nones, F.R.: Reuse water network synthesis by modified PSO approach. Chem. Eng. J., pp. 183, 198–211 (2012)
6. Maldonado, Yazmin, Castillo, Oscar, Melin, Patricia: Particle swarm optimization of interval type-2 fuzzy systems for FPGA applications. Appl. Soft Comput. **13**, 496–508 (2013)
7. Tanabe, R., Fukunaga, A.: Evaluation of a randomized parameter setting strategy for island-model evolutionary algorithms. In: Proceedings of IEEE CEC, June 20–23, Cancún, México (2013)
8. Sun, S., Li, J.: A two-swarm cooperative particle swarms optimization. Swarm Evolut. Comput. **15**, 1–18 (2014)
9. Wang, H., Zhao, X., Wang, K., Xia, K., Tu, X.: Cooperative velocity updating model based particle swarm optimization. Appl. Intell., **40**, 322–342 (2014)
10. Parsopoulos, K.E., Vrahatis, M.N.: UPSO: a unified particle swarm optimization scheme. In: Proceedings of International Conference of Computational Methods in Sciences and Engineering (ICCMSE). Lecture Series on Computer and Computational Sciences, vol. 1, pp. 868–873 VSP International Science Publishers, Zeist (2004)
11. Liang, J.J., Qin, A.K.: Student member comprehensive learning particle swarm optimizer for global optimization of multimodal functions. IEEE Trans. Evol. Comput. **10**(3), 281–295 (2005)
12. Niu, B., Zhu, Y., He, X., Wu, H.: MCPSO: a multi-swarm cooperative particle swarm optimizer. Appl. Math. Comput., **185**, 1050–1062 (2007)
13. Joanna, Y.-T., Tung, S.L., Chiu, H.C.: Adaptive fuzzy particle swarm optimization for global optimization of multimodal functions. Inf. Sci., **181**(20), 4539–4549 (2011)
14. Chatterjee, S., Goswami, D., Mukherjee, S., Das, S.: Behavioral analysis of the leader particle during stagnation in a particle swarm optimization algorithm. Inf. Sci. **279**(20), 18–36 (2014)
15. Mahmoodabadi, M.J., Mottaghi, S.Z., Bagheri, A.: HEPSO: high exploration particle swarm optimization. Inf. Sci., **273**(20), 101–111 (2014)
16. Davoodi, E., Hagh, M.T., Zadeh, S.G.: A hybrid improved quantum-behaved particle swarm optimization–simplex method (IQPSOS) to solve power system load flow problems. Appl. Soft Comput., **21**, 171–179 (2014)
17. Sadeghi, J., Sadeghi, S., Niaki, S.T.A.: Optimizing a hybrid vendor-managed inventory and transportation problem with fuzzy demand: an improved particle swarm optimization algorithm. Inf. Sci. **272**, 126–144 (2014)
18. Ganapathy, K., Vaidehi, V., Kannan, B., Murugan, H.: Hierarchical particle swarm optimization with ortho-cyclic circles. Expert Syst. Appli., **41**(7), 3460–3476 (2014)
19. Mahmoodabadi, M.J., Momennejad, S., Bagheri, A.: Online optimal decoupled sliding mode control based on moving least squares and particle swarm optimization. Inf. Sci. **268**, 342–356 (2014)
20. Beheshti, Z., Shamsuddin, SMH.: CAPSO: centripetal accelerated particle swarm optimization. Inf. Sci., **258**, 54–79 (2014)
21. Cagnina, L., Errecalde, M., Ingaramo, D., Rosso, P.: An efficient particle swarm optimization approach to cluster short texts. Inf. Sci., **265**, 36–49 (2014)
22. Boubaker, S., Djemai, M., Manamanni, N., M'Sahli, F.: Active modes and switching instants identification for linear switched systems based on discrete particle swarm optimization. Appl. Soft Comput. Part C, **14**, 482–488 (2014)
23. Kundu, R., Das, S., Mukherjee, R., Debchoudhury, S.: An improved particle swarm optimizer with difference mean based perturbation. Neurocomputing, **129**, 315–333 (2014)

Short Term Prediction of Continuous Time Series Based on Extreme Learning Machine

Hongbo Wang, Peng Song, Chengyao Wang and Xuyan Tu

Abstract Extreme Learning Machine (ELM) is a popular tool of machine learning, which has been used in many fields. Time series prediction is usually a complex problem without related parameters or features. In this paper, a prediction method for continuous time series based on the theory of extreme learning machines is proposed, which focus on short term prediction of continuous time series. Firstly, the ST-ELMpredicting model is constructed. Then the ways of training and predicting is analyzed. ST-ELM uses time series and predicted value to adjust itself. Mackey-Glass and Lorenz time series have been used as example for demonstration. It is showed this method can predict continuous time series timely and accurately without related parameters or features of time series.

Keywords Extreme learning machine (ELM) · Time series prediction · Machine learning

1 Introduction

Time series prediction is a common problem, and has a lot of application areas like signal processing, pattern recognition, econometrics, mathematical finance, weather forecasting [1]. But time series analysis is a complex problem which makes linear prediction methods useless. Hence, some nonlinear prediction methods have been

H. Wang (✉) · P. Song · C. Wang · X. Tu
Department of Computer Science and Technology, School of Computer and Communication Engineering, University of Science and Technology Beijing, Beijing, China
e-mail: foreverwhb@ustb.edu.dn; foreverwhb@126.com

P. Song
e-mail: songpeng@188.com

C. Wang
e-mail: wangchengyao@ustb.edu.dn

X. Tu
e-mail: tuxuyan@126.com

J. Cao et al. (eds.), *Proceedings of ELM-2016*, Proceedings in Adaptation, Learning and Optimization 9, DOI 10.1007/978-3-319-57421-9_10

proposed including ARMA model [2], stationary process [3] and so on. Neural network [4] has been used in the area of time series prediction successfully. Neural network use descent method to get solution by adjusting network weights. It needs training a lot to update weights according to the errors. But the neural network has its disadvantages, for example, the multiple local minima problem, the over fitting etc., which make it difficult for some practical application.

The researchers also use support vector machine [5, 6] regression to solve the problem of time series prediction, which also achieved some results.

ELM trains easily when we have inputs and outputs, we can test its accuracy with train data and test data. And it is worth mentioning that we don't need to adjusting its parameters while training and predicting. In this paper, we use ELM in short term time series prediction, which is timely, effective and simple.

2 Methodology

2.1 Time Series

Recording the process of the development of random events according to the order of time, which constitutes a time series. Time series analysis comprises methods for analyzing time series data in order to extract meaningful statistics and other characteristics of the data. Time series forecasting is the use of a model to predict future values based on previously observed values.

From a mathematical point of view, the time series can be expressed as:

$$\{x_t, t = 1, 2, \ldots, N\} \tag{1}$$

Here t represents a moment or a time, x_t represents the value of t time (moment). The time series prediction is mainly based on the principle of continuity. The continuity principle means that the development of objective things has a regular continuity, and the development of objective things has a regular continuity, and the development of things is based on the inherent law of it. As long as the law depends on the conditions of the time series do not have qualitative change, then the basic development trend of things in the future will continue to go on. A basic method is arithmetic mean method, which is predicting next value based on the simple arithmetic mean of the history data. It can be computed as follows:

$$x_{n+1} = \frac{1}{n} \sum_{i=1}^{n} . \tag{2}$$

2.2 Extreme Learning Machine

Extreme Learning Machine (ELM) is a simple and effective algorithm for training the single-hidden layer feed-forward neural networks (SLFN) proposed by Huang et al. in 2004 [7]. ELM randomly generated input layer weights and hidden layer node bias, calculating the weights of the output layer. Huang proved that ELM has uniform approximation ability as SLFN [8].

There are P group samples, each with a N column input, expressed as:

$$x_t = \{x_{i1}, x_{i2}, \ldots, x_{iN}, i = 1, 2, \ldots, P\} \quad (3)$$

M column target output, expressed as:

$$t_i = \{t_{i1}, t_{i2}, \ldots, t_{iM}, i = 1, 2, \ldots, P\} \quad (4)$$

The number of nodes in the input layer and the number of nodes in the output layer are determined by the training samples. The hidden layer node number L is given by manmade, generally $L \leq P$, activation function is set to $g(x)$. Its network structure is shown in the following figure (Fig. 1).

In fact, in the ELM algorithm, the input layer to the hidden layer is a random mapping, which maps the training set of samples from the original space to a feature space. The dimension of feature space is determined by the number of hidden layer nodes. In general, the dimension of the feature space is higher than that of the original space. Compared with other SLFN training algorithms, the advantage of ELM does not need to adjust the weight parameters and has a very fast learning speed and a very good generalization ability.

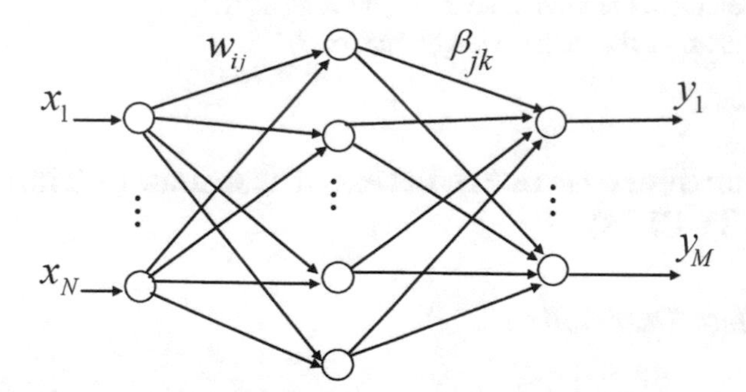

Fig. 1 Extreme Learning Machine Networks. w_{ij} indicates the connection weights of the i input layer nodes to the j hidden layer nodes, the matrix W is the input weight matrix, which is initialized random. β_{jk} indicates the connection weights of the j hidden layer nodes to the node of the k output layer, and the matrix B is the output weight matrix, which is obtained in the training process. For each hidden layer node, there is a corresponding threshold value b_j, which is initialized random

The training model can be expressed as:

$$t_l = \sum_{j=1}^{L} \beta_j g_j(w_j x + b_j), l = 1, 2 \ldots, P \tag{5}$$

These p equations can be expressed as a form of matrix multiplication: The training model can be expressed as:

$$HB = T \tag{6}$$

$$H = \begin{bmatrix} g(w_1 x_1 + b_1) & \ldots & g(w_L x_1 + b_L) \\ \vdots & \ddots & \vdots \\ g(w_1 x_p + b_1) & \ldots & g(w_L x_P + b_L) \end{bmatrix}_{P \times L}, B = \begin{bmatrix} \beta_1^T \\ \vdots \\ \beta_L^T \end{bmatrix}_{L \times M}, T = \begin{bmatrix} t_1^T \\ \vdots \\ t_P^T \end{bmatrix}_{P \times M}$$

H represents the hidden layer output matrix of ELM. Huang proves that if the activation function is infinite differentiable, the W and B do not need to be adjusted. We only need to solve the output weight matrix B, which can meet the target output with a minimum error approximation. The solution of B generally expressed as:

$$B = H^{\dagger} T \tag{7}$$

$H^{\dagger}$is the pseudo inverse (Moore-Penrose) of the hidden layer output matrix H. Thus, ELM can be summarized as follows:

Given a training set (x_t, t_i), which $(i = 1, \ldots, P)$, $xi = [x_{i1}, \ldots, x_{iN}] \in R^n$, $ti = [t_{i1}, \ldots, t_{iM}] \in R^m$, activation function $g(x)$, input nodes number N, output nodes number M, hidden nodes number $L(L \leq P)$.

- Step1: Assign arbitrary input weight matrix W, and bias matrix B.
- Step2: Calculate the hidden layer output matrix H.
- Step3: Calculate the output weight matrix B.

3 ELM for Short Term Prediction of Continuous Time Series(ST-ELM)

3.1 Problem Description

A continuous time series is an array of values, which samples usually comes from a dynamic systems output. It is assumed that neither of state of dynamic system is measurable nor the equation describing is known. If the dynamic system is deterministic, we can try to predict the time series based on ELM. The purpose is to predict values only use time series itself.

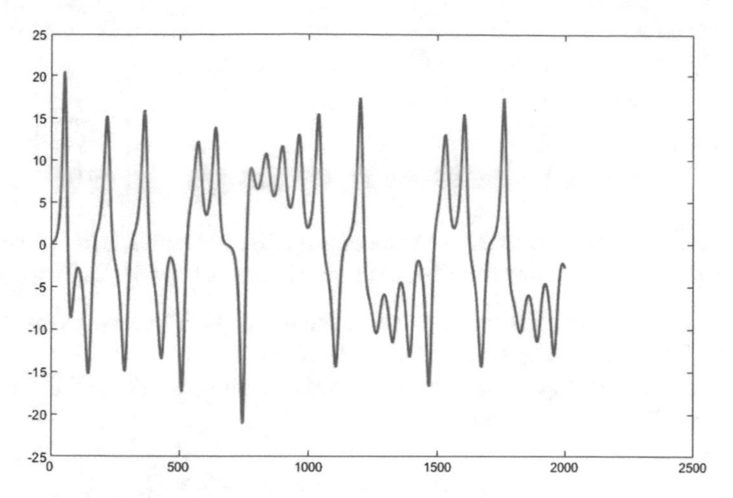

Fig. 2 A Continuous time series

We assume that we have a time series, which sampled every T and we do not know its features, expressed as $x_{(T)}, x_{2(T)}, \ldots, x_{NT}$. Then we already know N data, and we need predict $\{x_{(N+1)T}, \ldots, x_{(N+Q)T}\}$. Like formula (2), there are many one-step prediction methods. However, it provides insufficient information. Thus we use ELM to get Q steps to obtain $\{x_{(N+1)T}, x_{(N+Q)T}\}$.

For example, Fig. 2 is a continuous times series, which shows we already know 2000 values. The object is to predict next 500 values using these 2000 values.

And, a time series is usually continuous, which means it generates data all the time. In some applications the prediction results are only useful in a short time, for we can obtain the real data timely. So we need to determine predicted length Q, history values for learning in order to obtain predicted values.

3.2 ST-ELM Model for Time Series Prediction

Consider a given set of N data $P = \{x_t, t = 1, 2, \ldots, N\}$, predicted length Q, input length L. Split P to build inputs and outputs of ELM in order to train, $PH = \{x_1, x_2, \ldots x_{N-L}\}$, length $N - L$, and $PL = \{x_{N-L+1}, x_{N-L+2}, \ldots, x_N\}$, length L. Then we set the training set as:

$$PI = \begin{bmatrix} x_1 & \ldots & x_L \\ \vdots & \ddots & \vdots \\ x_{N-L} & \ldots & x_N \end{bmatrix}_{(N-L)\times L} \tag{8}$$

And:

$$H = \begin{bmatrix} g(w_1x_1 + b_1) & \dots & g(w_Lx_1 + b_L) \\ \vdots & \ddots & \vdots \\ g(w_1x_{N-L} + b_1) & \dots & g(w_Lx_{N-L} + b_L) \end{bmatrix}_{(N-L)\times L}, PL = \begin{bmatrix} x_{N-L+1} \\ \vdots \\ x_N \end{bmatrix}_{L\times 1}$$

Thus, for ELM, we have PI as $(N - L)$ inputs from PH. Because it is one dimensional time series. The actual output of ST-ELM is one. Here is how ST-ELM works:

- Step 1: Use PI as inputs, use PL as outputs to train, get H and calculate B.
- Step 2: Use PL as inputs, get $PreH$, then calculate $x_{n+1} = PreH * B$.
- Step 3: Use PI_2 as inputs, use PL_2 as outputs to train, get H_2 and calculate B_2. Which H_2, PI_2, PL_2 are as follows:

$$H_2 = \begin{bmatrix} g(w_1x_1 + b_1) & \dots & g(w_Lx_1 + b_L) \\ \vdots & \ddots & \vdots \\ g(w_1x_{N-L+1} + b_1) & \dots & g(w_Lx_{N-L+1} + b_L) \end{bmatrix}_{(N-L+1)\times L},$$

$$PL_2 = \begin{bmatrix} x_{N-L+1} \\ \vdots \\ x_{N+1} \end{bmatrix}_{(L+1)\times 1}, PI_2 = \begin{bmatrix} x_1 & \dots & x_{L+1} \\ \vdots & \ddots & \vdots \\ x_{N-L+1} & \dots & x_{N+1} \end{bmatrix}_{(N-L+1)\times L}$$

- Step 4: Use PL_2 as inputs, get $PreH_2$, then calculate $x_{n+2} = PreH_2 * B_2$.
- ...
- Step 2 * Q: Use PL_Q as inputs, get $PreH_Q$, then calculate $x_{n+Q} = PreH_Q * B_Q$. Then, we get a predicted series:

$$\{x_t, t = N + 1, N + 2, \dots, N + Q\}. \tag{9}$$

3.3 ST-ELM Model Optimization

Since ST-ELM use predicted values as inputs, so it is less useful when Q becomes longer. Thus the method is useful for a short term prediction. As mentioned in Sect. 3.1, we can obtain the real data in a short time. Then we can replace the predicted values with real values to make it more accurate. We can see that with the data set getting larger, we need larger quantity calculation. So we can set learning length (History Length) as HL. After once or several predictions, especially when we get real values of the predicted values. We can abandon the beginning data.

For example, when we know $\{x_t, t = N + 1, N + 2, \dots, N + Q\}$then we replaced the predicted data with it, and abandon $\{x_t, t = 1, 2, \dots, Q\}$. Thus, we set $HL = N$. In continuous time series, ST-ELM can keep running other than becoming slower and less effective.

4 Experimental Results and Discussions

4.1 For Mackey-Glass Time Series

In this section, we use data sets generated by the Mackey-Glass showing the effectiveness of using ST-ELM for time series prediction. We generate data by numerically integrating the Mackey-Glass time delay differential equation [9].

$$\frac{dx(t)}{dt} = -hx(t) + \frac{gx(t-\tau)}{1 + x^{10}(t-\tau)} \tag{10}$$

When $\tau < 17$, Eq. (10) generate a chaotic time series prediction. We use parameter $g = 0.2, h = 0.1, \tau = 18$ and initial condition $y = 0.72$.The values are shown in Fig. 3.

Now we use $N = 1000, L = 15, Q = 100$, hidden layers = 20, $g(x)$ uses Sigmoid. The predicted, desired values, RMSE of Mackey-Glass series as follows in Figs. 4 and 5.

4.2 Optimized ST-ELM

In continuous time series, usually we can obtain data instantly. In this section, we update the predicted value with real value to train ST-ELM.

For Mackey-Glass time series above with same parameters. For example, when we predicting 1002th value, we use real 1001th value instead of predicted 1001th value. The predicted, desired values, RMSE as follows in Figs. 6 and 7:

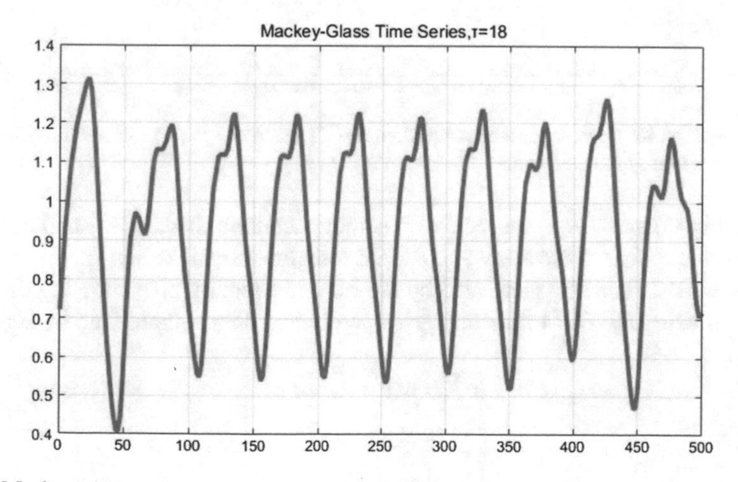

Fig. 3 Mackey-glass time series τ =18

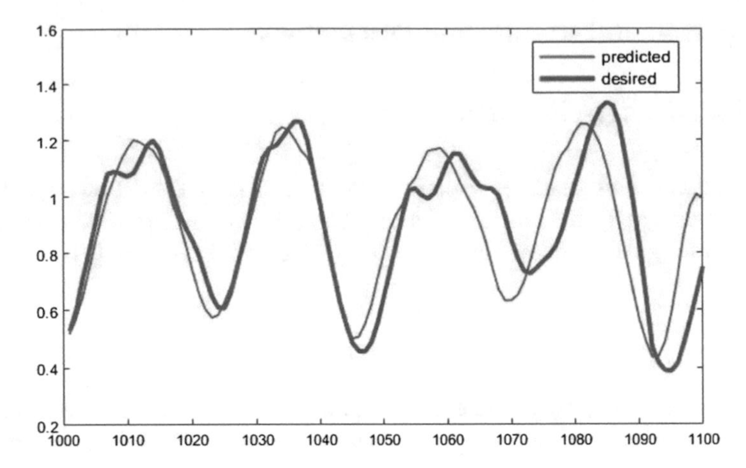

Fig. 4 The predicted and desired values. It shows ST-ELM performs well at the beginning. when predicted length gets longer, the predicted trend moves away from desired values gradually. It is because we use predicted values as inputs, which lends additional uncertainly

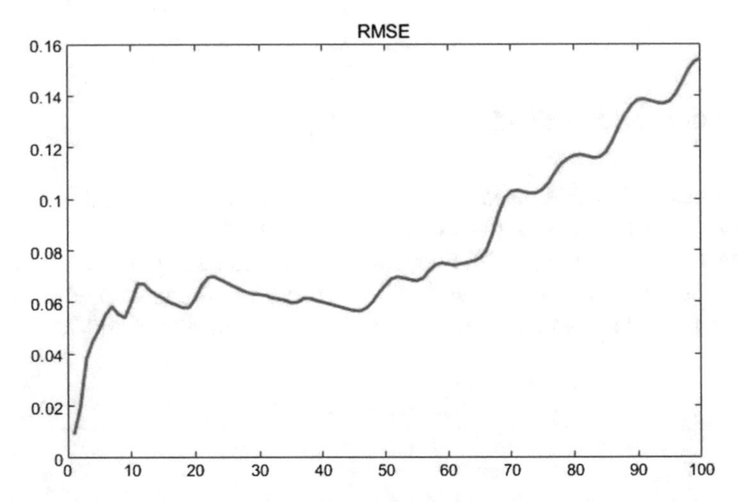

Fig. 5 RMSE,which appearances the same when RMSE getting larger. RMSE remains a low value before about fiftieth predicted value

From Figs. 6 and 7, we can see that it is more accurate than Figs. 4 and 5. Prediction does not appear move away and RMSE remains a small value.

Then, we use data sets generated by the Lorenz equation [10] showing the effectiveness of using ST-ELM for time series prediction.We generate data by equation:

$$\begin{cases} \frac{dx}{dt} = \sigma(y - x) \\ \frac{dy}{dt} = x(\rho - z) - y \\ \frac{dz}{dt} = xy - \beta z \end{cases} \tag{11}$$

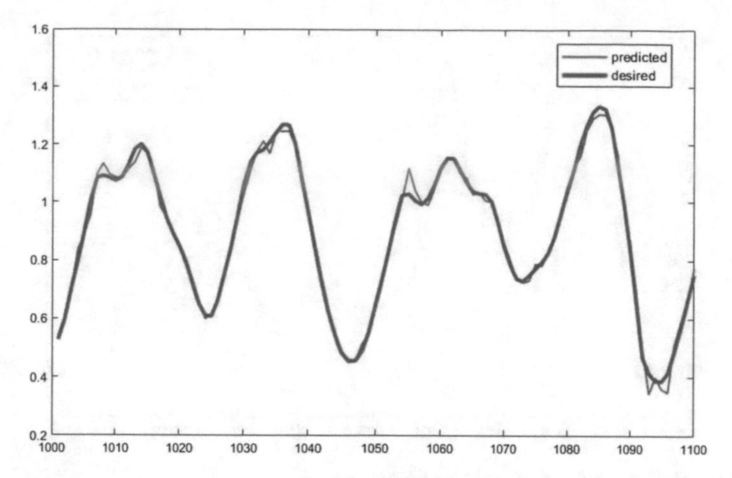

Fig. 6 The predicted and desired values of Mackey-Glass

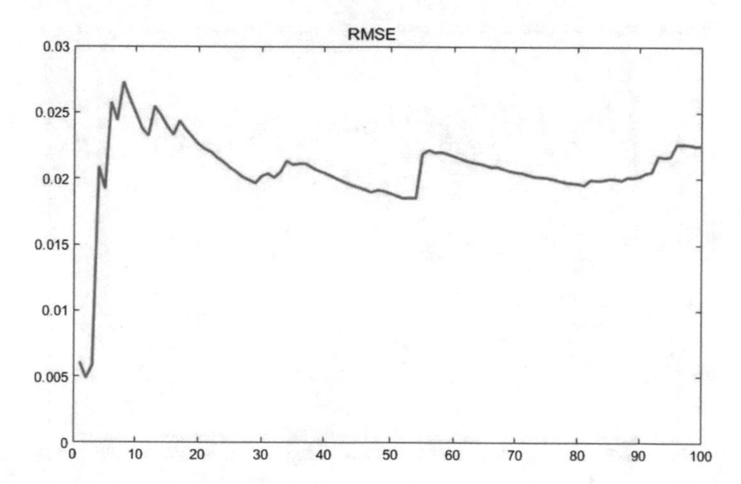

Fig. 7 RMSE of Mackey-Glass

When $\rho = 28$, Eq. (11) generate a chaotic time series prediction. We use common parameter $\beta = 8/3, \rho = 28, \sigma = 10$ and initial condition $x_0 = y_0 = z_0 = 0.1$. Figure 2 of Sect. 3.1 shows x component of Lorenz times series.

Now we use its x component, $N = 2000, L = 2, Q = 500$, hidden layers $=$ 20, $g(x)$ uses Sigmoid. the predicted, desired values and RMSE of Lorenz time series as follows in Fig. 8.

Figures 8 and 9 shows that ST-ELM is also useful for Lorenz time series with high accuracy.

Since the Mackey-Glass time series and Lorenz time series is nonlinear and chaotic, which has no clearly period and will not diverge or converge. So the prediction of these series is a benchmark problem.

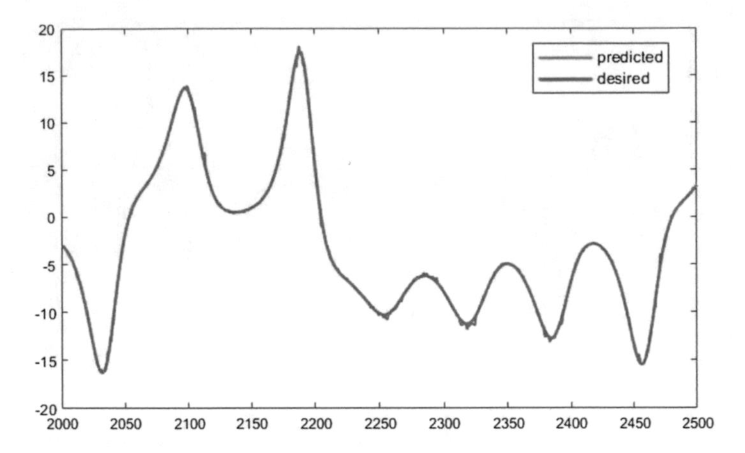

Fig. 8 The predicted and desired values of Lorenz

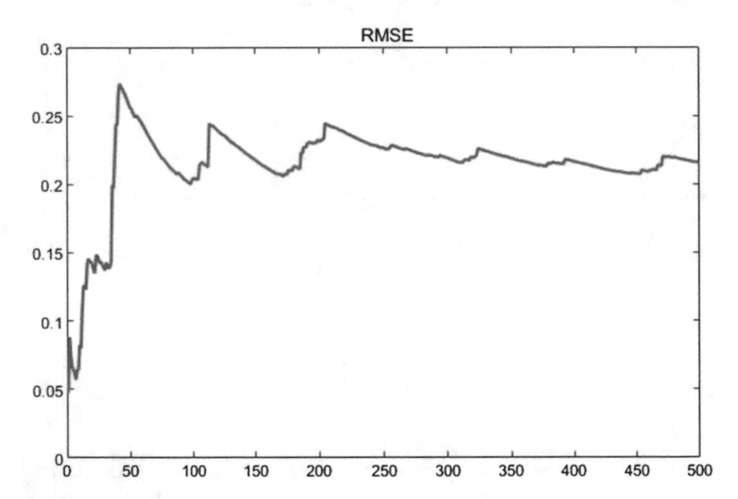

Fig. 9 RMSE of Lorenz

4.3 For Signal Generated by sinc

In this section, we use data sets generated by function *sinc*. We generate data by equation:

$$sinc(x) = \frac{sin(x)}{x} \tag{12}$$

It is similar with sample1 presented by Huang, and we use the same values as the sample, but we only use the outputs (time series of *sinc*) to predict next values. In sample1, we have 5000 values of *sinc*, now we use Optimized ST-ELM to predict the signal.

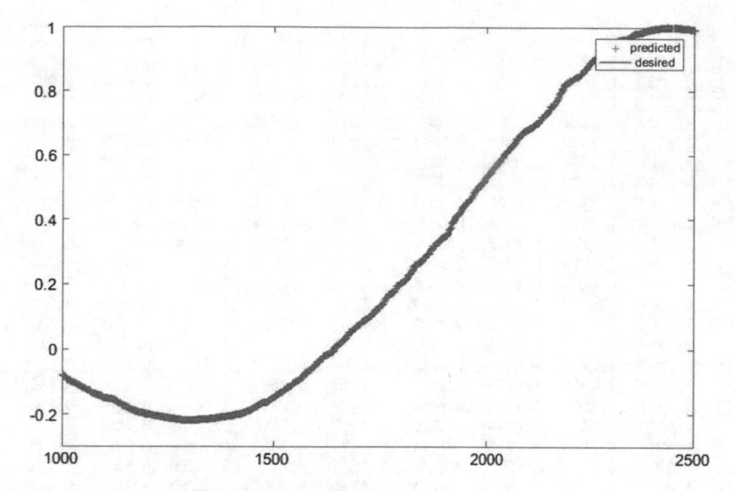

Fig. 10 The predicted and desired values of signal

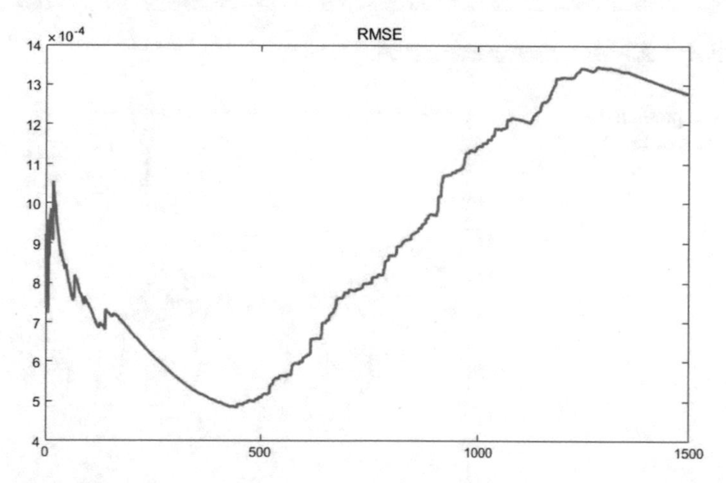

Fig. 11 RMSE of signal

Now we use $N = 1000, L = 15, Q = 1500$, hidden layers = 20, $g(x)$ uses Sigmoid, the predicted, desired values, RMSE of Signal (*sinc*) as follows in Figs. 10 and 11.

5 Application of Traffic Flow Prediction in Metro

Traffic flow prediction is a fundamental problem in transportation modeling and management. In this section we use ST-ELM to predict traffic flow in Beijing Metro. we use data of the website of Beijing subway from 2013 to 2015, which consists

million	line1	line2	line5	line6	line7	line8	line9	line10	line13	line15
2013-1-1	1.0405	0.8723	0.6143	0.2527	0.1245	0.1383	0.7711	0.5053	0.1052	0.2476
2013-1-2	0.9087	0.8046	0.5658	0.2382	0.1105	0.1373	0.7541	0.4662	0.0929	0.2153
2013-1-3	0.8746	0.8267	0.5665	0.231	0.1027	0.1599	0.7626	0.4811	0.0935	0.2147
2013-1-4	1.2829	1.2226	0.9021	0.3577	0.1636	0.1857	1.3256	0.7855	0.1174	0.3187
2013-1-5	1.2077	1.1496	0.8623	0.3518	0.1598	0.1802	1.2842	0.7477	0.1111	0.3025
2013-1-6	1.2036	1.1598	0.8585	0.3569	0.1653	0.1867	1.287	0.7476	0.1118	0.2994
2013-1-7	1.2649	1.2336	0.9186	0.3805	0.169	0.1936	1.358	0.7859	0.1181	0.3145
2013-1-8	1.2782	1.2482	0.9222	0.3873	0.1658	0.1977	1.3718	0.7856	0.1171	0.3136
2013-1-9	1.2747	1.2427	0.9159	0.3896	0.1666	0.1991	1.3728	0.778	0.1162	0.3128
2013-1-10	1.286	1.26	0.9229	0.395	0.1701	0.2061	1.3826	0.7868	0.1179	0.3136
2013-1-11	1.3341	1.314	0.9487	0.4137	0.1759	0.2251	1.4421	0.8221	0.1233	0.3211
2013-1-12	1.0063	0.9268	0.6681	0.3051	0.1306	0.1829	0.9811	0.5525	0.0996	0.2433
2013-1-13	0.9042	0.8547	0.5931	0.2627	0.11	0.1694	0.8562	0.4961	0.0986	0.2239
2013-1-14	1.2863	1.2785	0.9296	0.4025	0.1734	0.2213	1.3986	0.798	0.1213	0.3123
2013-1-15	1.2934	1.2623	0.9271	0.4047	0.1743	0.2196	1.3958	0.7839	0.1177	0.3096
2013-1-16	1.3043	1.2898	0.9335	0.4127	0.1769	0.2232	1.4111	0.7944	0.1195	0.3105
2013-1-17	1.3038	1.2915	0.9134	0.4233	0.1793	0.2286	1.4154	0.8005	0.1196	0.3093
2013-1-18	1.3462	1.3508	0.9489	0.4402	0.182	0.246	1.4647	0.8194	0.1237	0.3104
2013-1-19	1.0213	0.9646	0.6774	0.3273	0.1365	0.2001	1.008	0.5687	0.1005	0.2376
2013-1-20	0.8274	0.8874	0.6113	0.301	0.1172	0.2058	0.9138	0.5017	0.0993	0.2181
2013-1-21	1.2916	1.2775	0.929	0.4265	0.1666	0.2351	1.4252	0.7877	0.1221	0.3123
2013-1-22	1.3011	1.2714	0.9268	0.4277	0.1733	0.2267	1.4183	0.7684	0.1184	0.316
2013-1-23	1.2765	1.2443	0.9089	0.4169	0.1709	0.2264	1.3993	0.7507	0.1158	0.3025
2013-1-24	1.2706	1.2425	0.9087	0.4202	0.1727	0.2236	1.3875	0.7462	0.1156	0.2978

Fig. 12 Part of daily passenger volume of Beijing subway

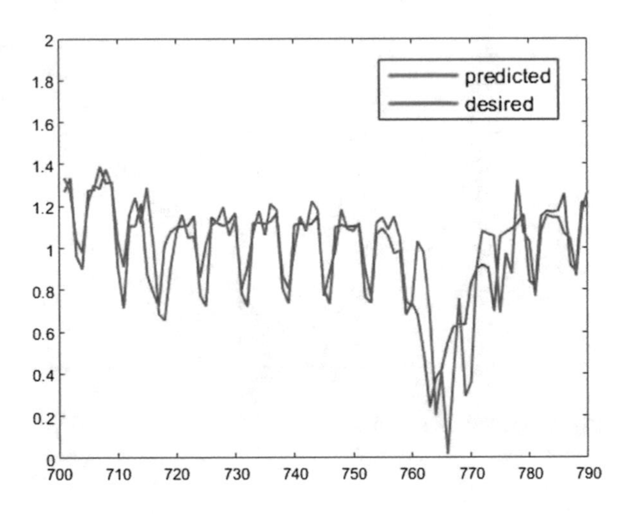

Fig. 13 The predicted and desired values of line 1

of daily passenger volume. Figure 12 shows part of the daily passenger volume of Beijing subway. We use the ST-ELM to predict last 90 days's passenger volume.

Figures 13, 14, 15 and 16 show the predicted and desired values of each line, Fig. 17 shows the RMSE of each prediction. We can see that the accuracy is obtainable.

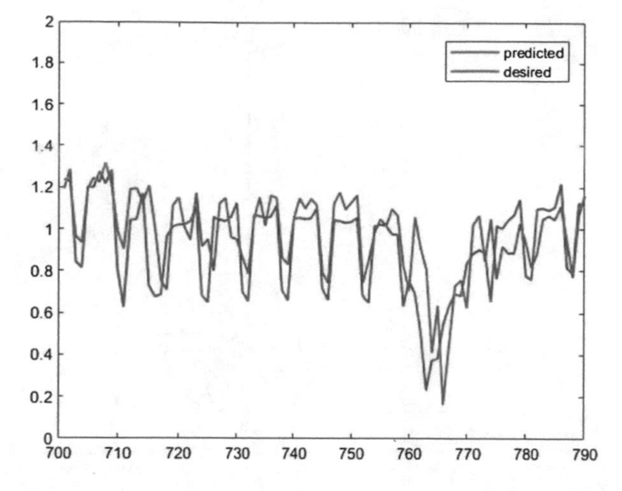

Fig. 14 The predicted and desired values of line 2

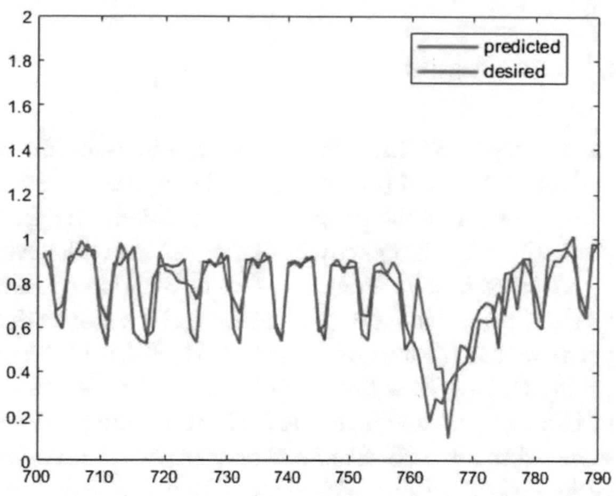

Fig. 15 The predicted and desired values of line 5

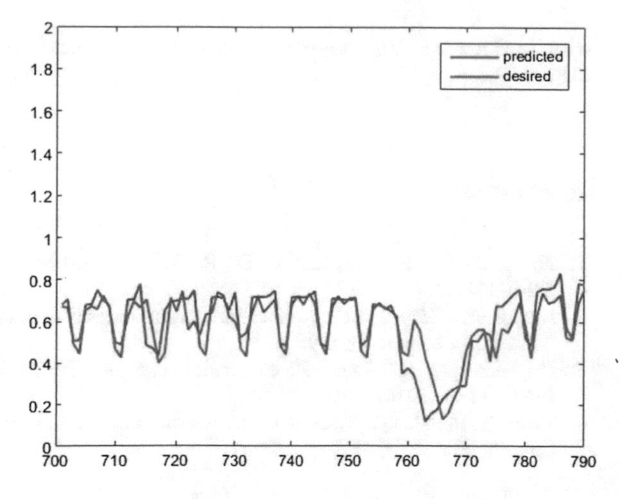

Fig. 16 The predicted and desired values of line 6

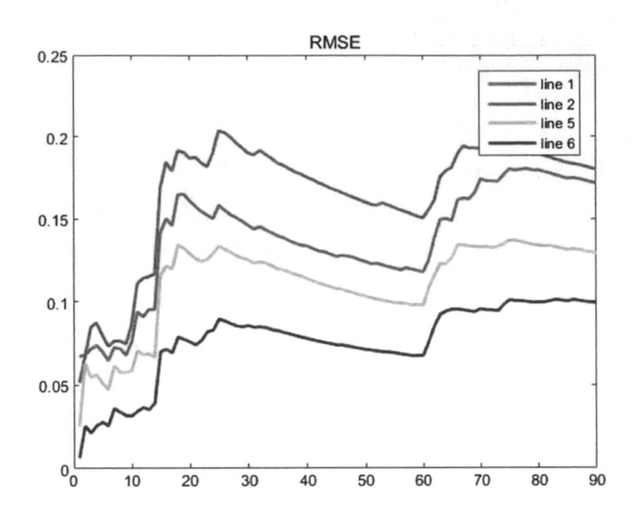

Fig. 17 RMSE of line 1, 2, 5, 6

6 Conclusions

In this paper, ST-ELM is used for continuous time series prediction. The Lorenz and Mackey-Glass time series has been used as examples for demonstration. The results show that the prediction method using ST-ELM is suitable for a short term prediction. And if updates the history values with desired values, it is more accurate.

ST-ELM is quite accurate and do not need to know models or other information of time series, and input weight matrix W, bias matrix B are assigned arbitrarily. In this paper we use random values in $[-1, 1]$. Hidden layers is set to 20, input number L is 15 in Mackey-Glass time series and 2 in Lorenz time series. Although the number of hidden layers and inputs do lead to different accuracy for different time series. We believed that the ST-ELM can be used to other continuous especially chaotic time series.At last we applied into traffic flow prediction in Metro and have some results.

Acknowledgements The Project Supported by the National Natural Science Foundation of China (61572074).

References

1. Zissis, D., Xidias, E.K., Lekkas, D.: Real-time vessel behavior prediction. Evol. Syst. **7**(1), 29–40 (2016)
2. Huang, W.J., Miles, M.D.: Application of time series amra model in long term forecasting. Sci. Sin. **25**(22), 1030–1033 (1980)
3. Furstenberg, H.: Stationary Processes and Prediction Theory (AM-44), vol. 44. Princeton University Press (2016)
4. Virili, F., Freisleben, B.: Neural network model selection for financial time series prediction. Comput. Stat. **16**(3), 451–463 (2001)

5. Mellit, A., Pavan, M.A., Benghanem, M.: Least squares support vector machine for short-term prediction of meteorological time series. Theor. Appl. Climatol. **111**(1–2), 297–307 (2013)
6. Ye, M.: Prediction of chaotic time series using LS-SVM with simulated annealing algorithms. In: International Symposium on Neural Networks, pp. 127–134. Springer (2007)
7. Huang, G.-B., Zhu, Q.-Y., Siew, C.-K.: Extreme learning machine: a new learning scheme of feedforward neural networks. In: 2004 IEEE International Joint Conference on Neural Networks, 2004. Proceedings, vol. 2, pp. 985–990. IEEE (2004)
8. Huang, G.-B., Chen, L., Siew, C.K., et al.: Universal approximation using incremental constructive feedforward networks with random hidden nodes. IEEE Trans. Neural Netw. **17**(4), 879–892 (2006)
9. Mackey, M.C., Glass, L., et al.: Oscillation and chaos in physiological control systems. Science **197**(4300), 287–289 (1977)
10. Lorenz, E.N.: Deterministic nonperiodic flow. J. Atmos. Sci. **20**(2), 130–141 (1963)

Learning Flow Characteristics Distributions with ELM for Distributed Denial of Service Detection and Mitigation

Aapo Kalliola, Yoan Miche, Ian Oliver, Silke Holtmanns, Buse Atli, Amaury Lendasse, Kaj-Mikael Bjork, Anton Akusok and Tuomas Aura

Abstract We present a methodology for modeling the distributions of network flow statistics for the specific purpose of network anomaly detection, in the form of Distributed Denial of Service attacks. The proposed methodology offers to model (using Extreme Learning Machines, ELM), at the IP subnetwork level (or all the way down to the single IP level, if computations allow), the usual distributions of certain network flow characteristics (or statistics), and then to use a One-Class classifier in the detection of abnormal joint flow statistics. The methodology makes use of the original ELM for its good performance to computational time ratio, but also because of the needs in this methodology to have simple update rules for making the model evolve in time, as new traffic and hosts come in.

1 Introduction

Distributed denial-of-service (DDoS) attacks are a present and increasing threat [1] to the availability of networks and internet services. Solutions such as global scale distribution of a service can be effective in mitigating large DDoS attacks, but there are many use cases, such as the telco cloud, where a network or service needs to be protected on a local scale. Defences against DDoS attacks are commonly largely based on signature based detection and mitigation. While this can be effective against known attack patterns, it is vastly preferable for the defence mechanism to autonomously learn to differentiate between normal and attack traffic patterns for the purpose of attack-time traffic filtering.

A. Kalliola (✉) · Y. Miche · I. Oliver · S. Holtmanns · B. Atli
Bell Labs, Nokia, Finland
e-mail: aapo.kalliola@nokia.com

A. Kalliola · B. Atli · T. Aura
Aalto University, Espoo, Finland

K.-M. Bjork · A. Akusok
Arcada University of Applied Sciences, Helsinki, Finland

A. Lendasse
The University of Iowa, Iowa, USA

J. Cao et al. (eds.), *Proceedings of ELM-2016*, Proceedings in Adaptation, Learning and Optimization 9, DOI 10.1007/978-3-319-57421-9_11

In this paper, we propose to model the traffic entering a network as distributions of traffic features tied to traffic source addresses or subnetworks. While existing research [2] has demonstrated the feasibility of simple source-based hierarchical clustering for DDoS mitigation, a more comprehensive analysis methodology is likely to provide further insight against a wider range of DDoS attacks patterns and decrease the potential for false positives.

Our mitigation mechanism is designed for the protection of Cloud infrastructure, including Telco Cloud, against possible DDoS attacks from remote hosts while maintaining low levels of false negatives and false positives in malicious traffic detection and filtering.

In order to fulfil the scenario requirements the traffic analytics needs to output a means of traffic filtering which is directly applicable for line-rate traffic filtering, preferably on existing network elements. For this purpose we aggregate the traffic patterns to sub-networks, which effectively enables easily deployable traffic control on software defined networking hardware with limited flow entry budgets.

In the following, we first describe the problem of network traffic analysis, handling and filtering, for protecting said networks, in Sect. 2. We then move on to the description of the proposed methodology in Sect. 3, building upon the strategies in [2] to create models of the typical traffic for several key indicators. In Sect. 4, we explicitly detail how the model is used, and precisely what computations are required, when the model is actually used for prediction, while the following Sect. 5 is about techniques enabling partial re-training of the model, and update mechanisms with the lowest possible computational cost, and therefore minimal model use disruption. We finally propose a brief state of the art analysis in the related work Sect. 6.

2 Network Traffic

Data traffic in networks comprises of data packets transmitted between communications endpoints. These transmissions form network traffic flows. The extraction of features from these data packets and flows is the first step in our detection and mitigation mechanism. While the available traffic network traffic features have been previously surveyed [3], we herein highlight some element which are potentially useful for our network traffic analysis.

As previously mentioned, from the analysis perspective network traffic falls into two categories: data packets and traffic flows. The core difference is that packet analysis deals with the raw data of a data packet, while flow analysis uses data which has been somehow aggregated as input. Both packet and flow analysis can also be performed on input which is not complete, but rather sampled from the live network traffic.

Individual IP data packets have multiple relevant features in their headers which need to be considered. Source and destination addresses, when not forged, identify the communications endpoints, TTL conveys information about the length of the path the packet has traversed and protocol identifier gives some constraints on what kind of a packet sequence is natural for this type of communication. Depending on the protocol, protocol headers may also provide further insight into the communication. In addition to header values, a sequence of packets also provides indirect features such as packet inter-arrival times and packet fragmentation levels.

A network traffic flow is a unidirectional sequence of data packets with a set of common features, such as IP source and destination address and port number. In modern flow export definitions, e.g. in IPFIX [4], the selection of these features is very flexible and can also contain information beyond the content of the data packet, for instance next-hop IP addresses or physical network port indexes.

The main difference between packet and flow analysis is in the level of available data. With packet analysis we have potential access to the complete communications content, while flow records provide metadata of communications. Commonly monitored statistics include e.g. endpoint addresses, flow duration, traffic packet count, flow duration and traffic protocol.

Deep packet inspection (DPI) and application layer analysis of network traffic is commonly done by firewalls and intrusion detection systems (IDSs). These approaches have benefits in detecting certain attacks, for instance non-bandwidth-intensive attacks [5] such as Slowloris.

However, volumetric DDoS attacks still form the vast majority of DDoS attacks on the internet [1]. Thus, analysing and mitigating volumetric attacks will be our target scenario within the scope of this paper. This type of an attack aims to exhaust the target's capacity to handle incoming traffic by either overloading the serving endpoint or by congesting the capacity of network links connecting the endpoint to the wider network.

In order to achieve meaningful differentiation between normal and anomalous traffic, these traffic classes must have some differentiating features. In real world some attack types, e.g. SYN floods, can be relatively easy to separate from real traffic, while others, such as heavily distributed HTTP valid request floods, can be more difficult to identify.

We base our work in this paper on the realistic assumption that *normal traffic and attack traffic are not in all aspects identical*. Since an attacker does not typically have detailed knowledge of the normal traffic pattern of the target, it is a near certainty that there are distinguishing features between normal and attack traffic. In our approach we do not manually predefine any features as more or less significant than others, but rather build distributions of all monitored features and use the differences in the distributions over time as the basis for attack detection and mitigation.

Attack detection and traffic analysis do not automatically provide useful information for attack mitigation. In our work we have considered the available set of traffic forwarding capabilities provided by routing hardware, and use the lowest common denominator, i.e. subnet based rules, as the vehicle for deploying our mitigation on network devices. This approach has the benefit of being widely applicable and extremely lightweight in terms of traffic forwarding performance impact.

3 Methodology

3.1 Initial Training and Construction of the Model

In this initial training of the model components, we take a certain time window and keep it fixed during the training. In this sense, the model obtained at the end of this construction phase, reflects the behavior of the IP addresses and sub networks for this very specific time frame (Fig. 1).

In Sect. 5, we propose mechanisms that allow new information to be included in the model, without the need to recompute all the model components.

Let us denote by $T_{1;N} = [t_1; t_N]$ the time period spanning from t_1 to t_N, sampled uniformly at the time intervals $\{t_i\}_{1 \leq i \leq N}$.

Let us then assume that we have built, over this period of time $T_{1;N}$, the Hierarchical tree holding the structure of the IP addresses present in the data file D used for the training. This data D is, e.g. a pcap file that was recorded on a network while there were no suspected attacks or misbehaviour. An example depiction of that tree is proposed on Fig. 2. Note that on this illustration, the leaves of the tree are not necessarily single IP addresses, but more likely, whole ranges/subnetworks. This can be decided and controlled to avoid obtaining a tree that would be too large to handle, computationally.

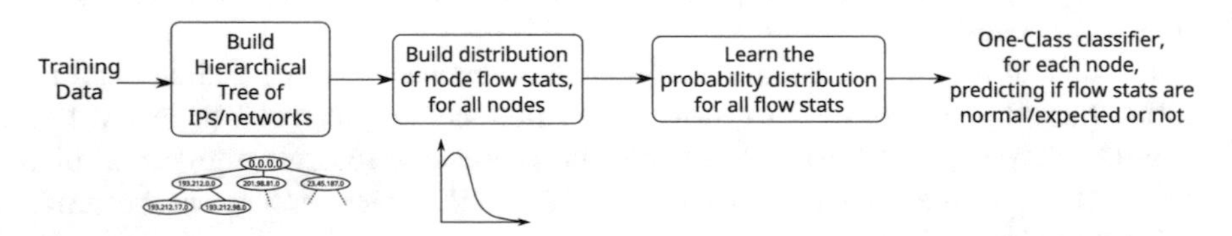

Fig. 1 High level description of the methodology: The training data is first analyzed and processed into the IP/subnetwork tree it contains; each of the vertex in this tree/graph holds the time series for each flow statistic that is recorded; the probability distributions of these flow statistics are then modeled, and a final model (for each graph vertex) learns the "normal" situation in terms of the several probability distributions of the flow statistics

Fig. 2 Example of a hierarchical tree holding the IP structure encountered so far in the training data. The *dashed lines* illustrate the tree nodes continuing, but not depicted here for clarity. Note that the leaves of the tree do not have to be single IP addresses, but can be whole subnetworks/IP ranges

We denote by

$$G\left(T_{1;N}\right) = \left(V\left(T_{1;N}\right), E\left(T_{1;N}\right)\right), \tag{1}$$

the temporal graph holding this tree structure devised on the data file D. And we note K the number of vertices present in this graph (K is of course also dependent on $T_{1;N}$, but we omit the notation for clarity).

Note that once the following training procedure is finished, new incoming data might alter this graph (by adding new nodes to it, e.g.). We discuss this in Sect. 5.1 later, and for now consider the graph "static" over the considered time period.

Assuming that we record d flow statistics (TTL, average packet size, packet timings…) at a certain time t, denoted as

$$\left\{s_i^j(t)\right\}_{1 \le j \le d} \tag{2}$$

at each vertex V_i of the graph (thus representing either a single IP or a whole subnet of them), we then have that the structure $V\left(T_{1;N}\right)$ holding the data related to all vertices, is a tensor

$$\mathbf{V}\left(T_{1;N}\right) = \begin{bmatrix} \mathbf{V}_1^1\left(T_{1;N}\right) & \cdots & \mathbf{V}_1^d\left(T_{1;N}\right) \\ \vdots & \vdots & \vdots \\ \mathbf{V}_K^1\left(T_{1;N}\right) & \cdots & \mathbf{V}_K^d\left(T_{1;N}\right) \end{bmatrix}, \tag{3}$$

with $\mathbf{V}_i^j\left(T_{1;N}\right) \in \mathbb{R}^N$ the vector holding the values for flow statistic j at vertex (i.e. node/IP subnet) i.

With N large enough (meaning we have a sufficient number of flow statistics values evolving over time), we can form

$$f_{\mathbf{V}_i^j}(x), \tag{4}$$

the distribution of the values of V_i^j over the time period $T_{1;N}$, that is, the distribution of the values of flow statistic j at vertex i.

We then model this distribution using a Universal Function Approximator [6], in this case the Extreme Learning Machine [7]. This allows to have a model of the distribution with a small computational time spent on the training, from which we can extract probabilities of occurrence of unseen flow statistics events.

Thus, we create the approximation function $\tilde{f}_{\mathbf{V}_i^j}(x)$ to the real distribution as

$$\tilde{f}_{\mathbf{V}_i^j}(x) = \mathrm{ELM}\left(f_{\mathbf{V}_i^j}\right)(x), \tag{5}$$

where the ELM($\cdot$) notation means that the distribution is learned by the means of an ELM [7]. We do not mention the normalization of the distribution $\tilde{f}_{\mathbf{V}_i^j}$ by the normalization factor, here, but it is performed so as to make sure that the values predicted by $\tilde{f}_{\mathbf{V}_i^j}$ are in fact probabilities. In effect, $\tilde{f}_{\mathbf{V}_i^j}(x)$ is approximating the probability of the value x to happen for the statistic j, at vertex i.

We can then form the matrix of functions $\mathbf{V}_{\tilde{f}}$ as

$$\mathbf{V}_{\tilde{f}} = \begin{bmatrix} \tilde{f}_{V_1^1} & \cdots & \tilde{f}_{V_1^d} \\ \vdots & & \vdots \\ \tilde{f}_{V_K^1} & \cdots & \tilde{f}_{V_K^d} \end{bmatrix}, \tag{6}$$

which is in effect a matrix of functions.

More importantly, we can define, for a specific vertex i, the matrix of values

$$\tilde{\mathbf{V}}_i = \begin{bmatrix} \tilde{f}_{V_i^1}(s_i^1(t_1)) & \cdots & \tilde{f}_{V_i^d}(s_i^d(t_1)) \\ \vdots & & \vdots \\ \tilde{f}_{V_i^1}(s_i^1(t_N)) & \cdots & \tilde{f}_{V_i^d}(s_i^d(t_N)) \end{bmatrix}, \tag{7}$$

which holds the estimates of the probabilities of the statistics s_i^j for a specific vertex i.

For each vertex i, i.e. subnetwork, we want to have a model that is capable of recognising the “normality”; that is to say, a model that learns that the probabilities values from $\tilde{\mathbf{V}}_i$ are the normal situation for vertex i.

We thus propose to learn, using a One Class classifier (the OC-ELM [8], in this work, further presented in the following Sect. 3.3), the normal situation as that represented in $\tilde{\mathbf{V}}_i$. We then have, for vertex i, a model $\mathcal{M}_i(\mathbf{x})$ that takes as argument $\mathbf{x}$, the probabilities of the different statistics it has been trained on, and returns whether the given probabilities taken together, represent a normal situation, or an outlier.

3.2 About Extreme Learning Machines

In this work, we propose to use Extreme Learning Machines (ELM) [7, 9] as the learning tool to create a model of the distributions $f_{\mathbf{V}_i^j}(x)$. There are three main reasons for using this specific Machine Learning technique. First, it is among the techniques with the best performance/computational time ratio, as the model is mathematically simple and involves a minimal amount of computations. Second, we need a model that can create a model of the distributions in reasonable time from large amounts of data, if such a need arises (such as in the cases of real-time processing and modeling). Third, the theory behind ELM (and behind single layer feed-forward neural networks in general) states that it is a universal function approximator (per the universal approximation theorem [6]), and can therefore fit any continuous function, to a $\varepsilon > 0$.

The Extreme Learning Machine algorithm was originally proposed by Guang-Bin Huang et al. in [9] (and further developed, e.g. in [10–12] and analysed in [13]). It uses the structure of a Single Layer Feed-forward Neural Network (SLFN). The main concept behind ELM is the replacement of a computationally costly procedure of training the hidden layer, by its random initialisation. Then an output weights matrix between the hidden representation of inputs and the outputs remains to be found. The ELM is proven to be a universal approximator given enough hidden neurons [9]. It works as following:

Consider a set of N distinct samples $(\mathbf{x}_i, \mathbf{y}_i)$ with $\mathbf{x}_i \in \mathbb{R}^d$ and $\mathbf{y}_i \in \mathbb{R}^c$. Then a SLFN with n hidden neurons is modelled as $\sum_{j=1}^{n} \beta_j \phi(\mathbf{w}_j \mathbf{x}_i + b_j)$, $i \in [1, N]$, with $\phi : \mathbb{R} \rightarrow \mathbb{R}$ being the activation function, $\mathbf{w}_j$ the input weights, b_j the biases and β_j the output weights.

In case the SLFN would perfectly approximate the data, the errors between the estimated outputs $\hat{\mathbf{y}}_i$ and the actual outputs $\mathbf{y}_i$ are zero, and the relation between inputs, weights and outputs is then $\sum_{j=1}^{n} \beta_j \phi(\mathbf{w}_j \mathbf{x}_i + b_j) = \mathbf{y}_i$, $i \in [1, N]$ which can be written compactly as $\mathbf{H}\boldsymbol{\beta} = \mathbf{Y}$, with $\boldsymbol{\beta} = (\beta_1^T \dots \beta_n^T)^T$, $\mathbf{Y} = (\mathbf{y}_1^T \dots \mathbf{y}_N^T)^T$.

Solving the output weights $\boldsymbol{\beta}$ from the hidden layer representation of inputs $\mathbf{H}$ and true outputs $\mathbf{Y}$ is achieved using the Moore-Penrose generalised inverse of the matrix $\mathbf{H}$, denoted as $\mathbf{H}^\dagger$ [14]. The training of ELM requires no iterations, and the most computationally costly part is the calculation of a pseudo-inverse of the matrix $\mathbf{H}$, which makes ELM an extremely fast Machine Learning method.

Therefore, using ELM, we propose to learn the $f_{\mathbf{V}_i^j}(x)$, in order to build a (hopefully only interpolated) model of the real underlying distribution.

3.3 About the OC-ELM

The main idea behind the One-Class ELM (OC-ELM), as proposed in [8], is to use the distance between the predicted one-class output of a classical (although regularised) ELM, and the real output. In this method, we are thus effectively checking whether a sample lies within a ball of a certain radius θ centred on the real output value (which is unique for all samples, as we are working with a one-class assumption.

In effect, given the same notations as in Sect. 3.2 for the ELM, we here have the extra information that $\mathbf{y}_i = y = 1, \forall i$ (the actual value used for the $\mathbf{y}_i$ is not important, but choosing 1 makes computations simpler and normalisation implicit). We then have to define the distance

$$d_{\text{ELM}}(\mathbf{x}_i) = \left|\left|\mathbf{H}(\mathbf{x}_i)\boldsymbol{\beta} - y\right|\right|, \quad (8)$$

where the notation $\mathbf{H}(\mathbf{x}_i)$ conveniently denotes the output of the hidden layer of the ELM for the specific sample $\mathbf{x}_i$.

A key point of the OC-ELM in [8] is then to determine the optimum radius θ of the ball around y to achieve the best classification rate, while avoiding the trivial case where θ is too large (and therefore does not generalise well). The authors propose to reject a portion of the data, considered somewhat as "outliers" by the OC-ELM model, so as to obtain the most compact ball possible. Future samples outside of this ball are then considered as not belonging to the one class.

The decision function for the OC-ELM is then straightforward to derive once θ has been determined, and the authors define

$$\begin{aligned} C_{\mathrm{ELM}}(\mathbf{z}) &= \operatorname{sign}\left(\theta - d_{\mathrm{ELM}}(\mathbf{z})\right) \\ &= \begin{cases} +1 & \mathbf{z} \text{ belongs to the one class.} \\ -1 & \mathbf{z} \text{ is an outlier.} \end{cases}, \end{aligned} \tag{9}$$

for a new sample $\mathbf{z}$. For a thorough exploration of the proper determination of the key parameter θ for the OC-ELM, we refer the reader to the original paper [8].

4 Using the Methodology on Test Data

Using the devised methodology on test data is actually trivial, but can be separated in two distinct cases: (i) The test data is a single moment t in time, i.e. we want to know whether the network behaviour is normal for a specific moment in time, as described in Sect. 4.1. (ii) The test data is a time window $T_{1;N}$ (which can in effect be a rolling window) over which we want to know about the network behaviour, as described in Sect. 4.2.

4.1 Testing the Model on a Single Moment

In this case, we assume that we are collecting the statistics for a certain vertex i, at a moment t in time. Testing the model on such data then requires the following steps:

1. Gather all statistics $\mathbf{s}_i(t) = \left(s_i^1(t), \ldots, s_i^d(t)\right)$ for vertex i;
2. Use the functions in $\mathbf{V}_{\tilde{f}}$ to predict the probabilities $\mathbf{x}_i(t)$ of the statistics values $\mathbf{s}_i(t)$;
3. Calculate $\mathcal{M}_i(\mathbf{x}_i(t))$ to estimate whether the statistics are considered anomalous by the model, based on their probabilities;
4. Act based on the output of the model $\mathcal{M}_i$: if the statistics are considered abnormal by the model, there is anomalous behaviour at vertex i, and the host should, e.g. be blocked or restricted.

Testing the model on a single time instant, here, poses some questions. First, the computational load, even if the computations only involve using ELM models on a very limited set of values (for a given vertex), is large: This test potentially has to be run over the whole set of vertices of the graph, which may become large as time grows. More hosts end up connecting to the monitored cloud, and the graph holding the IP addresses structure G, grows large. To address this issue, we discuss current research ideas, in Sect. 5.4 about forgetting factors and globally scaling this methodology to large amounts of connecting hosts.

In practice, the approach of testing on a single time instant makes the methodology prone to false positives, as the combination of probabilities of the statistics for this specific instant, is likely to contain small anomalies—which would be better "ironed out" by a sliding time window.

4.2 Testing the Model on a Time Window

In this case, we consider that we collect the same statistics for the same vertex i, but over a certain time period $T_{k;l} = (t_k, \dots, t_l)$ (as before, uniformly sampled over this time period).

To keep things simple, we propose in this paper, to simply compute the average of each single statistic s_i^j over the time period. This case then reduces to the previous one.

The advantage in using a time window is twofold: First, the averaging ensures, as mentioned before, that single traffic anomalies get averaged out in the process, thus lowering false positives in practice; Second, this process avoids running the testing too often and therefore lowers the computational complexity, in testing many vertices for anomalies.

As with any model trained on a specific time window, though, there comes the problem of obsolescence of the model, after a certain amount of time has passed: new hosts are contacting the cloud, for which we have no previous data to predict behavior, and it is also very possible that former hosts have different traffic patterns over time (some subnetworks/hosts are more active during certain periods of the day, e.g.). For this purpose, we detail in the next Sect. 5, how to perform continuous learning of the model, first by adding new data to existing vertices of the graph—i.e. continuing to learn the possibly changing behaviour of some known hosts—, and second, by adding new vertices to the whole graph.

5 Continuous Learning of the Model

In this section, we detail two cases which use essentially the same principles to make the graph, and the model underlying it, evolve with new data. First, by adding new data to existing vertices, and then by adding new vertices to the graph. Since the computational costs of this whole methodology are a concern for close to real-time usage of the model, we want to have these update mechanisms as low cost as possible, and thus propose to update the model, to avoid heavy re-computations.

5.1 Adding Data to Existing Vertices

In this case, the new data to be added to an existing vertex V_i is coming in the form of new statistics values $\left\{ s_i^j(t) \right\}_{1 \leq j \leq d}$, for various values of t. For the sake of covering real-life cases, we would like to consider two cases regarding the t values, here: First in Sect. 5.1.1, the expected case in which the time instants at which the statistics are gathered are all the same and are uniformly sampled over a certain time period; Second in Sect. 5.1.2, the real-life case in which some statistics are actually collected at slightly different time instants and for which some might actually lack a value (because of data collection errors, e.g.).

5.1.1 New Data over Uniformly Sampled Time Instants

This case is rather trivial, thanks to the existing OS-ELM [15]. Here we assume that new incoming data for a vertex V_i is essentially dense and uniformly sampled. In this sense, we basically have new statistics $\left\{ s_i^j(t) \right\}_{1 \leq j \leq d}$ for t over a time window $T_{k;l} = \left(t_k, \ldots, t_l\right)$. The problem then reduces to that of extending the training data used in the ELM for the construction of $\tilde{f}_{\mathbf{V}_i^j}(x)$, which can in effect be solved by using the Online-Sequential ELM [15] (OS-ELM, more details given in Sect. 5.3) algorithm approaches. The OS-ELM algorithm allows to add new data to the training set without full retraining of the formerly calculated distribution $\tilde{f}_{\mathbf{V}_i^j}(x)$. We then need to also update the One-Class classifier model $\mathcal{M}_i(\mathbf{x})$ accordingly. Interestingly, while the OC-ELM differs slightly from the original ELM, the update equations and approaches of the OS-ELM can be applied directly to it, thus allowing to update at low cost the One-Class classifier used. One could argue that the determination of an updated θ value that is optimal for the whole data (former and new together), is required. But in practice, if the amounts of new data are not very large, and if the new data behaves in previously know ways, the update of θ is not critically required for every update. It might be required after a certain amounts of updates, though, or if the amount of detected outliers by the OC-ELM starts to grow large.

5.1.2 New Data with Non-uniform Sampling

This case deals with the possibility of getting new data that is not uniformly sampled over time, for the various statistics that are collected. In effect, this means that the tensor $\mathbf{V}(T_{1;N} + T_{k;l})$ (where the notation $T_{1;N} + T_{k;l}$ means concatenating these two time windows together), is potentially sparse (as well as not uniformly sampled), with no values in some places. Since all the computations performed in this methodology are assuming dense matrices, we propose to fall back from this case to the previous one (of dense tensor) by drawing the missing statistic values from the estimated distributions. In effect, we fill the tensor $\mathbf{V}(T_{1;N} + T_{k;l})$ in a missing statistic value $s_i^j(t)$ (for a time t for which we do not have that statistic), by drawing from the associated estimated probability distribution $\tilde{f}_{\mathbf{V}_i^j}(x)$. The previously sparse tensor is then brought to a full tensor, and we are in the same case as in the previous Sect. 5.1.1, and can use the OS-ELM to update the models.

5.2 Adding a New Vertex to the Graph

Adding a new vertex to the graph is straightforward, in terms of the computations to be made (assuming the parent vertices of this new vertex are updated using the aforementioned processes from Sect. 5.1), as this is essentially what is initially done for all the vertices in the training stage.

The steps for this process come down to:

1. Add the row corresponding to the new vertex V_{new} data in $\mathbf{V}$ (note that the time window for the new row will be different from the initial $T_{1;N}$ one used for the initial training, but this will only cause the tensor $\mathbf{V}$ to be sparse and has no other drawback);
2. Compute (as in the training phase) the estimated functions $\tilde{f}_{V^j_{\text{new}}}(x), \forall j \in [\![1, d]\!]$;
3. Use an OC-ELM to build the One-Class classifier $\mathcal{M}_{\text{new}}(\mathbf{x})$ for this new vertex.

Thus, we can add new IP sub-networks to the graph, without disrupting the entire structure with re-computations. This process can effectively be done "in parallel" and then added to the full model.

5.3 About the OS-ELM

In [15], a sequential learning modification of the original ELM is proposed, to be able to update the ELM model without complete re-computation. Taking the same notations as in Sect. 3.2, i.e. given a training set comprised of N^0 samples $\mathbf{X}^0 = \left(\mathbf{x}_1^0, \dots, \mathbf{x}_{N^0}^0\right)^T \in \mathbb{R}^{N^0 \times d}$ and targets $\mathbf{Y}^0 = \left(\mathbf{y}_1^0, \dots, \mathbf{y}_{N^0}^0\right) \in \mathbb{R}^{N^0 \times c}$, and an ELM model trained on this data, with of n neurons, we have (assuming that $\mathbf{H}^0$, the hidden layer output matrix, is full rank) the matrix $\boldsymbol{\beta}^0 = \left(\mathbf{H}^{0T}\mathbf{H}^0\right)^{-1} \mathbf{H}^{0T}\mathbf{Y}^0$ holding the output weights of the current model. Now, given a new chunk of data $\mathbf{X}^1, \mathbf{Y}^1$, with N^1 number of samples in it (which will be concatenated to the end of $\mathbf{X}^0$ and $\mathbf{Y}^0$), we now need to find the new $\boldsymbol{\beta}$ such that

$$\left\| \begin{bmatrix} \mathbf{H}^0 \\ \mathbf{H}^1 \end{bmatrix} \boldsymbol{\beta} - \begin{bmatrix} \mathbf{Y}^0 \\ \mathbf{Y}^1 \end{bmatrix} \right\| \tag{10}$$

is minimized. Omitting most of the derivations (which can be found in the original paper [15]), we can get the updated output weights $\boldsymbol{\beta}^1$ directly as

$$\boldsymbol{\beta}^1 = \boldsymbol{\beta}^0 + \left(\mathbf{H}^{0T}\mathbf{H}^0 + \mathbf{H}^{1T}\mathbf{H}^1\right)^{-1} \mathbf{H}^{1T} \left(\mathbf{Y}^1 - \mathbf{H}^1\boldsymbol{\beta}^0\right). \tag{11}$$

Note that most of the elements in this equation are already calculated (or have very low computational cost). It is of course possible to derive the general recursive formula for the general case, and we refer the reader to the original paper for it.

In this work, the OS-ELM is also potentially used for the special case of adding only one sample to the training set (the case where $N^1 = 1$), which happens to have a special form (using the Sherman-Morrison formula [16]). Again, please see the original paper for the derivation of this special case.

5.4 Future Work: Scaling and Forgetting Factor for Continuous Learning

From the design of the graph tree holding the IP sub-networks, it is easily seen that this approach cannot scale to large amounts of IP addresses and retain all the information it has ever seen. The tree would rather quickly become too large and saturate the memory. For this reason, we talk of sub-networks in this paper, instead of single IP addresses, as the leaves of the tree. We thus limit ourselves to aggregated IP networks so as to keep the size of the tree reasonable to handle.

Future work on this very specific problem will be centred around incorporating a forgetting factor in the computations.

A forgetting factor in this context would have two major advantages: limit the size of the tree by forgetting sub-networks that have not contacted the currently monitored host for a long time; and make sure that the density estimates only take into account the most recent data. Or at least, make sure that the most recent data is taken as the most relevant data for the estimations: sub-networks that have been encountered recently are most likely to connect again in the near future.

Our goal is to implement a variable forgetting factor (not a constant or linear one) which takes possibly into account the variability over time of some sub-networks connecting to the monitored host. This is likely to include some sort of seasonality-like behaviour.

Finally, while a preliminary evaluation of this method is performed internally, on private data, an extended version of this work will feature testing on publicly available data, such as the ISCX IDS dataset [17].

6 Related Work

Network traffic flows and the characteristics of anomalies in network traffic have been a subject of extensive previous research, e.g. Barford and Plonka [18] have performed time series analysis to flow data. IP source address based clustering and subnetwork based filtering has been shown to be a viable approach to DDoS mitigation by Kalliola et al. [2]. Compared to previous work, the method proposed in our paper is expected to more extensively capture differentiating traffic features, leading to lower false positives and negatives in detection and in traffic priorization and filtering.

Machine learning in the context of DDoS detection and filtering has been explored e.g. by Seufert and O'Brien [19], wherein the authors use an artificial neural network (ANN) with traffic features from different layers for anomaly detection. Berral et al. [20] have proposed a distributed mechanism for mitigating flooding DDoS attacks. In their mechanism many elements in the network share traffic views for the purpose of detecting and filtering attack traffic before it reached the destination network.

More recently, ELM based approaches specifically have been used for Intrusion Detection testing, as for example in [21], where the authors propose to use the ELM (both in binary classification—normal versus abnormal traffic, and multi-class classification aimed at detecting the type of attack). In [22], the authors also propose to use a variation of the ELM (Weighted ELM) to perform multi-class classification on existing data sets (NSL-KDD, e.g.). Similarly, in [23], a modification of the ELM in the form of a Kernel ELM, with Multiple Kernel Boosting and Ensemble approach, is applied to the NSL-KDD cup data set (among many others), and the authors note the excellent performance of such model, with the low computational cost of the ELM associated to it.

The major difference between this existing body of work, and the approach proposed in this paper is that we offer here a methodology tailored to the problem of learning, in time, what the traffic is like, in terms of the measured statistics of it; while the existing literature mostly concentrates on improving the performances of a model over a certain data set, we also propose in this paper, the detailed creation of said data set, by transforming the network flow data into the statistics that are used to build the tree model.

7 Conclusion

In this paper we have detailed a method for modelling network traffic as traffic feature distributions relating to a hierarchical network tree. By gathering traffic feature statistics over certain periods of time, the proposed methodology enables the creation of probability distribution of these statistics. One-class classification is then used to learn what is considered as the normal behaviour of the network, in terms of the probability values of the various statistics. Major deviations, probabilistically speaking, from this learned "normality", will trigger the detection of an anomaly.

While a detailed evaluation of the method is still in progress, our proposed approach shows promise in capturing essential characteristics of network traffic for normal versus attack traffic differentiation, and for providing a useful output from this differentiation for traffic filtering deployment on real-life network elements. In addition, the method takes into account the crucial timeliness of the network data, by summarising the input data values (in the form of the observed network statistics) into modelled distributions of values. In future work, the time aspect of this analysis will be more prevalent, by using forgetting factors to achieve a complex weighting scheme for the statistics distributions, in time.

References

1. Akamai: Akamai's [State of the Internet]/Security Q1/2016 Report. http://www.akamai.com/StateOfTheInternet (2016)
2. Kalliola, A., Lee, K., Lee, H., Aura, T.: Flooding DDoS mitigation and traffic management with software defined networking. In: 2015 IEEE 4th International Conference on Cloud Networking (CloudNet), pp. 248–254, Oct 2015
3. Iglesias, F., Zseby, T.: Analysis of network traffic features for anomaly detection. Mach. Learn. **101**(1), 59–84 (2015)
4. Claise, B., Trammell, B.: Specification of the IP flow information export (IPFIX) protocol for the exchange of flow information. RFC **7011** (2015)
5. Cambiaso, E., Papaleo, G., Aiello, M.: Taxonomy of Slow DoS Attacks to Web Applications, pp. 195–204. Springer (2012)
6. Cybenko, G.: Approximations by superpositions of sigmoidal functions. Math. Control Signals Syst. **2**(4), 303–314 (1989)
7. Guangbin, H., Chen, L., Siew, C.-K., Huang, G.-B., Lei, C., Siew, C.-K.: Universal approximation using incremental constructive feedforward neural networks with random hidden nodes. IEEE Trans. Neural Netw. **17**(4), 879–892 (2006)
8. Leng, Q., Qi, H., Miao, J., Zhu, W., Su, G.: One-class classification with extreme learning machine. Math. Prob. Eng. **2015**(Article ID 412957), 1–11 (2015)
9. Huang, G.-B, Zhu, Q.-Y., Siew, C.-K.: Extreme learning machine: theory and applications. Neurocomputing **70**(1), 489–501 (2006)
10. Miche, Y., Sorjamaa, A., Bas, P., Simula, O., Jutten, C., Lendasse, A.: OP-ELM: optimally-pruned extreme learning machine. IEEE Trans. Neural Netw. **21**(1), 158–162 (2010)
11. Miche, Y., van Heeswijk, M., Bas, P., Simula, O., Lendasse, A.: TROP-ELM: a double-regularized ELM using LARS and Tikhonov regularization. Neurocomputing **74**(16), 2413–2421 (2011)
12. Van Heeswijk, M., Miche, Y., Oja, E., Lendasse, A.: GPU-accelerated and parallelized ELM ensembles for large-scale regression. Neurocomputing **74**(16), 2430–2437 (2011)
13. Cambria, E., Huang, G.-B, Kasun, L.L.C., Zhou, H., Vong, C.M., Lin, J., Yin, J., Cai, Z., Liu, Q., Li, K., Leung, V.C.M., Liang F., Ong, Y.-S., Lim, M.-H., Anton A., Amaury L., Francesco C., Rui N., Yoan M., Paolo G., Rodolfo Z., Sergio D., Xuefeng Y., Kezhi M., Oh, B.-S., Jehyoung J. Toh, K.-A., Teoh, A.B.J., Kim, J., Yu, H., Chen, Y., Liu, J.: Extreme learning machines [trends and controversies]. IEEE Intell. Syst. **28**(6), 30–59 (2013)
14. Radhakrishna C.R., Mitra, S.K.: Generalized Inverse of Matrices and Its Applications. Wiley (1972)

15. Liang, N.Y., Huang, G.B., Saratchandran, P., Sundararajan, N.: A fast and accurate online sequential learning algorithm for feedforward networks. IEEE Trans. Neural Netw. **17**(6), 1411–1423 (2006)
16. Golub, G.H., Van Loan, C.F.: Matrix Computations. The Johns Hopkins University Press (2013)
17. Shiravi, A., Shiravi, H., Tavallaee, M., Ghorbani, A.A.: Toward developing a systematic approach to generate benchmark datasets for intrusion detection. Comput. Secur. **31**(3), 357–374 (2012)
18. Barford, P., Plonka, D.: Characteristics of network traffic flow anomalies. In: Proceedings of the 1st ACM SIGCOMM Workshop on Internet Measurement, IMW '01, pp. 69–73. ACM, New York, NY, USA (2001)
19. Seufert, S., O'Brien, D.: Machine learning for automatic defence against distributed denial of service attacks. In: 2007 IEEE International Conference on Communications, pp. 1217–1222, June 2007
20. Berral, J.L., Poggi, N., Alonso, J., Gavaldà, R., Torres, J., Parashar, M.: Adaptive distributed mechanism against flooding network attacks based on machine learning. In: Proceedings of the 1st ACM Workshop on Workshop on AISec, AISec '08, pp. 43–50. ACM, New York, NY, USA, (2008)
21. Cheng, C., Tay, W.P., Huang, G.B.: Extreme learning machines for intrusion detection. In: The 2012 International Joint Conference on Neural Networks (IJCNN), pp. 1–8, June 2012
22. Srimuang, W., Intarasothonchun, S.: Classification model of network intrusion using weighted extreme learning machine. In: 2015 12th International Joint Conference on Computer Science and Software Engineering (JCSSE), pp. 190–194, July 2015
23. Fossaceca, John M., Mazzuchi, T.A., Sarkani, S.: Mark-ELM: Application of a novel multiple kernel learning framework for improving the robustness of network intrusion detection. Expert Syst. Appl. **42**(8), 4062–4080 (2015)

Discovering Emergence and Bidding Behaviour in Competitive Electricity Market Using Agent-Based Simulation

Ly-Fie Sugianto and Zhigang Liao

Abstract The aim of this paper is to explore the implication of multi agent interaction, learning and competing in a repetitive trading environment. Using the complex systems paradigm, the study attempts to observe the behavior of the agents and the emergence phenomena resulting from the multi agent interaction. Using Q-learning, generator agents can rapidly learn the market mechanism and auction rules as they seek to maximize their revenue by modifying their bidding strategies. In this paper, we experiment with different pricing rule to observe its impact on agents' behavior. The paper also describes the types of agents in each domain, together with the properties, relationships, processes and events associated with the agents. *Emergence* from this study includes collusion and capacity withholding to inflate price. The *emergence* is evidence that we can gain new knowledge from the Sciences of the Artificial.

Keywords Agent-based model · Artificial intelligence · Complex systems · Q-learning

1 Introduction

Introducing competition in the electricity trading implies the importance of market mechanism to ensure a contestable market for new entrants and incumbents. Factors, such as demand, transmission constraints, types of power stations, regional settings and policy governing the market, are influencing the complex interactions among market participants. For generators, simply bidding at the marginal cost may not recover its fixed or stranded asset costs. Thus, there is a need for more sophisticated bidding strategy for generators to compete in the market environment. A heuristic technique for optimal bidding requires an assessment of market power

---

L.-F. Sugianto (✉) · Z. Liao
Monash Business School, Monash University, Caulfield East, Melbourne, VIC 3145, Australia
e-mail: Lyfie.sugianto@monash.edu

J. Cao et al. (eds.), *Proceedings of ELM-2016*, Proceedings in Adaptation, Learning and Optimization 9, DOI 10.1007/978-3-319-57421-9_12

and how each player exercises its market power in order to maximize its profit. The optimum bid corresponding to maximum profit depends on market conditions, such as the market power and the generation cost. In practice, the formulation must also account for complex factors, such as load forecast, network congestion, gaming strategies of competitors, generation cost and physical constraints of generators.

Given that the complexity of the relationship among many variables cannot be modeled accurately, a computerized simulation platform is deemed suitable to study the dynamics of a competitive market. The simulation platform also provides an empirical method to observe electricity price fluctuation. Running the simulation over a considerable time horizon allows us to answer important questions: Does the electricity price fall to the marginal cost level? Does the electricity price hit the price cap? If so, under what circumstances would such price spike occur? Which pricing rule leads to the most economical electricity price for the consumer? What is the role of the short-run repetitive trading of electricity in establishing a sustainable supply in this sector?

This paper presents an agent based simulation to analyze the impact of different pricing rules on the electricity price. The proposed simulation is aimed to assist strategic and operational challenges in the electricity supply industry, namely making strategic bids based on available generation capacity to ensure positive earning. In the longer term, generator companies can utilize the simulation as a decision support tool to make strategic investment to start a new plant.

The remainder of this paper has been structured as follows. Section 2 provides literature reviews on the pricing rules and the use of computer simulation and specifically the agent-based model, in studying the electricity market model. Section 3 presents the agent-based model, highlighting Q-learning as the learning mechanism to build the agents' intelligence for strategic bidding. This section extends our previous work on the use of agent-based simulation reported in [1]. The main distinction of the simulation scenarios presented in this paper is that we enrich the comparison of not only Uniform and Vickrey pricing rules, but also includes the Pay-as-bid pricing rule. Section 4 presents case examples to compare the impact of different pricing rules on the agents' bidding behavior when maximum capacity bid is enforced as well as when variable bidding quantity is allowed. Lastly, the paper provides analysis on the results, including on the agent's learning mechanism, and ends with a concluding remark.

2 Literature Review

2.1 Pricing Rule

Research in the Economics discipline reported that there is no significant difference in market performance between the Uniform and Pay-as-Bid pricing rule. The theory underpinning the research in [2, 3] is the Revenue Equivalence

Theorem (RET). The theorem, as cited in [4] stated that "*Suppose bidders have independent and identically distributed valuations and are risk neutral. Then any symmetric and increasing equilibrium of a direct revelation auction that assigns the item to the highest bidder such that the expected payment of bidder with value zero is zero, yields the same expected revenue*." However, RET only applies where a single unit of an indivisible good is being auctioned, in a single-period setting. In contrast, in a deregulated electricity market, auctions are normally happened between asymmetric bidders where multi-unit, multi-period bidding occurs. In order to apply the RET to a deregulated electricity market, it is necessary to assume that all generators bid with identical strategies and are risk neutral. This assumption is unrealistic considering that in electricity market, market power and price volatility are inevitable. In a deregulated electricity market, instead of being risk neutral and bidding identically, generator behaves differently to maximize its expected profit according to its understanding of the market environment [5–7].

Study in [8] stated that the Uniform pricing rule, which has been widely employed in many electricity markets around the world, causes poor performance in terms of market price, generator's revenue and total dispatch cost. It was recommended that the Pay-as-Bid rule should be introduced to replace Uniform pricing rule [9]. However, other contradictory results have also been reported by in the literature in terms of the impact of pricing rules on generator's revenue, market price and price volatility brought by these two pricing rules, such as [10–13]. While the argument over comparing these two pricing rules has lasted decades in the electricity market establishment process, little attention has been paid to the Vickrey pricing rule [14], despite being perceived to be highly favorable by economists [2].

Another commonly used assumption in previous studies is that, generators were often assumed to submit bid quantity at their maximum capacity and such bid quantity was either accepted completely or not at all. This assumption is inaccurate because generators can actually supply its bid quantity based on its own interest and such bid quantity can be partial accepted. Therefore, in this paper, partial accepted bid quantity has been modelled in the design of the simulated market in our study.

2.2 Computer Simulation

There have been a number of studies on competitive electricity markets employing agent-based models, as reported in [15–18]. Generator agents reported in these studies can adapt their bidding strategies, based on the success or failure experience of previous trading.

A popular learning algorithm called Q-Learning is employed as the learning method for the agents. Q-learning is a Reinforcement Learning technique that works by learning an action-value function that gives the expected utility of taking a given action in a given state and following a fixed policy thereafter. The entities modeled with Q-Learning can behave in such a way that they are able to use their past experience to improve their behavior [19]. It was proposed by Watkins [20] for solving the Markovian Decision problems with incomplete information. The main advantage of Q-Learning algorithm is that it is model-free and can be used on-line to find an optimal result based on the direct interaction with the environment. This feature makes it suitable for decision making problems in repeated games with unknown component [21].

3 Agent-Based Model for a Competitive Electricity Market

Figure 1 illustrates the proposed electricity market model with Q-Learning based generator. In our study, we employed four generator agents that bid into the market. Each generator agent is characterized by the ramp rate of the power plant and cost function characterizing the type of power plant.

In a competitive electricity market, generators are required to submit their bids for every trading period. The bid price that a generator can offer is between its marginal cost and the predefined price cap. The bid quantity that a generator can offer is between its minimum stable load and its max capacity. These settings can be described as:

$$mc_g \leq P_g \leq P_{cap} \quad (1)$$

$$minQ_g \leq Q_g \leq maxQ_g \quad (2)$$

where mc_g is the marginal cost of a competing generator; P_g and Q_g are the bid price and bid quantity submitted by a competing generator; P_{cap} is the predefined price cap; $maxQ_g$ is the maximum quantity that can be offered by a competing generator.

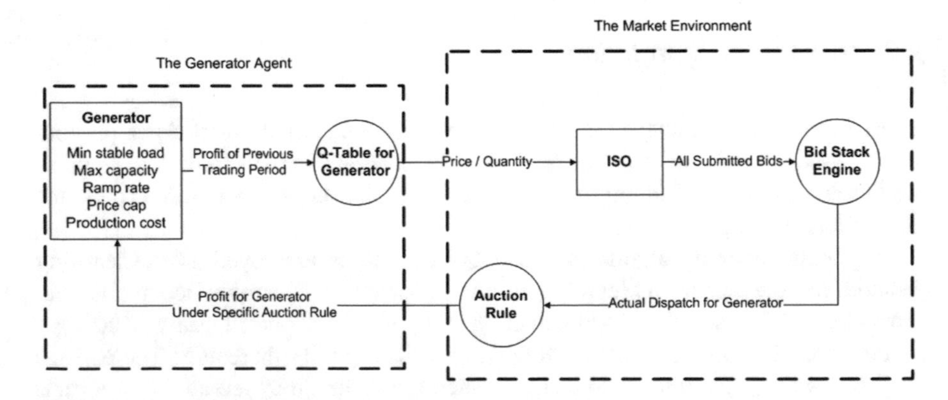

Fig. 1 Electricity market model with Q-Learning based generator agents

Once the bids are collected, the ISO schedules the actual dispatch starting from the generator offering the lowest bid price. If the demand is not fulfilled, the ISO then schedule the dispatch from the generator offering the second lowest bid price. This process is terminated until all the demand is fulfilled:

$$\sum_1^g Q_g = Demand \tag{3}$$

Generators competing for dispatch adopt a profit maximizing strategy. When the Uniform pricing rule is employed, the single market clearing price is the last bid price. Hence, the generator's profit is:

$$Profit_g = (MCP - mc_g) \times Dispatch_g \tag{4}$$

where MCP is the market clearing price; $Dispatch_g$ is the actual dispatch of a competing generator. When Vickrey pricing rule is employed, the generator will be paid based on the opportunity cost its presence introduces to all the other generators. Therefore, the generator's profit is calculated as:

$$U_{total} = \sum_1^g P_g * Dispatch_g \tag{5}$$

$$U_{total'} = \sum_1^{g'} P_g * Dispatch_g \tag{6}$$

$$U_{other} = U_{total} - P_g * Dispatch_g \tag{7}$$

$$U'_{other} = U_{total'} \tag{8}$$

$$R_g = U'_{other} - U_{other} \tag{9}$$

$$Profit_g = R_g - mc_g * Dispatch_g \tag{10}$$

where U_{total} is the total utility received by all the generators when generator g is competing for dispatch; U_{other} is the total utility received by all the other generators when generator g is competing for dispatch; $U_{total'}$ (or U'_{other}) is the total utility received by all the generators when generator g is excluded from competition; R_g is the revenue received by generator g.

The *state* describes the current situation of each generator. In this paper, it is defined as the bid price and quantity pair each generator submitted in previous trading period. For each generator, there are (N – 1) main intervals within each interval there are (M – 1) sub-intervals defined. (N – 1) is the number of quantity selections that can be chosen by a generator with each selection represents a bid quantity range; (M – *1*) is the number of price selections that can be chosen by a generator with each selection represents a bid price range. N and M is calculated as follow:

$$N = maxQ_g - minQ_g \quad (11)$$

$$M = P_{cap} - 0 \quad (12)$$

The increments for both main interval and sub-interval in each main interval are set to 1. Therefore, there are $(N - 1)$ times $(M - 1)$ sub-intervals in total with each representing a selected bid price and quantity pair.

The definition of *action* is similar to the definition of *state* except that the calculation of M and N are modified as follow:

$$N = maxQ_g - minQ_g. \quad (13)$$

$$M = P_{cap} - mc_g \quad (14)$$

Hence, taking an action is to locate the corresponding interval according to generator's bid price and quantity pair.

In Q-Learning algorithm, determining the action based on Q-values is referred to as an *exploitation*. Determining the action without sufficient information is an *exploration*. In this paper, the Simulated Annealing (SA)-Q-Learning algorithm [22] is adopted as the action selection policy for generator agent:

(1) Randomly select an action a_r, where $a_r \in$ A.
(2) Adopt greedy approach: select action a_p, where $a_p \in$ A.
(3) Generate a random number rand between 0 and 1.
(4) Select the final action based on the following calculation:

$$a = \begin{cases} a_p, & \text{rand} \geq \exp[\frac{Q(s, a_r) - Q(s, a_p)}{\text{temperature}}] \\ a_r, & \text{otherwise} \end{cases} \quad (15)$$

(5) Update the temperature parameter based on pre-defined temperature dropping function.

Assume T_n is the temperature in the n*th* iteration, then $T_n = \beta T_{n-1}$, where n is natural number and β is a constant number close to 1 to ensure a slow decay of the temperature in the algorithm.

In the beginning of the simulation process, generator agents will undertake exploratory actions. As the iteration procedure progresses, the Q-values will be updated accordingly. These updates decrease the tendency of exploratory action and increase the chances of taking exploitative actions. In this paper, T_1 is set to 100000 and β is defined as 0.999. This is decided by using trial-and-error to ensure that the temperature variable is not dropping too fast so that the generators can have sufficient exploration actions to understand the market more comprehensively.

The bid quantity of each generator in each trading period is constrained by its ramp rate, which is:

$$Q_g \leq Q_g^{'} + R_g \tag{16}$$

$$Q_g \geq Q_g^{'} - R_g \tag{17}$$

where Q_g and $Q_g^{'}$ are the bid quantity in current and previous trading period accordingly, R_g is the ramp rate of the generator.

For each trading period, there are four successive stages: (1) agents submit bids, select *action* and next *state*; (2) ISO receive bids and determine economic dispatch schedule; (3) agents receive their profit based on pricing rule; (4) agents update their knowledge (Q-table) based on their recent trading experience following:

$$Q(s,a) \leftarrow (1-\alpha)Q(s,a) + \alpha[r + \gamma \max Q(s^{'},a^{'})] \text{where } s \leftarrow s^{'} \tag{18}$$

In the Q-Learning algorithm, α is the learning rate with value between 0 and 1; γ is the preference for employing immediate reward. The learning rate in this paper is designed to be state action dependent as in [19, 21], which is inversely proportional to the visited number of a particular state-action pair up to the present trading period. The learning rate for a particular state-action pair will be smaller as the number of visits of such state-action pair increase. So the learning rate α for a state-action pair k is calculated as $\alpha_k = 1/N_k$, where N_k is the number of times that this pair k have been taken by a generator.

The discount factor γ in this paper is assigned with the value of 0.1. This value leads to a short reaction time for the generator agents to respond to the change in the market due to their *recent-reward pursuing* which makes their behavior easier to be captured and analyzed.

4 Case Examples

Three case studies are presented in this paper. Each case includes four competing generator agents with similar attributes, namely the same minimum stable load, ramp rate and production cost. The aggregated supply capacity of the four generators is fixed. In the three scenarios, the maximum generating capacity of generator agent 1 is set to: 25% of aggregated supply capacity in case 1, 50% in case 2, and 75% in case 3. The total demand is set to approximately 60% of aggregated supply capacity. These three scenarios are setup to emulate situations where (1) generator agent 1 has no dominant position, (2) generator agent 1 has a dominant position but its generating capacity cannot fulfill the demand, and (3) generator agent 1 has such a dominant position that it can fulfill the demand using its generating capacity. An additional backup generator with extremely large capacity and high price (reserve capacity price) is set for each case to ensure the demand can still be fulfilled even if the total supplied quantity is less than the demand. However, the involvement of this backup generator will lead to extremely high market price. In each case, two scenarios are tested. In the first scenario, the agent must submit bid that offers the maximum generating capacity; while in the second scenario, generator agent can submit bid quantity following their profit maximizing strategy. In addition, these scenarios are tested for different pricing rule, namely economic dispatch based on (1) Uniform pricing, (2) Pay-as-bid pricing, and (3) Vickrey pricing.

During each simulation process, the total dispatch cost is calculated by taking the average of the total dispatch costs in the last 2000 trading periods when the generator agents have learnt sufficiently. The total dispatch cost under Uniform pricing rule in Case 1, Scenario 1 is used to normalized the other total dispatch cost by a factor k for comparison and analysis. Thus, the value for the total dispatch cost under Uniform pricing rule is 1. Our intent is that through these case studies, the effect of different pricing rules and supply quantity variation on total dispatch cost can be examined.

4.1 Discussion of Case 1

As can be seen in Table 1, in Scenario 1, the total dispatch costs are similar under all pricing rules. This is because, all the generators have the same attributes and only one variable which is the bid price. Therefore, the chance of exhibiting similar bidding behavior among all four generators is relatively high. This situation somewhat satisfies the requirements over which the Revenue Equivalent Theorem (RET) [2, 3] holds supporting the same total dispatch cost. In short, RET holds in Scenario 1.

In Scenario 2, bid quantity has been added as another variable in the simulation. The approach for each generator to develop their bidding strategy becomes somewhat unpredictable. This indicates a reduced possibility of exhibiting similar bidding strategies among generators. As a result, the degree of satisfaction of the requirements over which the RET holds is weakened. Under this circumstance, it is observed that the Vickrey pricing rule produces a slightly higher dispatch cost.

In comparing Scenario 1 and 2, it is noted that allowing variable bid quantity leads to an increase in the total dispatch cost. With variable quantity bidding, there are more occurrence of balanced supply-to-demand since all the generators are

Table 1 Simulation results

Total dispatch cost	Scenario 1: Maximum capacity			Scenario 2: Variable quantity		
	Uniform	Pay-as-bid	Vickrey	Uniform	Pay-as-bid	Vickrey
Case 1	1	1	1.1	1.65	1.58	2.28
Case 2	6.05	2.19	2.69	9.84	1.24	2.85
Case 3	6.05	5.01	6.58	11.01	4.50	7.06

allowed to supply bid quantity at their own preference. However, since all the generators have similar market share and there is no dominating player in the market, the near-free competition existing in the market leads to a merely small increase in the total cost.

4.2 Discussion of Case 2

In Case 2 when bidding at maximum capacity is required, it is observed that the total cost under Uniform pricing rule is much higher than under the Pay-as-bid and the Vickrey pricing rules. This is because the supply capacity is on a par with the aggregated supply capacity of the other three competing generators. This cooperative behavior among the generator agents is unfavorable to the first generator agent. As a result, in order to get higher profit, it is observed that the first generator has to bid at a much higher price to recover its loss in dispatch. This high price, which becomes the single market clearing price, leads to a high total dispatch cost.

Under Pay-as-bid pricing rule, the profit of a generator agent relies only on its own bid price and dispatch quantity. The bidding behavior of any generator agent does not affect its competitors. The bidding behavior of the generator agents is somewhat dissociated from the market dynamics; hence, providing minimal feedback to the learning process. It is also observed that the time required for a generator agent to complete its learning process has significantly increased. Two distinctive characteristics notable with this market setting are: (1) generator agents tend to bid with risk averse; and (2) the low volatility in bid price leads to low volatility in generators' profit.

Under Vickrey pricing rule, due to the large generating capacity of generator agent 1, even if it submits the highest bid price among all the competing generators, at least 10% supply will need to be dispatched from its capacity. When its revenue is calculated by excluding generator agent 1, such supply shortage will have to be fulfilled by the backup generator. This may still lead to large revenue as calculated by (9) due to the extremely high reserve capacity price. Generator 1 may bid at a relatively high price without worrying about sufficient dispatch quantity. However, the small generator has to ensure enough dispatch quantity so that the U_{other} does not equal to U_{total} which otherwise leads to a zero revenue. This requirement can only be reached by reducing the bid price since neither of the small generators can incur a supply shortage due to its small capacity. Hence, it is observed that all the three small generators' bid prices are close to their production cost. As a result, only a small amount of dispatch will be paid at a very high price which dramatically lowers the total dispatch cost comparing to it under Uniform pricing rule.

In Scenario 2, when bid quantity can vary, under Uniform pricing rule, the large generator can strategically limit its maximum supply quantity comparable to the other three small generators. The difference in the bidding behavior of the large generator and the three small generators' is marginal, which results in a highly reduced profit difference when compared with Scenario 1. The large generator behaves as if it were a small generator in an attempt to break the collusion, causing a supply capacity withholding. This leads to extremely high market clearing price and total dispatch cost.

Under Pay-as-bid and Vickrey pricing rules, allowing variable bid quantity also increases the chance of supply capacity withholding phenomenon. However, unlike Uniform pricing, only the generator agent 1 has been observed to withhold supply capacity. For a reason similar to Scenario 1, only a small portion of dispatch is paid at a high price which therefore retains a comparatively low total dispatch cost comparing to it under Uniform pricing rule.

4.3 Discussion of Case 3

The last case study depicts a situation where a large generator has a clearly dominant position in the market. All the other three agents have very limited supply capacity to fulfill the market demand. In such situation, generator agent 1 can easily overcome the unfavorable collusion among the three small generators and gain the highest revenue. More importantly, its dominant position makes it the market price setter controlling the clearing price as the cap price in scenario 1 and the reserved capacity price in Scenario 2. Accordingly, the total dispatch costs under these two scenarios are both very high; and the total cost is even higher when bid quantity can vary.

Under Pay-as-bid and Vickrey pricing rules, the situation is similar as in Case 2. However, due to the increased size of generator agent 1, more shortage needs to be fulfilled if generator agent 1 is excluded. As a result, revenue gained by generator 1 increases substantially. Likewise, total dispatch cost increases significantly.

A notable observation from Case 3 compared to Case 2 for variable quantity bidding scenario (Scenario 2) is that the total dispatch cost under Uniform pricing rule is higher than the other pricing rules. When maximum capacity is supplied, under Uniform pricing rule, the price resolves to the cap price. However, when Vickrey pricing is adopted, the shortage caused by excluding the large generator from the competition will need to be fulfilled by the backup generator. This, dramatically increases the total dispatch cost.

5 Analysis

5.1 Electricity Market as a Complex System

The Cynefin framework [23] is a useful typology to describe problems and life phenomena. The model introduces four domains in which every contextual problem may be best approached using different practices: (1) in simple domain, the approach is to sense, categorize and respond—leading to *best practice*; (2) in complicated domain, the approach is to sense, analyze and respond—leading to *good practice*; (3) in complex domain, the approach is to probe, sense and respond—leading to *emergence*; (4) in chaotic domain, the approach is to act, sense and respond—leading to *novelty*.

Although there is no consensus in formalizing the problem definition in the complex domain, typically complex problems can be characterized by four attributes: (1) interaction, interconnection, heterogeneity and tension. We submit that the competitive electricity market exhibits these attributes. First, the high level of interaction is evident in the repetitive nature of trading in the electricity market. In the Australian market, there are 48 trading intervals every day. Second, the interconnection is evident as the *spot price* and the electricity dispatch and schedule of each generator depends on each other's quantity-price bids. Third, the heterogeneity in the context of this problem is inherent in nature, as generators are distinctive in its capacity and in its mode of operation. There is a diverse range of power plants generated using coal, gas, hydro and many others—leading to different operational costs, thus may be bidding differently when competing in the market. Fourth, the tension in this problem exists among the profit maximizing objective of each competing generators and operational boundaries (or limitations) that govern the trading, may it be internal in nature, such as operating characteristics of the power plants, or external in nature, such as trading policy and price cap.

Complexity in the electricity market is a result of the highly dynamic and inherent non-linearity in the market environment. It is also a result of interconnected components (coupling) with high level of interactions among these components. With the increase in the number of interactions, it becomes more difficult for us to understand the system as a whole. And with the increase in the number interconnected components, it becomes more difficult for us to identify and isolate causal relationships in the system.

In terms of identifying emergence as findings of our studies, we noted four phenomena in agents' bidding behavior, namely capacity withholding, collusion, risk taking and risk averse biddings.

5.2 Agents' Learning

In our study, we employed Q-learning as the mechanism for generator agents to learn from repetitive trading experience in order for them to bid strategically to maximize their profit. In the mathematical model, the three parameters that influence the learning process are the immediate reward r_t, learning rate α and discount factor γ.

In theory, Q-learning algorithm for a single agent will lead to a convergence [24, 25]. In a more dynamic environment with multi-agent settings, a stable optimal state may not exist. Under this circumstances, it is crucial to find the appropriate learning rate α and discount factor γ.

The agents' learning in our study is tightly coupled with the pricing rule employed in the competitive market environment. Specifically, the immediate reward r_t has been defined as the profit the generator agent receives in the dispatch interval t. Given that the purpose of the Q-learning algorithm is to obtain the maximum overall reward, the purpose of the Q-learning based generator agent in this platform is to discover the best action at each state so as to maximize its overall profit.

As for the learning rate for the generator agents, in our study, we defined the learning rate to be state and action dependent following [26]. In other words, the learning rate is inversely proportional to the visited number of a particular state-action pair up to the present state. In effect, the agent will learn more if it is performing a new situation and rarely taken any action. When a particular action of a state has been undertaken several times, implying that the agent has gained some experience to deal with that particular state, there are fewer new information that an agent ought to learn in that particular state. To implement this, we keep a copy of the Q-table for each generator agent and populate the Q-table with the number visits of each state-action pair.

In Q-learning algorithm, a discount factor models the relevance of recent vs past reward. Setting the discount factor to 0 implies that an agent considers only the recent reward while setting it to 1 highlights the importance of long-term reward. In the highly interdependent and dynamic electricity market, the concept of *long-term* becomes elusive, as there is neither single convergence state that a generator agent can reach nor a terminating point in a repetitive trading of the electricity market. In other words, the learning process of a generator agent is never ending and continuous. In addition to this elusive long-term concept, the dispatch information is updated and overridden every trading day so as the generator agent can compete effectively.

6 Conclusion

Using agent based model in our studies gives us significant advantages over other traditional methods. In particular, by using an agent-based model, we can create hypothetical situations with more potency than existing ones. As a result, we are able to examine a wide range of scenarios. This approach enables a more systematic and imaginative thinking rather than constraining us with limited scenarios. Agent-based model allows us a relatively rapid development and testing of alternatives in a cost effective manner. Likewise, this method provides us with a controlled environment for experimentation. However, to effectively develop and use agent based systems, two complementary skills are required, namely modeling and interpreting (to discover emergence). Likewise, we found the complex systems paradigm as a useful framework guiding us to focus our observation and guide us in understanding our findings in a meaningful way.

References

1. Liao, Z., Sugianto, L.F.: A comparative study on pricing rules and its effect on total dispatch cost. In: Dagli, C.H. (ed.) Complex Adaptive Systems, Procedia Computer Science, vol. 1. Elsevier (2011)
2. Fabra, N., et al.: Designing electricity auctions. Rand J. Econ. **37**, 23–24 (2006)
3. Hinz, J.: A revenue-equivalence theorem for electricity auctions. J. Appl. Probab. **41**, 299–312 (2004)
4. Chekuri, C., Pellizzoni, R.: Revenue Equivalence Theorem. http://www.cs.uiuc.edu/homes/chekuri/teaching/spring2008/Lectures/scribed/Notes20.pdf (2008)
5. Bunn, D.W., Oliveira, F.S.: Evaluating individual market power in electricity markets via agent-based simulation. Ann. Oper. Res. **121**, 57–77 (2003)
6. Mount, T.: Market power and price volatility in restructured markets for electricity. Decis. Support Syst. **30**, 311–325 (2001)
7. Reinisch, W., Tezuka, T.: Market power and trading strategies on the electricity market: a market design view. IEEE Trans. Power Syst. **21**, 1180–1190 (2006)
8. Fabra, N., et al.: Modeling electricity auctions. Electricity J. **15**, 72–81 (2002)
9. Kahn, A., et al.: Uniform pricing or Pay-as-bid pricing: a dilemma for California and beyond. Electr. J. **14**, 70–79 (2001)
10. Federico, G., Rahman, D.: Bidding in an electricity Pay-as-bid auction. J. Regul. Econ. **24**, 175–211 (2004)
11. Rassenti, S., et al.: Discriminatory price auctions in electricity markets: low volatility at the expense of high price levels. J. Regul. Econ. **23**, 109–123 (2003)
12. Son, Y.S., et al.: Short-term electricity market auction game analysis: uniform and Pay-as-bid pricing. IEEE Trans. Power Syst. **19**, 1990–1998 (2004)
13. Maskin, E., Riley, J.: Asymmetric Auctions. Rev. Econ. Stud. **67**, 413–438 (2000)
14. Vickrey-Clarke-Groves auction. http://en.wikipedia.org/wiki/Vickrey-Clarke-Groves_auction
15. Bunn, D.W., Oliveira, F.S.: Agent-based simulation—an application to the new electricity trading arrangements of England and Wales. IEEE Trans. Evol. Comput. **5**, 493–503 (2001)
16. Bunn, D.W., Oliveira, F.S.: Evaluating individual market power in electricity markets via agent-based simulation. Ann. Oper. Res. **121**, 57–77 (2003)
17. Watanabe, I.: Agent-based simulation model of electricity market with stochastic unit commitment. In: Proceedings of the 8th International Conference on Probabilistic Method Applied to Power Systems. Iowa State University, Ames, Iowa, pp. 403–408 (2004)
18. Nicolaisen, J., et al.: Market power and efficiency in a computational electricity market with discriminatory double-auction pricing. IEEE Trans. Evol. Comput. **5**, 504–523 (2001)
19. Krause, T., et al.: Nash equilibria and reinforcement learning for active decision maker modelling in power markets. In: The Proceedings of the 6th IAEE Conference–Modeling in Energy Economics, Zurich (2004)
20. Watkins, C.J.C.H.: Learning from Delayed Rewards. Ph.D. thesis, Cambridge University (2002)
21. Xiong, G., et al.: Multi-agent based experiments on uniform price and Pay-as-bid electricity auction markets. In: Proceedings of the 2004 IEEE International Conference on Electric Utility Deregulation, Restructuring and Power Technologies, pp. 72–76 (2004)
22. Guo, M. et al.: A new Q-Learning algorithm based on the metropolis criterion. IEEE Trans. Syst. Man Cybern. **34** (2004)
23. Snowden, D.: Complex acts of knowing: paradox and descriptive self-awareness. J. Knowl. Manage. **6**(2) (2002)
24. Watkins, C., Dayan, P.: Technical note: Q-Learning. Mach. Learn. **8**, 279–292 (1992)

25. Littman, M., Szepesvari, C.: A generalized reinforcement learning model: convergence and applications. In: Proceedings of the 13th International Conference on Machine Learning (1996)
26. Nie, J., Haykin, S.: A dynamic channel assignment policy through Q-learning. IEEE Trans. Neural Networks **10**, 1443–1455 (1999)

Multi-kernel Transfer Extreme Learning Classification

Xiaodong Li, Weijie Mao, Wei Jiang and Ye Yao

Abstract In this paper, a novel transfer extreme learning machine (TELM) algorithm based on multi-kernel (MK) framework has been proposed for classification. In this case, the problem is transformed into a semi-supervised learning problem, which allows multi-kernel extreme learning machine (MK-TELM) classifiers to be trained for the data categorization. Compared with many popular algorithms, the proposed method, named as MK-TELM, shows its satisfactorily experimental results on the variety of data sets, which highlights the robustness and effectiveness for classification applications.

Keywords Extreme learning machine · Transfer learning (TL) · Multiple kernel learning

1 Introduction

In the past couple of decades, neural networks (NN), as powerful intelligence tools, have been extensively studied and successfully applied to deal with various problems [1]. The famous error back-propagation (EBP) adopts gradient methods to

X. Li (✉) · Y. Yao
School of Computer Science and Techonogy, Hangzhou Dianzi University,
Hangzhou 310018, People's Republic of China
e-mail: hzxiaodong22@163.com

W. Mao · W. Jiang
Department of Control Science and Engineering, Zhejiang University,
Hangzhou 310028, People's Republic of China
e-mail: wjmao@iipc.zju.edu.cn

W. Jiang
e-mail: jiangwei@iipc.zju.edu.cn

J. Cao et al. (eds.), *Proceedings of ELM-2016*, Proceedings in Adaptation, Learning and Optimization 9, DOI 10.1007/978-3-319-57421-9_13

optimize the weights in the network [2]. Due to SVMs' simplicity and relatively stable generalization performance, it has been widely applied to various domains [3]. Recently, a new learning algorithm, i.e., extreme learning machine (ELM) was proposed by Huang et al. [4]. Compared with EBP neural network and SVM, the ELM only update the output weights between the hidden layer and the output layer, while the parameters, i.e., the input weights and biases of the hidden layer, are randomly generated, and has better generalization performance at a much faster learning speed.

Moreover, ELMs provide a universal model which include but not limit to neural network, SVM, and regularized network. Kim et al. [5] introduced a variable projection method to lower the dimension of the parameter space. Wang et al. [6] made a proper selection of the input weights and bias of ELM in order to improve the performance of ELM. Li et al. [7] proposed a structure-adjustable online ELM learning method, which can adjust the number of hidden layer RBF nodes. A pruned ELM (PELM) was proposed by Rong et al. [8] for classification problem. Zong et al. [9] put forward the weighted extreme learning machine for imbalance learning. The kernel trick applied to ELM was introduced in previous work. Liu et al. [10] designed sparse, non-sparse, and radius-incorporated MK-ELM algorithms. Li et al. [11] proposed the issue of multiple kernel learning for ELM by formulating it as a semi-infinite linear programming (SILP). Zeng et al. [12] studied and analyzed from the optimization point of view. Peng et al. [13] proposed a discriminative graph regularized Extreme Learning Machine (GELM), in which the constraint imposed on output weights enforces the output of samples from the same class to be similar.

Though conventional ELMs have become popular in a broad range of domains, they are mainly applied in supervised learning such as classification and regression problem that greatly restricts their applicability. Transfer learning is contributed to sharing and transferring knowledge between related domains, especially under such conditions as different distributions and feature representation. According to the relationship between the source and target domains, classified transfer learning into three kinds of knowledge transfer: parameter-based transfer, feature-based transfer and transfer instance-based knowledge [14–16]. Pan et al. [17] proposed a Q learning system for continuous spaces which is constructed as a regression problem for an ELM. Scardapane et al. [18] extend Extreme Learning Machine (ELM) theory to the transductive circumstance, known as the transductive ELM (TELM). Huang et al. [19] showed the general architecture of local receptive fields based ELM (ELM-LRF) and the proposed algorithm lowers the error rate compared with conventional deep learning solutions.

2 MK-TELM

2.1 Minimum Norm Least-Squares (LS) Solution of SLFNs

In kernel ELM algorithm, the input weights w_i and the hidden layer biases b_i not necessarily tuned and the hidden layer output matrix $\mathbf{H}$ can really remain unchanged once random values have been assigned to these parameters in the beginning of learning. For fixed input weights w_i and the hidden layer biases b_i, seen from Eq. (1), to train an SLFN is simply equivalent to finding a least-squares solution β of the linear system $\mathbf{H}\boldsymbol{\beta} = \mathbf{T}$:

$$\begin{aligned} &\left\| \mathrm{H}(\hat{w}_1, \ldots, \hat{w}_{\tilde{N}}, \hat{b}_1, \ldots, \hat{b}_{\tilde{N}})\beta - \mathbf{T} \right\| \\ &\quad = \min_{w_i, b_i, \beta} \left\| \mathrm{H}(w_1, \ldots, w_{\tilde{N}}, \hat{b}_1, \ldots, \hat{b}_{\tilde{N}})\beta - \mathbf{T} \right\| \end{aligned} \tag{1}$$

$$\begin{aligned} &\left\| \mathrm{H}(w_1, \ldots, w_{\tilde{N}}, b_1, \ldots, b_{\tilde{N}})\beta - \mathbf{T} \right\| \\ &\quad = \min_{\beta} \left\| \mathrm{H}(w_1, \ldots, w_{\tilde{N}}, b_1, \ldots, b_{\tilde{N}})\beta - \mathbf{T} \right\|. \end{aligned} \tag{2}$$

The smallest norm least squares solution of the above linear system is

$$\boldsymbol{\beta} = \mathbf{H}^{\dagger}\mathbf{T}, \tag{3}$$

where $\mathbf{H}^{\dagger}$ is the Moore-Penrose generalized inverse of matrix $\mathbf{H}$.

We will simply introduce the MK-ELM algorithm as follows [19].

In MK-ELM, the kernel $K(x, x')$ is actually an approximate convex linear combination of other single ELM kernels:

$$\begin{aligned} \max_{\alpha, t} \quad & -\frac{1}{2}t + \alpha^T y - \frac{1}{2C}\alpha^T \alpha \\ \text{s.t.} \quad & t \geq \alpha^T K_i \alpha, \quad i = 1, \ldots, p, \\ & \alpha^T \mathbf{1}_n = 0. \end{aligned} \tag{4}$$

$$\begin{aligned} \max_{\theta, u} \quad & \mathrm{u} \\ \text{s.t.} \quad & \theta_j \geq 0, \quad j = 1, \ldots, p+1 \\ & \sum_{j=1}^{p+1} \theta_j^2 \leq 1, \\ & \frac{1}{2}\sum_{j=1}^{p+1} \theta_j f_j(\beta_q) - \frac{1}{2}\sum_{q=1}^{k} \beta_q^T Y_q^{-1} \mathbf{1}_q \geq u, \quad \forall \beta_q, \ q = 1, \ldots, k \end{aligned} \tag{5}$$

$$f_j(\beta_q) = \sum_{q=1}^{k} \left(\frac{1}{2}\beta_q^T K_j \beta_q\right), \ j = 1, \ldots, p+1, \ q = 1, \ldots, k$$

2.2 Framework of MK-TELM

Let $D = \{x_1, \ldots, x_{l+u}\}$ define the entire data set. Without loss of generality, we assume the first l samples are labeled $\{(x_i, y_i)\}_{i=1}^{l}$ where $y_i \in \{-1, +1\}, i = 1, \ldots, l$ and followed by u unlabeled samples $\{x_i\}_{i=l+1}^{l+u}$. The unknown labels are binary entries of the vector $y_u = [y_{l+1}, \ldots, x_{l+u}]^T$. A collection of l kernel functions $K = \{\kappa_j : \chi \times \chi \rightarrow \mathbb{R}, j = 1, \ldots, l\}$.

2.2.1 Problem Setting

The aim of semi-supervised ELM is to learn an ELM that exploits the information conveyed by the unlabeled data [20]. The general picture is to determine a decision function able to classify the labeled data and to correctly predict the class of unlabeled samples while maximizing the margin. Generally speaking, Semi-Supervised ELM algorithms rely on the optimization of the following generic objective function:

$$\Omega(f) + C \sum_{i=1}^{l} V(y_i g(x_i)) + C^* \sum_{i=l+1}^{l+u} U(g(x_i)), \tag{6}$$

where the decision function is defined as $g(x) = f(x) = h(x)\beta$ with f. The first term in (6) represents the regularization term which aims at controlling the complexity of f. The two last terms are respectively the fitting errors for the labeled and unlabeled samples which are evaluated through the margin loss function V (labeled data) and U (unlabeled data). The regularization parameters C and C^* balance the importance of those errors in the optimization process.

Indeed, problem (4) can be seen equivalently as

$$\min_{f,b} \frac{1}{2} \|f\|_H^2 + C \sum_{i=1}^{l} V(y_i g(x_i)) + C^* \sum_{i=l+1}^{l+u} U(|g(x_i)|) \tag{7}$$

Given a set of m kernels κ_K, these methods aim at learning a linear combination of the kernels i.e. $\kappa(x_i, x_j) = \sum_k d_k \kappa_k(x_i, x_j)$ with $d_k \geq 0$. Those kernels can be defined according to some a priori knowledge. Based on Eq. (7), the following formal setup for MK-TELM problem:

$$\begin{aligned} &\min_{f_k, b, d \geq 0} \frac{1}{2} \sum_{k=1}^{m} \frac{a_k}{d_k} \|f_k\|_{H_k}^2 + C \sum_{i=1}^{l} V(y_i g(x_i)) + C^* \sum_{i=l+1}^{l+u} U(|g(x_i)|) \\ &s.t. \quad \|d\|_1 \leq 1, \ \frac{1}{u} \sum_{i=l+1}^{l+u} g(x_i) = \frac{1}{l} \sum_{i=1}^{l} y_i \end{aligned} \tag{8}$$

where a_k is a normalization term, usually set as the trace of the kernel matrix κ_k induced by κ_k.

2.3 Stochastic MK-TELM Algorithms

The idea of a stochastic MK-TELM approach is that we could try to avoid unnecessary costs of training classifiers with some kernels that have relatively poor classification performance for the classification task. To this purpose, we define a variable $S_t(j)$ as the kernel sampling probability, which indicates how likely a kernel κ_j will be sampled at the t-th trial. At the beginning of the MK-TELM algorithm, all $S_1(j)$ values are set to 1, which means that all M kernels will be definitely selected at the first trial.

For each trial, we choose a subset of kernels according to the kernel sampling probability S_t. The proposed MK-TELM algorithm will train classifiers only for those selected kernels. At the end of each trial, we update the kernel sampling probability according to its classification performance:

$$S_t(j)\beta^{\varepsilon_t^j} \rightarrow S_{t+1}(j) \tag{9}$$

where $\beta \in (0, 1)$ is a constant parameter introduced as a sampling decay factor for updating the kernel sampling probability, and ε_t^j is the misclassification rate of the kernel classifier with a selected kernel κ_j. The above formulation indicates the larger the misclassification rate, the more decay penalty will be applied to the kernel to reduce the chance of being sampled in the next trial. Finally, at the end of a trial, we normalize to ensure all kernel sampling weights which are in [0, 1]. This normalized step could affect the sampling weights of those kernels that are not selected.

Algorithm 1:

Step 1: labeled data: $\{X_l, Y_l\} = \{x_i, y_i\}_{i=1}^{l}$, unlabeled data: $X_u = \{x_i\}_{i=1}^{u}$, kernel functions: $\kappa_j(\cdot,\cdot): \chi \times \chi \rightarrow \mathbb{R}, j = 1, \quad, \mu$, initial estimation $D_1(i) = 1/l$, $i = 1, \cdots, l$, $S_1(j) = 1, j = 1, \cdots, u$, sampling decay rate $0 < \beta < 1$

Step 2: for $t = 1, \cdots, T$

do

Step3: for $\kappa_j \in K_t$

do

Step4: Train weaker classifier with kernel κ_j;

Step5: $f_t^j : \chi \rightarrow \{-1, +1\}$

Step6: Compute the training error over D_t

$$\varepsilon_t^j = \sum_{i=1}^{N} D_t(i)\ (f_t^j(x_i) \neq y_i)$$

Step7: Update $S_t(j)\beta^{\varepsilon_t^j} \rightarrow S_{t+1}(j)$

Step8: end for

Step9: Update $S_{t+1}(j)/Z^S \rightarrow S_{t+1}(j)$, $Z^S = \max(S_{t+1})$

Step10: $f_t = \arg\min_{\kappa_j \in K_t} \varepsilon_t^j = \arg\min_{\kappa_j \in K_t} \sum_{i=1}^{N} D_t(i)(f_t^j(x_i) \neq y_i)$

Step11: Compute the training error over D_t

$$\varepsilon_t = \sum_{i=1}^{N} D_t(i)(f_t(x_i) \neq y_i)$$

Step12: $\alpha_t = \frac{1}{2}\ln(\frac{1-\varepsilon_t}{\varepsilon_t})$

Step13: Update $D_t/Z_t \exp(-\alpha_t y_i f_t(x_i)) \rightarrow D_{t+1}(i)$

Step14: end for

Step15: Output $f(x)$

In the above algorithm, at each trial, we simply choose the best classifier among the l kernel classifiers as the classifier and simply abandon the other $l - 1$ classifiers which may make complementary contribution in improving the performance in some cases. Thus, we introduce another way to build the classifier by combining all these l classifiers, each of which is assigned with a weight.

3 Empirical Results Performance Evaluation of MK-TELM

3.1 Data Sets

In this section, in order to evaluate the properties of our framework, we perform the experiments on one none-text data set from the UCI machine learning repository and

Table 1 UCI data sets used in the experiments

Data set	Dimensionality	Sample size
Banana	2	4900
Titanic	3	2200
Waveform	21	5000
Image	18	2310
Heart	13	270
Diabetis	8	768
Flare Solar	9	1066
Splice	60	3175

Table 2 Top class and sub-class on the 20-Newsgroups dataset

Top class	Subclass number	Subclass	Sample size
Comp	1	comp.sys.ibm.pc.hardware	968
	2	comp.windows.x	978
	3	comp.sys.mac.hardware	956
Rec	4	rec.motorcycles	975
	5	rec.sport.baseball	982
	6	rec.sport.hockey	973
Sci	7	sci.electronics	976
	8	sci.med	985
	9	sci.space	987
Talk	10	talk.politics.guns	907
	11	talk.politics.misc	992
	12	talk.religion.misc	778

text data sets 20-Newsgroups repository (Table 1). The 20-NewsGroups data set contains 20,000 documents distributed evenly in 20 different newsgroups (Table 2).

3.2 *Classification Performance Assessment*

For the 20-NewsGroups categorization data, in each case the goal is to correctly discriminate between articles at the top level, e.g. “comp” articles versus “rec” articles, using different sets of sub-categories within each top-category for training and testing. For the UCI categorization data, the different attribute is to classify between dataset. The parameter C is chosen from the range {0.001, 0.01, 0.05, 0.1, 0.2, 0.5, 1, 2, 5, 10, 20, 50, 100, 500, 1 000, 2 000, 4 000, 8 000}. Because of the less training sample in target domain, the 12 different values of the parameter C_t are {0.0001, 0.001, 0.01, 0.05, 0.1, 0.2, 0.5, 1, 2, 5, 10, 100}. However, Tr-AdaBoost (Tr-SVM) parameter set according to [11].

Table 3 Different algorithms' performance on the UCI dataset

Dataset	Dataset		Accuracy (%)		
	Source sample (Positive class vs. negative class)	Target sample (Positive class vs. negative class)	Tr-AdaBoost (Tr-SVM)	ELM	Our method
Banana	460 versus 1700	390 versus 1700	81.52	82.72	87.41
Titanic	250 versus 683	250 versus 684	92.31	93.78	96.88
Waveform	520 versus 6535	520 versus 6534	90.93	92.62	96.95
Image	260 versus 722	260 versus 722	96.31	96.40	97.92
Heart	30 versus 85	30 versus 85	93.76	94.67	96.52
Diabetis	86 versus 241	86 versus 242	87.25	91.09	97.24
Flare Solar	120 versus 333	120 versus 333	89.69	89.91	93.82
Splice	368 versus 974	368 versus 975	87.78	90.42	96.54

For each dataset, we create a set of 18 base kernels, i.e.,

- Gaussian kernels with 15 different widths ($2^{-7}, 2^{-6}, \ldots, 2^{7}$) on all features.
- Polynomial kernels of degree 1 to 3 on all features.

For the implementation of our MK-TELM algorithms, by default, we set the total number of trials T to 100, the sampling ratio to 0.1, and the sampling decay factor β to 2^{-6} for stochastic MK-TELM algorithms. To avoid unstable results for stochastic MK-TELM algorithms, we run 8 times under each setting and report average performances. Finally, we give both classification accuracy and time cost for performance evaluation.

We compare the performance of MK-TELM with Tr-AdaBoost (Tr-SVM), ELM. As shown in Table 3, the MK-TELM method delivers more stable results across all the datasets and is highly competitive in most of the datasets. It obtains the best classification accuracy more than any other method. Hence, as discussed in the above section, MK-TELM possesses overall Tr-AdaBoost (Tr-SVM) advantages over other methods in the sense of classification accuracy.

From Table 4, MK-TELM is obviously superior than Tr-AdaBoost (Tr-SVM) in classification accuracy for almost all these datasets. The MK-TELM obtains the good performance compared with the traditional ELM, MK-ELM.

3.3 *Parameter Evaluation*

3.3.1 Assessment of Parameter

The first set of experiments is to verify the influence of the total number of trials T for the MK-TELM algorithms. In this set of experiments, we examine the experimental results by varying the parameter T from 20 to 200. Figure 1 shows the

Table 4 Different algorithms' performance on the 20-Newsgroups datasets

Task	Datset	Accuracy (%)			
	Source sample (Target sample)	Tr-AdaBoost (Tr-SVM)	ELM	MK-ELM	Our method
1	1 versus 5 (3 vs. 6)	88.65	96.35	96.85	97.48
2	1 versus 7 (3 vs. 8)	93.21	94.31	95.71	96.33
3	1 versus 11 (3 vs. 12)	86.74	88.29	89.32	90.60
4	5 versus 7 (6 vs. 8)	83.21	87.12	88.41	89.39
5	5 versus 6 (11 vs. 12)	79.62	86.27	86.61	89.81
6	7 versus 11 (8 vs. 12)	88.15	89.67	90.84	91.67

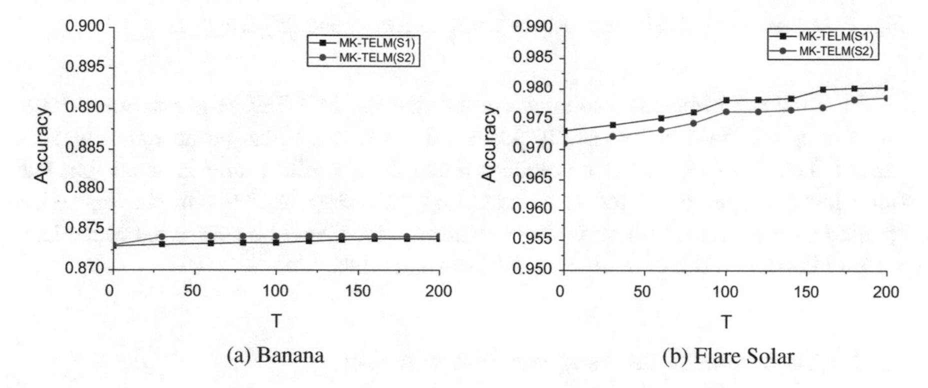

Fig. 1 Assessment of classification accuracy with respect to parameter T

evaluation results for the impact of the parameter T on the classification accuracy and learning time cost, respectively.

The empirical observations as above indicate that choosing a proper parameter T is essentially a weigh between classification accuracy and efficiency performances. However, it is not difficult to make a choice a proper T value that usually falls in between 20 and 200, which sometimes also relies on the empirical experience of efficiency and accuracy in an effective application.

3.3.2 Assessment of Sampling Ratio

Another parameter that may affect the MK-TELM algorithms is the sampling ratio, which depends on the proportion of training data examples sampled from the whole collection of training data at each trial. Figure 2 shows the evaluations of accuracy performance with respect to the sampling ratio by varying its value from 0.05 to 0.5.

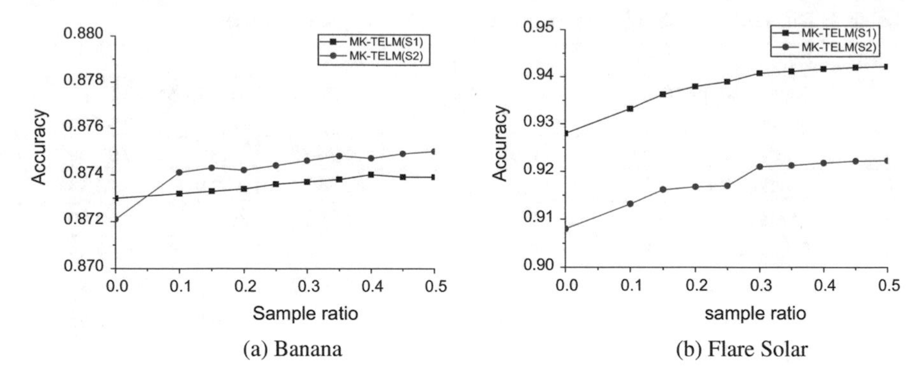

(a) Banana (b) Flare Solar

Fig. 2 Assessment of classification accuracy with respect to boosting sample ratio

From the experimental results, we found that the MK-TELM algorithms with a large sampling ratio value usually produced better classification accuracy performance. This is especially more evident when the sampling ratio is small. On the other hand, employing a too large sampling ratio may lead to sample too many training data examples for some large dataset, which may be somewhat redundant for building the basic classifiers at the boosting trials.

3.3.3 Assessment of the Sampling Decay Factor

The last set of experiments is to examine the effect of the sampling decay factor β for the two stochastic MK-TELM algorithms (S1 and S2). Figure 3 show that MK-TELM-S1 usually likes larger β, while MK-TELM-S2 prefers smaller β.

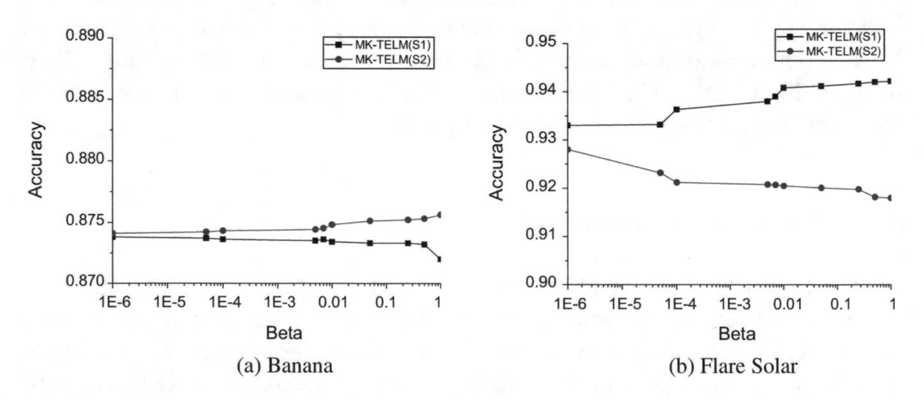

(a) Banana (b) Flare Solar

Fig. 3 Assessment of classification accuracy with respect to the sampling decay factor β

4 Conclusions and Future Research

We address the issue of transfer learning based on MK-ELM for classification in this paper. The results show that the proposed method using MK-TELM can effectively improve the classification by learning cross-domain knowledge and is robust to different sizes of training data.

Acknowledgements We appreciate the anonymous reviewers for the valuable comments. This work was supported by the National Natural Science Foundation of China (No. 61473252) and the National Natural Science Foundation of China (No. 61375049).

References

1. Jain, A.K., Mao, J.: Artificial neural networks: a tutorial. IEEE Comput. 31–44 (1996)
2. Hetch-Neilsen, R.J.: Theory of the backpropagation neural network. In: International Joint Conference on Neural Networks, pp. 593–605 (1989)
3. Cortes, C., Vapnik, V.N.: Support vector networks. Mach. Learn. **20**, 273–297 (1995)
4. Huang, G.B., Chen, L., Siew, C.K.: Universal approximation using incremental constructive feedforward networks with random hidden nodes. IEEE Trans. Neural Netw. **17**, 879–892 (2006)
5. Kim, C.T., Lee, J.J.: Training two-layered feedforward networks with variable projection method. IEEE Trans. Neural Netw. **19**, 371–375 (2008)
6. Wang, Y., Cao, F., Yuan, Y.: A study on effectiveness of extreme learning machine. Neurocomputing **74**, 2483–2490 (2011)
7. Li, G.H., Liu, M., Dong, M.Y.: A new online learning algorithm for structure-adjustable extreme learning machine. Comput. Math. Appl. **60**, 377–389 (2010)
8. Rong, H.J., Ong, Y.S., Tan, A.H., Zhu, Z.: A fast pruned-extreme learning machine for classification problem. Neurocomputing **72**, 359–366 (2008)
9. Zong, W.W., Huang, G.B., Chen, Y.: Weighted extreme learning machine for imbalance learning. Neurocomputing **101**, 229–242 (2013)
10. Liu, X.W., Wang, L., Huang, G.B., Zhang, J., Yin, J.P.: Multiple kernel extreme learning machine. Neurocomputing **149**, 253–264 (2015)
11. Li, X.D., Mao, W.J., Jiang, W.: Multiple-kernel-learning-based extreme learning machine for classification design. Neural Comput. Appl. **27**, 175–184 (2016)
12. Zeng, G.Q., Chen, J., Li, L.M., Chen, M.R., et al.: An improved multi-objective population-based extremal optimization algorithm with polynomial mutation. Inf. Sci. **330**, 49–73 (2016)
13. Peng, Y., Wang, S.H., Long, X.H., Lu, B.L.: Discriminative graph regularized extreme learning machine and its application to face recognition. Neurocomputing **149**, 340–353 (2015)
14. Pan, S.J., Yang, Q.: A survey on transfer learning. IEEE Trans. Knowl. Data Eng. **22**, 1345–1359 (2010)
15. Dai, W., Yang, Q., Xue, G., Yu, Y.: Boosting for transfer learning. In: Proceedings of the 24th International Conference on Machine Learning, pp. 193–200 (2007)
16. Lam, D., Wunsch, D.: Unsupervised feature learning classification using an extreme learning machine. In: Proceedings of the 2013 International Joint Conference on Neural Networks (IJCNN), pp. 1–5 (2013)
17. Pan, J., Wang, X., Cheng, Y., et al.: Multi-source transfer ELM-based Q learning. Neurocomputing **137**, 57–64 (2013)

18. Scardapane, S., Comminiello, D., Scarpiniti, M., et al.: A preliminary study on transductive extreme learning machines. In: Recent Advances of Neural Network Models and Applications, pp. 25–32. Springer International Publishing (2014)
19. Huang, G.B., Bai, Z., Kasun, L.L.C., et al.: Local receptive fields based extreme learning machine. IEEE Comput. Intell. Mag. **10**, 18–29 (2015)
20. Huang, G., Song, S., Gupta, J.N.D., et al.: Semi-supervised and unsupervised extreme learning machines. IEEE Trans. Cybern. **44**(12), 2405–2417 (2014)

Chinese Text Sentiment Classification Based on Extreme Learning Machine

Fangye Lin and Yuanlong Yu

Abstract With the rapid growth of the Web text data, mining and analyzing these text data, especially the online review data posted by the users, can greatly help better understand the usersconsuming habits and public opinions, it also plays an important role in decision-making for the enterprises and the government. But in the process of vectoring text, many current Chinese text sentiment classifications treat words as atomic units, there is no notion of similarity between words. In order to solve this problem, this paper imports word embedding to capturing both the semantic and syntactic information of words from a large unlabeled corpus. In the section of experiment, we toke the noun, verb, and adjectives as candidate set, used χ^2 statistic to reduce the number of dimensions. We mainly compared one-hot representation and word embedding as the expression of word to certain tasks, we also proposed the pooling method with word embedding to standardizing the vector, the ELM with kernels was adopted to analyze the text emotion tendentiousness. Finally the paper summarizes the current status, remaining challenges, and future directions in the field of sentiment classification.

Keywords Sentiment classification · Word embedding · Extreme learning machine

1 Introduction

With the rapid increase usage of the Internet, there are more and more subjective information appearing at the social medium, such as forum, community, blog and shopping websites. Both individual and organization became strongly relying on the review information obtained from the Internet to make their own decisions. However, due to the huge amount of information available on the Internet, one has to

F. Lin · Y. Yu (✉)
The College of Mathematics and Computer Science, Fuzhou University,
Fuzhou 350116, Fujian, China
e-mail: yu.yuanlong@fzu.edu.cn

F. Lin
e-mail: linfangye20@163.com

J. Cao et al. (eds.), *Proceedings of ELM-2016*, Proceedings in Adaptation, Learning and Optimization 9, DOI 10.1007/978-3-319-57421-9_14

search, check and judge each review one by one before the person or organization can make the final decision. In this situation, it will be very useful to first summarize the relevant huge amount of information, this summary will be valuable for both the customer and manufacturer. This kind of work is called opinion-based multi-document summarization. Furthermore, it will greatly enhance the customers efficiency to obtain the information if there is an automatic analysis of the original information, for example, which is positive attitude, which is negative attitude, and to what extent. This is called sentiment classification, which is a very important research topic in the field of natural language processing. While sentiment classification for English text has made a great progress, current research work on sentiment classification for Chinese text is still in its infancy. Due to the huge difference between English and Chinese in syntax, semantics and pragmatics etc., we face more problems in the processing of Chinese text.

In recent years, domestic scholars have done the relevant research according to the characteristics of emotion classification problem. Xu jun [1] used Naive Bayes and Maximum Entropy classification for the sentiment classification of Chinese news and reviews, the experimental results show that the methods they employed perform well. Moreover, they found that selecting the words with polarity as features, negation tagging and representing test documents as feature presence vectors can improve the performance of sentiment classification. ZHOU Jie [2] summarized the characteristics of netnews comments firstly, and selected different sets of feature, different feature dimensions, different feature-weight methods and parts of speech to construct classifiers, then made the comparison and analysis to the experimental results. The results of comparison showed that the features combining sentiment words and argument words perform well to those only employing sentiment words.

In this paper, sentiment classification for Chinese text will be seen as two classification problems, namely positive and negative tendencies. We selected 2607 negative reviews and 5149 positive reviews as the experimental data, and the rate of training sample and testing sample is 2:1. Finally the model of ELM with kernels learned the training sample set, and gave out the result of sentiment classification on the testing sample set. Section 2 will describe some basic models used in sentiment classification and some novel technologies which are proposed in recent years. Section 3 will be detailedly introduce the data processing and feature extraction, we will show the comparison of experimental results. The last section we will have the conclusion of the present stage of the work.

2 System Architecture

The proposed method consists of two modules as shown in Fig. 1: training and testing.

In the part of training phase to need to training the sentiment lexicons from training set, the training set which consists of positive samples and negative samples. We also need to training the word embedding from corpus. In the stage of testing, the

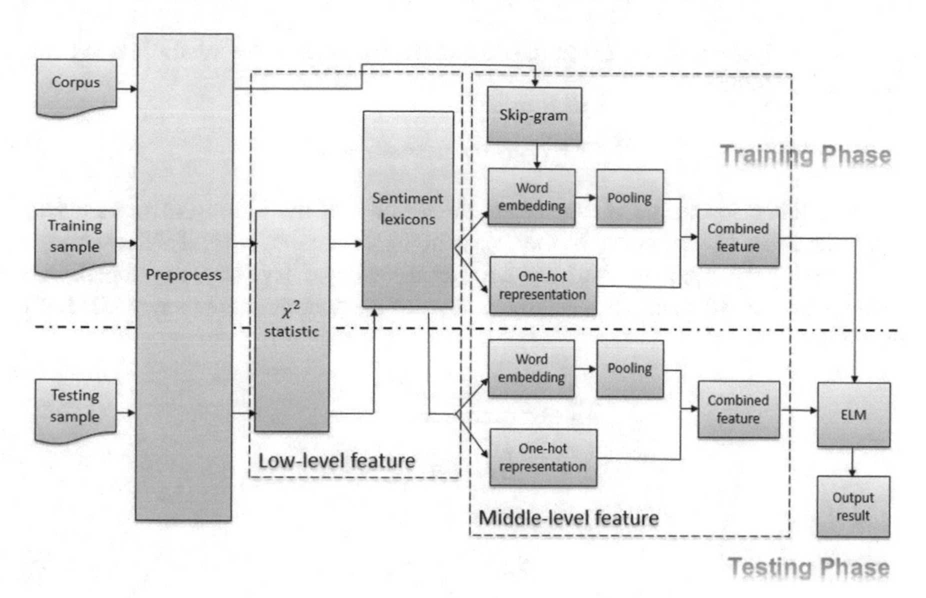

Fig. 1 Architecture of the proposed sentiment classification

testing samples are mapped to the sentiment lexicons after preprocess, all the data set are expressed as middle-level feature and put them into ELM for classifying.

2.1 Preprocess

Denoising: In the stage of experiment the tendency of the reviews should not be equivocal, so we check the whole corpus and delete some ambiguous content and repeated content.

Segmentation: In this paper, we use the Institute of Computing Technology, Chinese Lexical Analysis System (ICTCLAS) [3] as word segmentation tool. After this part we get a stream of words which bring the part-of-speech tagging (POS), we saved all these information by using hashmaps for the next step of study.

Filtering the Stop-Word: After comparison we choose the baidu stop-list to remove some words which have little contribution but occur frequently to the sentimental classification processing.

2.2 Low-Level Feature

Vector space model is an algebraic model for representing text documents (and any objects, in general) as vectors of identifiers. Each dimension corresponds to a separate term. If a term occurs in the document, its value in the vector is non-zero. forms are as follows:

$$D = \left(W_{term1}, W_{term2}, \ldots, W_{termn}\right) \tag{1}$$

Each dimension in the vector means the weights of the term in this document, and it describes the influence of the words in the document, several different ways of computing these values, also known as (term) weights, have been developed. One of the best known schemes is term frequency-inverse document frequency (TF-IDF) weighting.

$$W_{ik} = \frac{tf_{ik} * idf_k}{\sqrt{\sum_{k=1}^{t} (tf_{ik})^2 \left(idf_k\right)^2}} \tag{2}$$

$$tf_{ik} = \frac{n_{ik}}{\sum_{i=1}^{n} n_{ik}} \tag{3}$$

$$idf_k = \log \frac{|D|}{m + \left|\{k : t_i \in d_k\}\right|} \tag{4}$$

The $|D|$ is the total number of documents in the document set; the $|\{k : t_i \in d_k\}|$ is the number of documents containing the term t_i. If the term is not appeared in this document, the divisor is zero. tf_{ik} means the frequency of the term appear in this document set and idf_k means the words on the distribution of the documents in the collection of quantitative.

Considering the traditional vector space model needs more time expense as its vector dimension turns greater, we use the χ^2 statistic (CHI) [4] as feature selection method to reduce the number of dimensions. The χ^2 statistic measures the lack of independence between t and c and can be compared to the χ^2 statistic distribution with one degree of freedom to judge extremeness. Using the two-way contingency table of a term t and a category c, where A is the number of times t and c co-occur, B is the number of time the t occurs without c, C is the number of times c occurs without t, D is the number of times neither c nor t occurs, and N is the total number of documents, the term-goodness measure is defined to be:

$$\chi^2(t, c) = \frac{N \times (AD - CB)^2}{(A + C) \times (B + D) \times (A + B) \times (C + D)} \tag{5}$$

2.3 Middle-Level Feature

Many current natural language processing (NLP) systems and techniques treat words as atomic units, there is no notion of similarity between words. The most common expression of word is one-hot representation, in this method every word is expressed as a long vector, the dimension of this vector is the vocabulary size, only one dimension of the value is 1 and others value is 0, this dimension expresses current word and we use sparse coding to storage it, this choice has several good reasons like simplicity, robustness and the observation that simple models trained on huge amounts of data outperform complex systems trained on less data.

However, the simple techniques have some drawbacks in many tasks. For example, the amount of relevant in-domain data for automatic speech recognition is limited, the performance is usually dominated by the size of high quality transcribed speech data. In machine translation, the existing corpora for many languages contain only a few billions of words or less. Thus, there are situations where simple scaling up of the basic techniques will not result in any significant progress, and we have to focus on more advanced techniques.

With progress of machine learning techniques in recent years, it has become possible to train more complex models on much larger data set, and they typically outperform the simple models. Probably the most successful concept is to use distributed representations of words [5] -word embedding. It can capture both the semantic and syntactic information of words from a large unlabeled corpus and has attracted considerable attention from many researchers.

Mikolov [6] and his team proposed two novel model architectures for computing continuous vector representations of words from very large data sets. The two architectures are continuous bag-of-words (CBOW) model and skip-gram model. the models are shown in Fig. 2. We choose the Skip-gram model to train word embedding, it tries to maximize classification of a word based on another word in the same sentence. More precisely, we use each current word as an input to a log-linear classifier with continuous projection layer, and predict words within a certain range before and after the current word. The training complexity of this architecture is proportional to:

$$Q = C \times (D + D \times \log_2(V)) \tag{6}$$

where C is the maximum distance of the words. Thus, if we choose $C = 5$, for each training word we will select randomly a number R in range $< 1; C >$, and then use R words from history and R words from the future of the current word as correct labels. This will require us to do $R \times 2$ word classifications, with the current word as input, and each of the $R + R$ words as output. In the following experiments, we use $C = 5$.

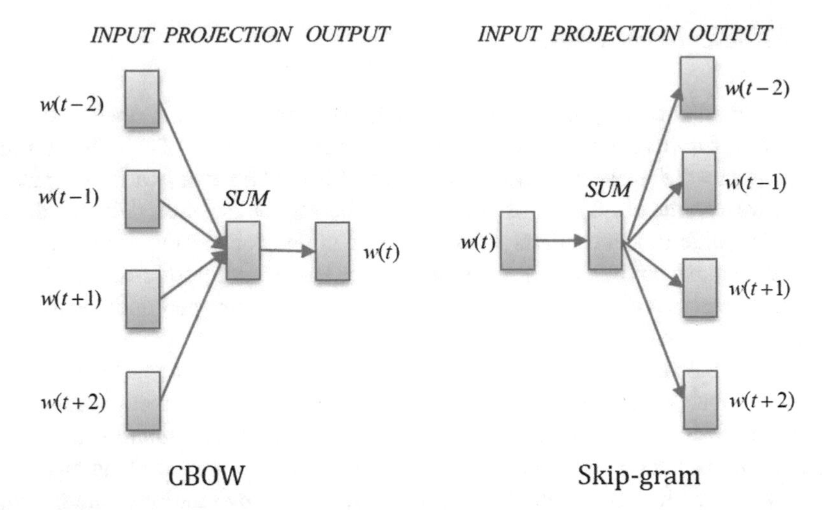

Fig. 2 The CBOW architecture predicts the current word based on the context, and the Skip-gram predicts surrounding words given the current word

2.4 Extreme Learning Machine

In the field of machine learning, a series of traditional machine learning algorithms were improved in order to satisfy the higher data processing needs. For instance, the model of Naive Bayes, Maximum entropy, Support vector machines (SVMs) [7] and so on which reduced the difficulty of solving a certain task. However, there are still some problems with those algorithms: (1) the speed of solution is slower than required for large data; (2) the model related to SVMs need to manual adjustment parameters (C, γ) frequently, they also repeat training in order to obtain the optimal solution with tedious time-consuming process and poor generalization ability.

Under the circumstances, extreme learning machine provides a new way to solve these problems. Extreme Learning Machine was first proposed by Huang [8] in 2006, ELM is generalized single-hidden layer feedforward networks (as illustrated in Fig. 3). 'Extreme' means it breaks limitations of traditional artificial learning methods and aims to work like the brain. Compare to the SVM algorithm, ELM may get better or similar predictive accuracy with less time.

The output function of ELM for generalized SLFNs (take one output node case as an example) is:

$$f(x) = \sum_{i=1}^{L} \beta_i G(a_i, b_i, x) = \beta \cdot h(x) \tag{7}$$

ELM can guarantee the regression prediction accuracy by minimizing the output error:

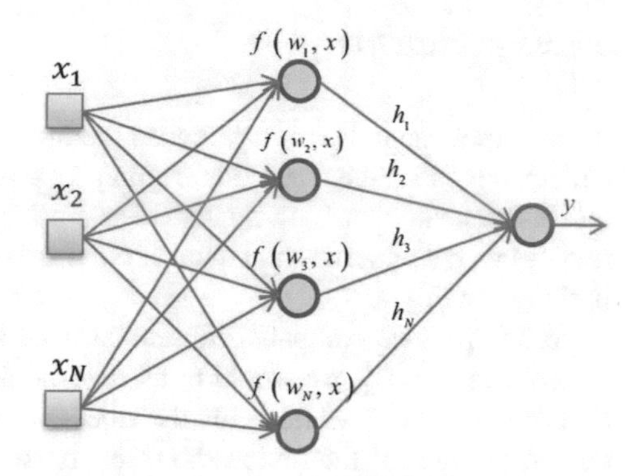

Fig. 3 Feedforward neural network with single hidden layer

$$\lim_{L\to\infty}(f(x)-f_o(x)) = \lim_{L\to\infty}(\sum_{i=1}^{L}\beta_i h_i(x)-f_o(x)) = 0 \tag{8}$$

where L is the number of hidden neurons h_i and $f_o(x)$ is the goal of the forecast function value.

At the same time, ELM is guaranteeing the generalization ability of the network by minimizing the output weights. In general, β is calculating with the least squares, the formula is:

$$\beta = H^{\dagger}O = H^{T}\left(HH^{T}\right)^{-1}O = H^{T}(\frac{1}{C}+HH^{T})^{-1}O \tag{9}$$

where H is the hidden-layer output matrix and $H^{\dagger}$ is the Moore-Penrose generalized inverse of matrix H. It can add a constant to get a better generalization ability according to ridge regression.

As for ELM with kernels, it obtains a better regression and classification accuracy by introducing kernel.

$$f(x) = h(x)\beta = \begin{pmatrix} K(x,x_1) \\ \ldots \\ K(x,x_N) \end{pmatrix}\left(\frac{1}{C}+\Omega_{ELM}\right)^{-1}O \tag{10}$$

$$\Omega_{(i,j)} = \exp(-\gamma(x_i-x_j)^2) \tag{11}$$

where Ω_{ELM} is the kernel function and N is the dimension of input.

3 Experiments

In the experimental part, we choose the hotel BBS reviews information which was collected by song-bo tan as our original corpus. After filtering some ambiguous content and repeated content. finally we select 2607 negative reviews and 5149 positive reviews as the experimental data, and the rate of training sample and testing sample is shown in Fig. 4.

In this part,we compare different parts of speech to build the suitable sentiment lexicon. Intuitively, we would think that the adjective directly determines the emotion of the text but we find only the adjective can not completely express the semantic information of the review.the result is shown in Table 1. If we only choose the adjective as candidate set, althrough the accuracy is 82.63% we only get 799 terms from training sample, too much information is lost. Considering there are not enough adverbs are trained in the word embedding,finally we choose the noun, verb, adjective to build the sentiment lexicon.

Meanwhile, we compare the different dimension of the vector and we find feature dimension has less influence on the accuracy of classification, The result is shown in Table 2. So we decide to choose the 3000 as the feature dimension, after this section every text is represented as a vector.

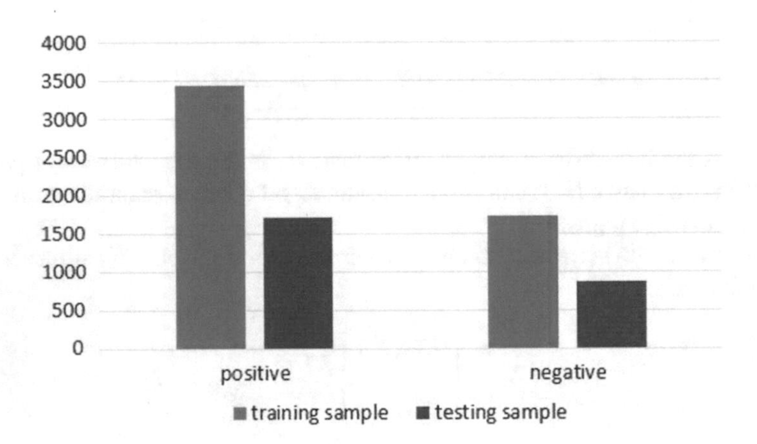

Fig. 4 The rate of training sample and testing sample

Table 1 Comparison between different parts of speech

Parts of speech	Accuracy (%)	TrainingTime (s)	TestingTime (s)
Noun	73.00	2.9317	1.6657
Noun, verb	75.42	3.3108	1.7063
Noun, verb, adjective	80.81	3.1306	1.9216
Noun, verb, adjective, adverb	80.97	2.8238	1.4792

Table 2 Comparison between different dimensions of vector

Dimension of vector	Accuracy (%)	TrainingTime (s)	TestingTime (s)
2000	80.78	2.3067	1.0994
3000	80.81	3.1306	1.9216
4000	80.70	3.1515	1.9125

Word embedding models capture useful information from unlabeled corpora, so we try to use word embeddings as the feature to improve the performance of certain tasks. The first task is the sentiment classification in which the word embedding is the only feature, the second task we use word embedding as an additional feature to achieve the sentiment classification.

According to experiment of Siwei Lai [9] which proves the simplest model, Skip-gram, is the best choice when using a 10M- or 100M- token corpus. We use the Skip-gram model to train the word embedding and the dimension of every term is 100, the word embeddings are used as feature to classify.

3.1 Pooling

In the section we use word embedding as feature, we meet a problem that we find it is hard to standardizing the vector, we hope to use the word embedding to reduce the dimension of vector and we can also remain the semantic information of every review at the same time. However, after preprocess every text are divided as diverse words and the number of words is different. At the first time we simply add the every word embedding which occurs in the text together Unfortunately we find the result is not good enough, the method of simple addition which loses the connection between words, so we propose the pooling method to standardizing the vector, we split the 3000 index of words into 10 parts and we add every word embedding which occurs in same part, if there is no words in this part, we use a zero vector of 100 dimension to express this part, finally we connect the vector according to the sequential order as the final expression of text. After comparison we find pooling method performs much better than simple addition method, the result is shown in Table 3.

Table 3 Comparison between pooling and no pooling

Method	Accuracy (%)	TrainingTime (s)	TestingTime (s)
Word embedding	78.51	1.5723	0.3145
Word embedding (pooling)	79.54	2.0491	0.7345

Table 4 Comparison between SVM and ELM with kernels

Classifier	Accuracy (%)	TrainingTime (s)	TestingTime (s)
SVM	79.5391	103.8211	69.9009
ELM with kernels	80.81	3.1306	1.9216

3.2 *Classification Result*

In this section we firstly compare SVM and ELM with kernels as the sentiment classifier in one-hot representation, the simulations for SVM and ELM with kernels algorithms are carried out in MATLAB environment running in a Core i7-4770, 3.40GHz CPU, 32G RAM. In Table 4. we find ELM with kernels performs better than SVM both in accuracy and saving times.

Then we compare the one-hot representation and word embedding with pooling, we try to connect the one-hot representation and word embedding as combined feature, it works but the promotion of accuracy is very tiny. The final result is shown in Table 5. The curve of training accuracy versus the kernel parameters is shown in Fig. 5.

Table 5 Comparison between one-hot representation and word embedding

	Accuracy (%)	TrainingTime (s)	TestingTime (s)
One-hot Representation	80.68	6.1624	3.8013
Word embedding (pooling)	79.86	3.4444	1.1522
One-hot + Word embedding (pooling)	80.89	5.5068	2.9478

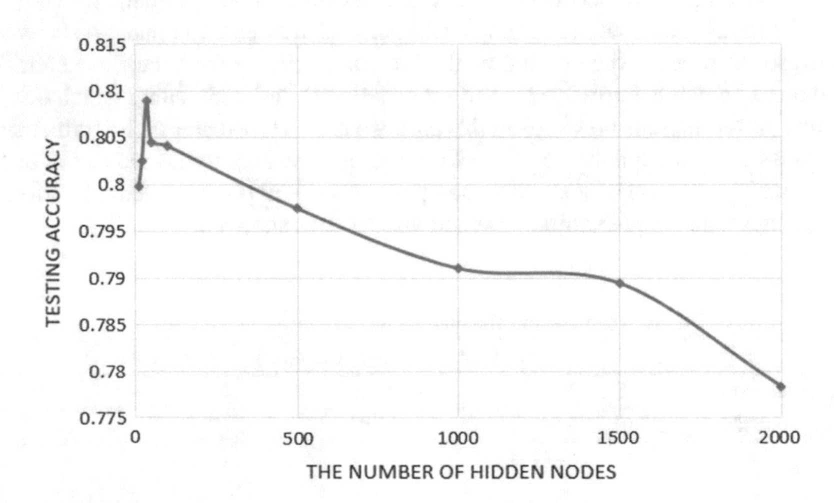

Fig. 5 The curve of testing accuracy versus the Kernel Parameters

4 Conclusion

This paper thought the analysis of the emotional polarity of text as two-classification problems. We used the VSM model to represents a document, and compared one-hot representation and word embedding in expressing words, ELM with kernel gave out the result of classification. Our main operation to the data set was cleaning, word segmentation, removing stop words, feature selection and classification. We found word embedding with pooling method has more advantages than one-hot representation in reducing the dimension of text vectoring, simultaneously it also captured both the semantic and syntactic information of words. In the part of classifier we found it took less time for ELM to training and testing the same data set than SVM. The further research we think is to design a better corpus for getting better word embeddings, we hope the word embedding can help to improve some certain tasks of sentiment classification.

References

1. Xu, J., Ding, Y.X., Wang, X.L.: Sentiment classification for Chinese news using machine learning methods. J. Chin. Inf. Process. **21**(6), 95–100 (2007)
2. Zhou, J., Lin, C., Bi-Cheng, L.I.: Research of sentiment classification for net news comments by machine learning. J. Comput. Appl. **30**(4), 1011–1014 (2010)
3. Zhang, H.P., Yu, H.K., Xiong, D.Y., Liu, Q.: HHMM-based Chinese lexical analyzer ICTCLAS. In: Proceedings of the Second SIGHAN Workshop on Chinese Language Processing, 17, pp. 184–187 (2003)
4. Yang, Y., Pedersen, J.O.: A comparative study on feature selection in text categorization. ICML **97**, 412–420 (1997)
5. Hinton, G.E., McClelland, J.L., Rumelhart, D.E.: Distributed representations. In: Parallel Distributed Processing: Explorations in the Microstructure of Cognition. MIT Press, Cambridge, MA (1986)
6. Mikolov, T., Chen, K., Corrado, G.: Efficient estimation of word representations in vector space. Comput. Sci. (2013)
7. Pang, B., Lee, L., Vaithyanathan, S.: Thumbs up?: sentiment classification using machine learning techniques. In: ACL-02 Conference on Empirical Methods in Natural Language Processing, pp. 79–86 (2002)
8. Huang, G.B., Zhu, Q.Y., Siew, C.K.: Extreme learning machine: a new learning scheme of feedforward neural networks. In: 2004 IEEE International Joint Conference on Neural Networks, 2004. Proceedings, vol. 2, pp. 985–990 (2004)
9. Lai, S., Liu, K., He, S., Zhao, J.: How to generate a good word embedding? Intell. Syst. IEEE (2015)

Incremental ELMVIS for Unsupervised Learning

Anton Akusok, Emil Eirola, Yoan Miche, Ian Oliver, Kaj-Mikael Björk, Andrey Gritsenko, Stephen Baek and Amaury Lendasse

Abstract An incremental version of the ELMVIS+ method is proposed in this paper. It iteratively selects a few best fitting data samples from a large pool, and adds them to the model. The method keeps high speed of ELMVIS+ while allowing for much larger possible sample pools due to lower memory requirements. The extension is useful for reaching a better local optimum with greedy optimization of ELMVIS, and the data structure can be specified in semi-supervised optimization. The major new application of incremental ELMVIS is not to visualization, but to a general dataset processing. The method is capable of learning dependencies from non-organized unsupervised data—either reconstructing a shuffled dataset, or learning dependencies in complex high-dimensional space. The results are interesting and promising, although there is space for improvements.

Keywords ELM · Visualization · Assignment problem · Unsupervised learning · Semi-supervised learning

---

A. Akusok (✉) · E. Eirola · A. Lendasse
Arcada University of Applied Sciences, Helsinki, Finland
e-mail: anton.akusok@arcada.fi; akusok.a@gmail.com

Y. Miche · I. Oliver
Nokia Solutions and Networks Group, Espoo, Finland

Y. Miche
Department of Computer Science, Aalto University, Espoo, Finland

K.-M. Björk
Risklab at Arcada University of Applied Sciences, Helsinki, Finland

A. Gritsenko · S. Baek · A. Lendasse
Department of Mechanical and Industrial Engineering, The University of Iowa, Iowa City, USA

A. Gritsenko · A. Lendasse
The Iowa Informatics Initiative, The University of Iowa, Iowa City, USA

J. Cao et al. (eds.), *Proceedings of ELM-2016*, Proceedings in Adaptation, Learning and Optimization 9, DOI 10.1007/978-3-319-57421-9_15

1 Introduction

The ELMVIS method [1] is an interesting Machine Learning method that optimize a cost function by changing assignment between two sets of samples, or by changing the order of samples in one set which is the same. The cost function is learned by an Extreme Learning Machine (ELM) [2–4], a fast method for training feed-forward neural networks with convenient mathematical properties [5, 6]. Such optimization problem is found in various applications like open-loop Traveling Salesman problem [7] or clustering [8] (mapping between samples and clusters), but not in Neural Networks. ELMVIS is unique in a sense that it combines the optimal assignment task with neural network optimization problem; the latter is optimized at each step of ELMVIS.

A recent advance in ELMVIS+ method [9] set its runtime speed comparable or faster than other state-of-the-art methods in visualization application. However there are unresolved problems like a greedy optimization leading to a local optimum. Also ELMVIS+ can be applied to a much wider range of problems than a simple visualization or a visualization accounting for the class information [10], which have not been tested or reported yet. This paper addresses the aforementioned drawbacks, and presents the most recent research advances in the family of ELMVIS methods.

The proposed incremental ELMVIS allows for iterative growth of dataset size and model complexity. Incremental ELMVIS learns an approximate global data structure with a few data samples and a simple ELM model, because at a very small scale global and local optimums are similar or the same. Then more data samples are added to the model, choosing the ones that better fit an existing ELM. After adding a batch of new samples, the current model is refined by ELMVIS+ method. This refinement keeps the global optimum due to the greedy optimization and only small changes. More neurons are added to ELM as the dataset size grows, to better separate the data.

Iterative ELMVIS is useful for semi-supervised learning, starting from the data with known outputs and adding more data with unknown outputs, simultaneously updating the model. It can even apply to completely unsupervised datasets, where it finds an input-output dependency, learns it with ELM model, and then simultaneously expands the supervised part of a dataset and updates an ELM model that encodes the input-output dependency.

The experiments have shown the ability of iterative ELMVIS to improve global optimum, successfully perform semi-supervised and unsupervised learning with complex tasks. Current version of the method is found limited to good separation between only two classes in data (which it learns first), ignoring samples of the additional classes until the first two ones are exhausted, and poorly fitting the additional classes into the learned two-class model. Solution to this problem will be considered in further works on the topic.

2 Methodology

An iterative extension of ELMVIS+ methodology is rather straight-forward, as explained below. ELMVIS methods start with a list of (visualization) samples and an unordered set of data samples; and it finds an optimal order of data samples in the set by a greedy search with changing positions of many samples at once (ELMVIS) or only two samples at once (ELMVIS+).

Iterative ELMVIS splits data samples into fixed, candidate and available ones. Fixed samples have their order fixed and cannot be moved by an iterative ELMVIS. Candidate samples are a small number of samples which are chosen from candidate+available ones to maximize the cost function. This cost function takes into account fixed and current candidate samples, but ignores the available samples. Once current candidate samples are chosen optimally, they are added to the fixed ones, and the method is repeated with a few more candidate samples—until the available data samples are exhausted.

2.1 Extreme Learning Machine

Extreme Learning Machine is a way of training feedforward neural networks [11] with a single hidden layer that features randomly assigned input weights [12], explicit non-iterative solution for output weights and an extreme computation speed and scalability [13]. This model is used as a non-linear cost function in all ELMVIS methods. This short summary introduces the notations to the reader.

The goal of ELM model is to approximate the projection function $\hat{\mathbf{y}}_i \approx f(\mathbf{x}_i)$ using a representative training dataset. As ELM is a deterministic model, the function $f()$ is assumed to be deterministic, and a noise ϵ is added to cover the deviation of true outputs $\mathbf{y}$ from the predictions by a deterministic function $f()$

$$\mathbf{y} = f(\mathbf{x}) + \epsilon \tag{1}$$

The Extreme Learning Machine [2] (ELM) is a neural network with d input, L hidden and c output neurons. The hidden layer weights $\mathbf{W}_{d \times L}$ and biases $\text{bias}_{1 \times L}$ are initialized randomly and are fixed. The hidden layer neurons apply a transformation function ϕ to their outputs that is usually a non-linear transformation function with bounded output like sigmoid or hyperbolic tangent.

The output of the hidden layer is denoted by $\mathbf{h}$ with an expression

$$\mathbf{h}_i = \phi(\mathbf{x}_i \mathbf{W} + \text{bias}) \tag{2}$$

where the function $\phi()$ is applied element-wise, and can also be gathered in a matrix $\mathbf{H}_{N \times L}$ for convenience.

The output layer of ELM poses a linear problem $\mathbf{H}\boldsymbol{\beta} = \mathbf{Y}$ with unknown output weights $\boldsymbol{\beta}_{L\times c}$. The solution is derived from the minimization of the sum of squared residuals $\mathbf{r}^2 = (\mathbf{y} - \hat{\mathbf{y}})^2$, that gives the ordinary least squares solution $\boldsymbol{\beta} = \mathbf{H}^{\dagger}\mathbf{Y}$, where $\mathbf{H}^{\dagger}$ is a Moore-Penrose pseudoinverse [14] of the matrix $\mathbf{H}$.

2.2 *ELMVIS+ Method*

ELMVIS+ method [15] approximates a relation between visualization (or input) space $\mathcal{V}$ and data space $\mathcal{X}$ by an ELM model. The task has N representative samples $\mathbf{v} \in \mathcal{V}$ and $\mathbf{x} \in \mathcal{X}$, but their order is unknown. The method assumes fixed order of samples $\mathbf{v}$ joined in matrix $\mathbf{V}$, and finds a suitable order of samples $\mathbf{x}$ joined in matrix $\mathbf{X}$ by exchanging pairs of rows in $\mathbf{X}$. Contrary to a common use of ELM, data samples $\mathbf{x}$ are the outputs of ELM and visualization coordinates $\mathbf{v}$ are the inputs (thus ELM predicts original data $\hat{\mathbf{x}}$). Visualization coordinates $\mathbf{V}$ are chosen arbitrary and fixed—they can be distributed randomly with normal or uniform distribution, or initialized on a regular grid.

The optimization criterion is a cosine similarity between $\mathbf{X}$ and $\hat{\mathbf{X}}$ predicted by ELM. A low error means that its possible to reconstruct data from the given visualization points, thus the visualization points keep information about the data. The reconstruction is approximated by the ELM model in ELMVIS.

There is an explicit formula for a change of error (negative cosine similarity) for swapping two rows in $\mathbf{X}$ and re-training ELM with this new dataset. The readers can refer to the original paper [15] for the full formula. It is based on the expression for the change on error Δ_E in case a row $\mathbf{x}_a$ in $\mathbf{X}$ is changed by $\delta \in \mathcal{X}$ amount:

$$\Delta_E = \sum_{j=1}^{d} \left(\mathbf{A}_{a,a}\delta_j^2 + 2\hat{\mathbf{x}}_{a,j}\delta_j \right) \tag{3}$$

$$\hat{\mathbf{X}} \leftarrow \hat{\mathbf{X}} - \mathbf{A}_{:,a} \times \delta \tag{4}$$

$$\mathbf{X}_{:,a} \leftarrow \mathbf{X}_{:,a} + \delta \tag{5}$$

2.3 *Incremental ELMVIS*

Incremental ELMVIS splits all data samples in three groups: fixed, candidate and available samples. The separation is maintained with two indexes: i_A is the number of fixed samples, and i_B is the number of fixed+candidate ones.

Incremental ELMVIS works similar to ELMVIS+. First, the initial numbers of fixed and candidate samples are given by i_A and i_B. Then swap indexes $a \in [i_A, i_B]$ and $b \in [i_B + 1, N]$ are selected randomly to replace one candidate sample with an available one. The change of error Δ_E is computed by the formula (3) for the change in the candidate row of the data matrix $\mathbf{X}$. Compared to ELMVIS+, the change in the available sample is ignored. Then for the negative Δ_E, an update step is performed for sample $\mathbf{x}_a$ as in Eq. (4) and for both $\mathbf{x}_a$ and $\mathbf{x}_b$ as in Eq. (5).

If there is no improvement during a large number of swaps, the current candidate samples are added to the fixed ones ($i_A \leftarrow i_B$), and k more samples are added as candidates ($i_B \leftarrow i_B + k$). Candidate samples are already initialized with the data samples at indexes $\mathbf{x}_{i_A}, \ldots, \mathbf{x}_{i_B}$. The method then repeats for another iteration. Iterations stop when no available samples are left.

In the original ELMVIS+, matrix $\mathbf{A}$ took the most space and limited the maximum amount of processed samples (its memory size is $\mathcal{O}(N^2)$). In incremental ELMVIS, only a $\mathbf{A}_{i_B \times i_B}$ part of the whole matrix $\mathbf{A}$ is needed. That relaxed memory requirements of the method, and allows to use a very large pool of available samples. The memory constraints of incremental ELMVIS apply only to the number of optimized data samples.

3 Experimental Results

Incremental ELMVIS method is developed for two main applications. The first one is achieving a better global optimum in ELMVIS+. The original EMLVIS+ is a good visualization method, however it has an unwanted feature: with a large number of neurons in ELM it fragments clusters in the visualized data. This happens with large amount of data and a complex ELM model. An iterative ELMVIS that starts with small amount of data and a simple model keeps all similar data together; then more data samples are gradually added while the total picture changes little due to local minimum in ELMVIS+ optimization.

The second application is finding unknown relations in datasets. This is an unsupervised learning field relevant to the current Big Data trends, when a large amount of interesting data is available—but there pre-processing like manual labeling or classification. It is possible to extract relations inside data automatically by iteratively growing an ELMVIS+ model between two sets of data samples (they don't have to be related to visualization). Results for both applications are presented below.

3.1 Better Optimum with ELMVIS

ELMVIS+ method is fast and works with large datasets, but it has a greedy optimization approach that leads to local optimality of the solution. Such local optimum is close to a global one for small datasets and simple ELM models, but with a large dataset and many neurons in ELM model the visualization data is split into multiple small clusters with local optimality, non-representative of a global picture.

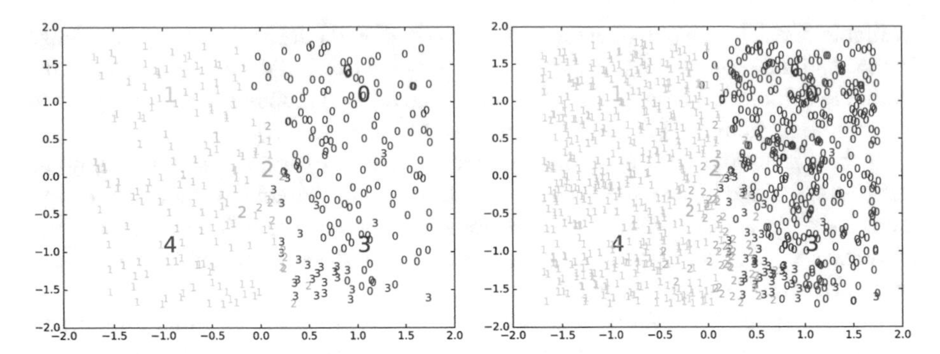

Fig. 1 Visualization of 300 (*left*) and 900 (*right*) MNIST digits on a two-dimensional space with incremental ELMVIS. The visualization is initialized with 5 digits shown in *larger font*. *Digit symbols* and *colors* are for the presentation purpose only; ELMVIS method does not have access to them and works with raw pixel data

A better global optimality is achievable with an incremental ELMVIS. This experiment uses MNIST digits [16] with their original 28×28 features (grayscale pixels), with 500 digits for each of the classes 0–4. It starts by seeding several cluster as shown on Fig. 1 by bold samples, and a simple ELM model. Then gradually added data fits into the existing model (Fig. 1, left). An ELM learns an easy separation between two clusters, and an incremental ELMVIS prefers to add samples of these clusters until they are available (Fig. 1, right).

The incremental ELMVIS learns a good model that separated between two different kinds of data. Then there is no data samples of these two types left, it begins adding more types, starting at the boundary (Fig. 2, left). These additional classes are mapped to a single area, although they go over the two previously learned cluster as there is no space left on the visualization (Fig. 2, right). The ELMVIS still ignores the last available class (digits 4) because it is not represented on the visualization space.

3.2 Better Optimum with Semi-supervised ELMVIS

In the previous experiment, there were no sharp borders between clusters because ELMVIS used all the visualization space to show only two clusters, and then has to map additional data clusters other them. Sharper borders can be obtained by running the original ELMVIS+ on the fixed set of data points after each iteration of the incremental ELMVIS. That will make space for more clusters by compacting the existing ones; while still preserving the global structure as ELMVIS+ optimization goes to the local optimum only.

In addition, a semi-supervised approach is tested where the clusters are initialized with a larger number of samples. This experiment uses 5 classes of digits with 1000

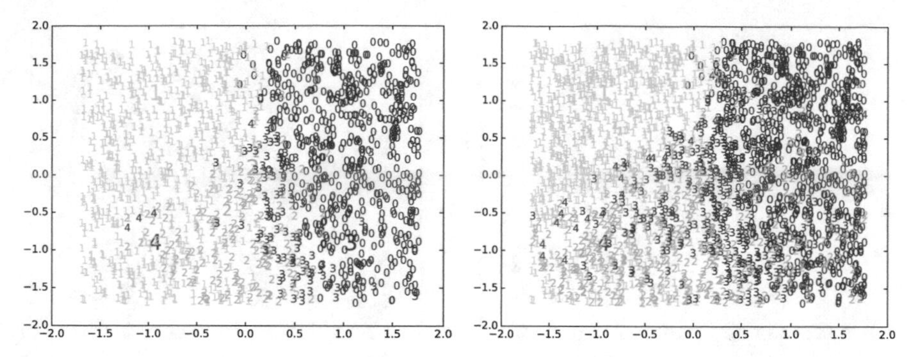

Fig. 2 Visualization of 1500 (*left*) and 2500 (*right*) MNIST digits on a two-dimensional space with incremental ELMVIS. There is only 500 samples of each class; ELMVIS has to add more classes then it exhausted zeros and ones in the available set. *Digit symbols* and *colors* are for the presentation purpose only; ELMVIS method does not have access to them and works with raw pixel data

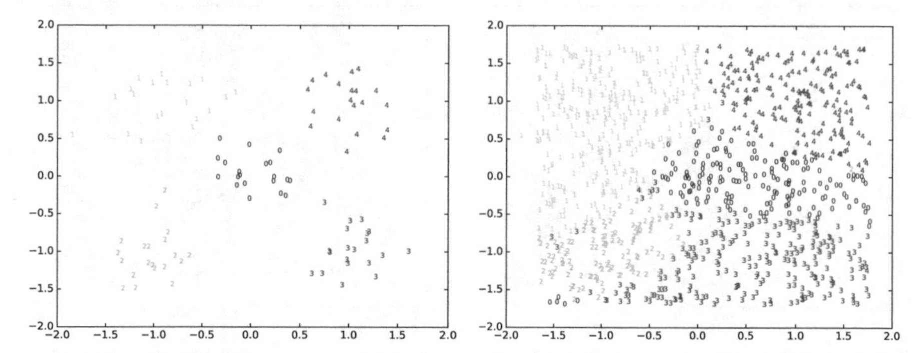

Fig. 3 Semi-supervised incremental ELMVIS initialized with 20 samples per class (*left*). An ELM learns all the clusters from a larger initialization set, and add new samples for all of them instead of just two (*right, 1000 samples mapped*). Classes of samples are used for initialization, but ELMVIS method does not have access to them and works with raw pixel data

samples per class, initialized with 20 samples per class as shown on Fig. 3 (left). A large initialization set forces ELM to learn all the classes, and add samples from all of them instead of only two (Fig. 3, right).

When the method need space to map more samples from a particular class, an additional ELMVIS+ step moves existing clusters to give that space while keeping sharp boundaries between the classes. The effect is shown on Fig. 4 where all clusters are moved to give more space for digits *2*. The global structure is well preserved, with large clusters keeping their place.

The added ELMVIS+ step refines the candidate samples placement within the fixed ones, that is necessary towards the end of visualization when there may be no samples of the desired class left, or no spaces left within the desired class area. An improvement of ELMVIS+ step fitting the candidate samples is presented on Fig. 5.

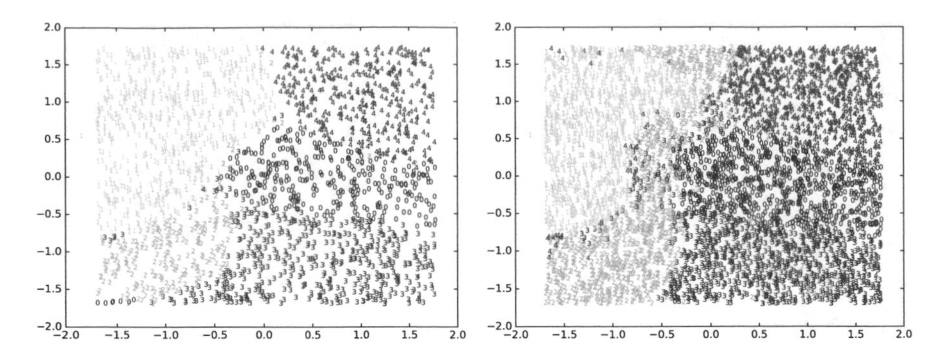

Fig. 4 An additional ELMVIS+ step moves existing clusters if more space is needed for the remaining data, without disturbing the global data structure. Here 2000 samples visualized on the *left* with under-represented class *2*; 5000 samples visualized on the *right* with other clusters moved to fit class *2* without overlap

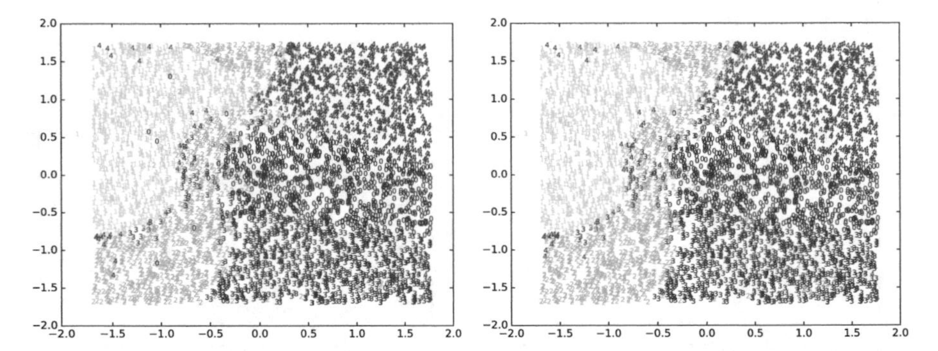

Fig. 5 The last step on an incremental ELMVIS (*left*), further refined by ELMVIS+ (*right*). Candidate samples of class *0* that are mis-placed due to the lack of space inside class *0* are correctly fitted by ELMVIS+

3.3 Data Structure Discovery with Unsupervised ELMVIS

The inputs to ELMVIS are not limited to visualization coordinates; they can be arbitrary data. Thus ELMVIS is a feasible method for finding structure in the data in an unsupervised manner. This experiments takes a number of MNIST digits in random (undefined) order, and maps them to the same number of classes (in zero/one encoding), or to the same number of different MNIST digits of the same classes.

First dataset has zero-one classes as inputs, and MNIST digits as outputs. However it cannot be used to train a supervised model because it is unknown which input corresponds to which output—a common situation in the analysis of large automatically acquired data corpora that has not been manually labeled. The goal of an incremental ELMVIS is to reconstruct the correct input-output pairing. The experiment uses 100 samples per class with 2 or 5 classes of digits. Note that the best pairing across all permutations of classes is reported, as in the unsupervised setup with equal

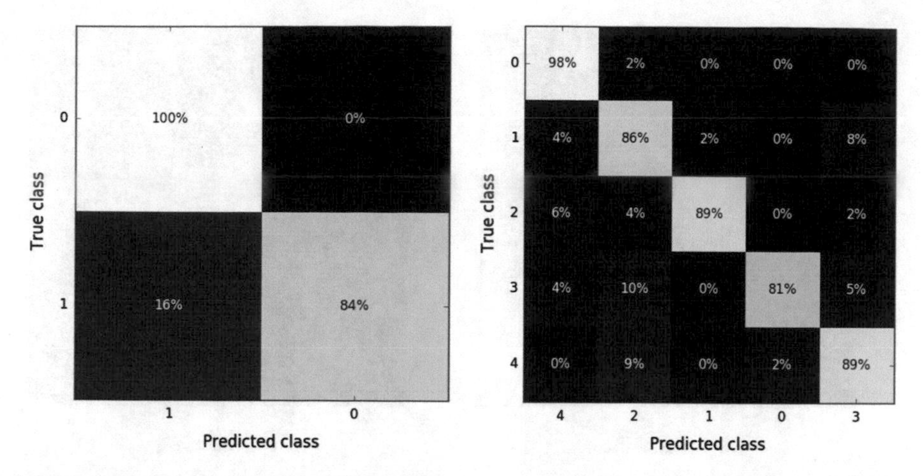

Fig. 6 Binary class to MNIST digit mapping with two or five classes, confusion matrices present correct percentages. The best mapping across all permutation of classes is shown, because it is not possible to specify matching of particular classes in an unsupervised method

amount of samples there is no way to tell ELMVIS which class should go to which digit.

The resulting confusion matrices are shown on Fig. 6. An incremental ELMVIS successfully paired classes with pictures of MNIST digits. The method mapped classes arbitrary (i.e. class $[1, 0, 0, 0, 0]$ is mapped to digit *4*), but this is to be expected from a purely unsupervised method.

Another experiment is performed in a similar setup, but instead of binary class representations the incremental ELMVIS tried to map MNIST digits to other MNIST digits (of the same classes, Fig. 7). The mapping is successful with two classes. With more than two classes the same feature always appears: two random classes are separated well while other classes are randomly mixed with them. This outcome is in line with the results observed in Sect. 3.1 where two classes are clearly separated at the beginning, followed by other classes mapped over them.

4 Conclusions

An iterative extension to the original ELMVIS+ method is proposed in this paper. It iteratively selects a small number of best fitting samples from all the available ones, and adds them to the model. It allows for a much larger set of potential samples than ELMVIS+ by limiting memory requirements to already fitted samples rather than to all available ones, keeping the high speed of the ELMVIS+ at the same time.

The method improves global structure of ELMVIS+ visualization by starting with a small dataset, and gradually adding more data or increasing the complexity of the model. It preserves the global structure, sharp boundaries between classes, and has

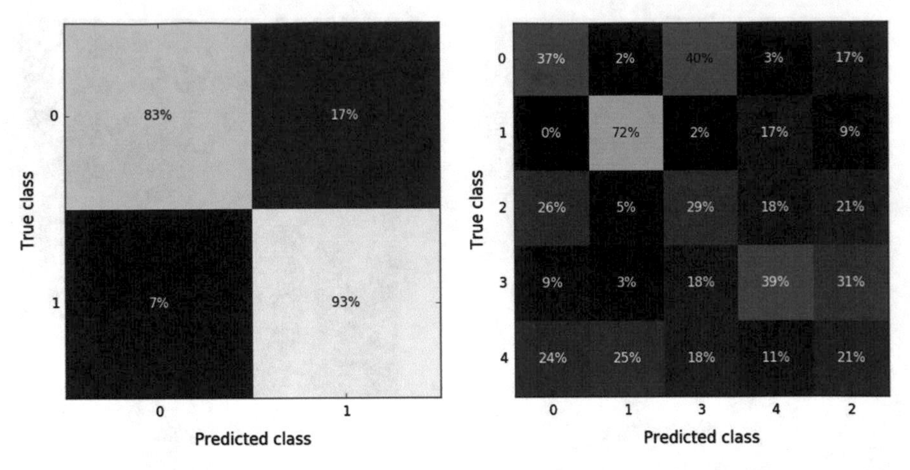

Fig. 7 MNIST digits to MNIST digits mapping, using different digits from the same 2 or 5 classes. Two classes are separated clearly, while any additional number of classes are mixed with them

a possible semi-supervised extension where the samples are mapped to the specified places on the visualization space.

The proposed method is capable of unsupervised data structure detection. It excels in reconstructing a randomly shuffled dataset with unknown pairing between inputs and outputs. It is also capable of finding a mapping between two complex data spaces as shown on MNIST digits example.

The methodology needs further investigation and improvement to counter the observed drawbacks, specifically the tendency of learning an easy model first leading to problems in incorporating more complex parts to the global picture.

References

1. Cambria, E., et al.: Extreme learning machines [Trends & Controversies]. IEEE Intell. Syst. **28**(6) 30–59 (2013)
2. Huang, G.B., Zhu, Q.Y., Siew, C.K.: Extreme learning machine: theory and applications. In: Neural Networks Selected Papers from the 7th Brazilian Symposium on Neural Networks (SBRN '04)7th Brazilian Symposium on Neural Networks, vol. 70(1–3), pp. 489–501 (2006)
3. Huang, G.B., Zhou, H., Ding, X., Zhang, R.: Extreme learning machine for regression and multiclass classification. IEEE Trans. Syst. Man Cybern. Part B: Cybern. **42**(2), 513–529 (2012). April
4. Huang, G.B.: What are extreme learning machines? Filling the gap between Frank Rosenblatt's dream and John von Neumann's puzzle. Cogn. Comput. **7**(3), 263–278 (2015)
5. Huang, G.B., Chen, L., Siew, C.K.: Universal approximation using incremental constructive feedforward networks with random hidden nodes. IEEE Trans. Neural Netw. **17**(4), 879–892 (2006). July
6. Miche, Y., Sorjamaa, A., Bas, P., Simula, O., Jutten, C., Lendasse, A.: OP-ELM: optimally-pruned extreme learning machine. IEEE Trans. Neural Netw. **21**(1), 158–162 (2010). January

7. Gutin, G., Punnen, A.P. (eds.): The Traveling Salesman Problem and its Variations. Combinatorial Optimization. Kluwer Academic, Dordrecht, London (2002)
8. Alpaydin, E.: Introduction to Machine Learning. MIT press (2014)
9. Akusok, A., Miche, Y., Björk, K.M., Nian, R., Lauren, P., Lendasse, A.: ELMVIS+: improved nonlinear visualization technique using cosine distance and extreme learning machines. In: Cao, J., Mao, K., Wu, J., Lendasse, A. (eds.) Proceedings of ELM-2015 Volume 2: Theory, Algorithms and Applications (II), pp. 357–369. Springer, Cham (2016)
10. Gritsenko, A., Akusok, A., Miche, Y., Björk, K.M., Baek, Stephen, Lendasse, A.: Combined Nonlinear Visualization and Classification: ELMVIS++C. Vancouver, Canada (2016)
11. Haykin, S.: Neural Networks: A Comprehensive Foundation, 2nd edn. Prentice Hall (1998)
12. Huang, G.B.: An insight into extreme learning machines: random neurons, random features and kernels. Cogn. Comput. **6**(3), 376–390 (2014)
13. Akusok, A., Björk, K.M., Miche, Y., Lendasse, A.: High-performance extreme learning machines: a complete toolbox for big data applications. IEEE Access **3**, 1011–1025 (2015). July
14. Rao, C.R., Mitra, S.K.: Generalized inverse of a matrix and its applications. In: Proceedings of the Sixth Berkeley Symposium on Mathematical Statistics and Probability, Volume 1: Theory of Statistics, pp. 601–620. University of California Press, Berkeley, CA (1972)
15. Akusok, A., Miche, Y., Björk, K.M., Nian, R., Lauren, P., Lendasse, A.: ELMVIS+: Fast nonlinear visualization technique based on cosine distance and extreme learning machines. Neurocomputing (2016)
16. Lecun, Y., Bottou, L., Bengio, Y., Haffner, P.: Gradient-based learning applied to document recognition. Proc. IEEE **86**(11), 2278–2324 (1998). November

Predicting Huntington's Disease: Extreme Learning Machine with Missing Values

Emil Eirola, Anton Akusok, Kaj-Mikael Björk, Hans Johnson and Amaury Lendasse

Abstract Problems with incomplete data and missing values are common and important in real-world machine learning scenarios, yet often underrepresented in the research field. Particularly data related to healthcare tends to feature missing values which must be handled properly, and ignoring any incomplete samples is not an acceptable solution. The Extreme Learning Machine has demonstrated excellent performance in a variety of machine learning tasks, including situations with missing values. In this paper, we present an application to predict the onset of Huntington's disease several years in advance based on data from MRI brain scans. Experimental results show that such prediction is indeed realistic with reasonable accuracy, provided the missing values are handled with care. In particular, Multiple Imputation ELM achieves exceptional prediction accuracy.

Keywords Extreme learning machine · Missing values · Multiple imputation · Huntington's disease · Prediction

1 Introduction

The prevalence of machine learning has been steadily increasing in the current information age. Engineering advances in processor performance and storage capacities have provided an opportunity to make practical use of computational statistics on a large scale. Simultaneously, the research community has contributed by devising

E. Eirola (✉) · A. Akusok
Arcada University of Applied Sciences, Helsinki, Finland
e-mail: emil.eirola@arcada.fi

K.-M. Björk
Risklab at Arcada University of Applied Sciences, Helsinki, Finland

H. Johnson
Department of Electrical Engineering, The University of Iowa, Iowa City, USA

A. Lendasse
Department of Mechanical and Industrial Engineering and The Iowa Informatics Initiative, The University of Iowa, Iowa City, USA

J. Cao et al. (eds.), *Proceedings of ELM-2016*, Proceedings in Adaptation, Learning and Optimization 9, DOI 10.1007/978-3-319-57421-9_16

new clever algorithms to maximize the amount of relevant information that can be extracted from data. While data sets are large at times, the more common situation is that the number of samples is limited by practical issues, meaning that all of the available data must be used as efficiently as possible in order to achieve the desired results.

One pertinent issue is incomplete data sets, where some samples have missing information [1]. Most methods in machine learning are based on the assumption that data is available as a fixed set of measurements for each sample. However, this is not always true in practice, as several samples may have incomplete records for any of a number of reasons. These could include measurement error, device malfunction, operator failure, non-response in a survey, etc. Simply discarding the samples or variables which have missing components often means throwing out a large part of data that could be useful for the model. It is relevant to look for better ways of dealing with missing values in such cases.

If the fraction of missing data is sufficiently small, a common pre-processing step is to perform imputation to fill in the missing values and proceed with conventional methods for further processing. Any errors introduced by inaccurate imputation may be considered insignificant in terms of the entire processing chain. With a larger proportion of measurements being missing, errors caused by the imputation are increasingly relevant as errors propagate in non-obvious ways, and the missing value imputation cannot be considered a separate step. Instead, the task should be seen from a holistic perspective, and the statistical properties of the missing data should be considered more carefully [2].

A particularly important area where incomplete data is commonplace is in healthcare, where varying procedures and equipment affect which data is available. Studies generally include a limited number of subjects, and often requires expensive equipment and highly trained professionals, meaning that discarding data samples with a few unknown values would not be cost-effective, and all the data must be used to its maximal potential.

Recently, significant results in machine learning have been achieved with methods based on the Extreme Learning Machine (ELM) [3]. Several modified approaches have been published with the goal of using datasets with missing values [4–8]. In this paper, we describe an application of ELM with multiple imputation [5] to predict the onset of Huntington's disease from early brain scans.

The structure of this paper is as follows: Sect. 2 describes the application scenario and data used. The modelling procedure is detailed in Sect. 3, and results are presented in Sect. 4.

2 Application: Predicting Onset of Huntington's Disease

Huntington's disease (HD) is an inherited condition caused by a genetic disorder. It affects muscle coordination and leads to mental decline and behavioral symptoms, and ultimately death. All patients with a sufficiently severe form of the disorder will eventually get the disease. Physical symptoms can begin at any age, but usually begin between 35 and 44 years of age. No cure is known, but therapy can considerably mitigate symptoms, especially if started at an early stage

While identifying the disease can be achieved early by testing for the genetic disorder, it is more difficult to predict how quickly symptoms will manifest, as the progression of the disease is not fully understood in detail [9–12]. It has been observed that subtle changes in brain structure can be identified several years before diagnosis [9]. The main research question here is to study how well MRI data allows to predict when symptoms will appear, up to 10 years in advance.

The data consists of a number of measurements related to the patients. Each sample corresponds to one session with a patient. For many patients, measurements (sessions) are available *before and after* they have been diagnosed with the disease, and this is crucial for studying the progression in detail. As most patients attended several sessions, it is important to consider that several samples are associated to the same patient. The sessions were planned to be conducted approximately once every 2 years, but in reality the data is available at very irregular intervals, differing for each patient.

There are a total of 3729 sessions and 1370 patients. There is a control group of 288 patients, which do not have the genetic disorder, and as such do not contract the disease.

The measurements (variables) consist of key metrics derived from an MRI scan, e.g., volume or length of specific structures. There are 561 variables in total. In addition, the data contains a target variable representing whether the physician diagnosed the patient with Huntington's disease or not.

The data has been collected at several different locations, by different people, on a variety of equipment. The varying procedures and equipment mean that many values are missing for a large number of patients. For each session, the number of available measurements varies from 95 to 561, and only 10% of sessions have no missing values. No measurement available for all sessions. Overall, 45% of values are missing in the data.

3 Model

The goal of the model is to predict the progression of the disease several years in advance. However, the data directly includes only the diagnosis at the time of the measurement session. Fortunately, the majority of the patients return for follow-up sessions, meaning that some information about the progression can be inferred.

Ideally, the output variable Y should contain the state of the patient up to 10 years in the future, but this information is not fully available. For example, consider the sample related to a visit in 2001, at which the patient does not show symptoms. Say the other available sessions for this patient are as follows:

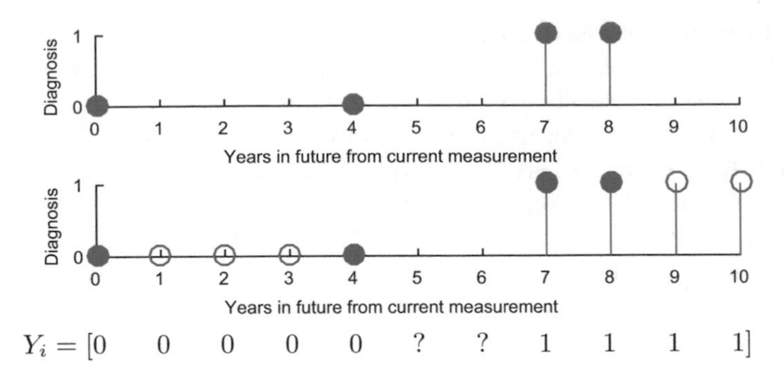

Fig. 1 Constructing the output as a prediction trajectory for a sample patient with data from three future sessions available. *Filled points* represent available data, and *non-filled points* inferred information. The state at 5 and 6 years remains unknown

- 2005 (not diagnosed)
- 2008 (diagnosed)
- 2009 (diagnosed)

By assuming the progression is monotonic (i.e., once diagnosed, a patient will never return to a non-diagnosed state), we can conclude that for the years 2001–2005 (0–4 years in the future) the patient should be considered as not diagnosed, and 2008 → (7+ years in the future) should be considered diagnosed. Information for 5–6 years is however still not available. We construct a *prediction trajectory*, such that the output is 0 for years with no diagnosis, 1 for years with diagnosis, and missing values when the state is unknown (see Fig. 1).

As such, the output is an 11-dimensional vector, and there are missing values in both input and output variables of the data. Note that several other types of particular situations occur in the data:

Diagnosed at current visit	$Y_i = [\,1\,1\,1\,1\,1\,1\,1\,1\,1\,1\,1\,]$
Infrequent visits	$Y_i = [\,0\,?\,?\,?\,?\,?\,?\,1\,1\,1\,1\,]$
Single visit, not diagnosed	$Y_i = [\,0\,?\,?\,?\,?\,?\,?\,?\,?\,?\,?\,]$
Not diagnosed after several visits	$Y_i = [\,0\,0\,0\,0\,0\,?\,?\,?\,?\,?\,?\,]$
Control group subject	$Y_i = [\,0\,0\,0\,0\,0\,0\,0\,0\,0\,0\,0\,]$

3.1 Extreme Learning Machine

The prediction model is realised using the Extreme Learning Machine (ELM) [3], which is a single hidden-layer feed-forward neural network where *only* the output weights β_k are optimised, and all the weights w_{kj} between the input and hidden layer are assigned randomly. With input vectors $\boldsymbol{x}_i$ and the targets collected as a vector $\boldsymbol{y}$, it can be written as

$$\mathbf{H}\boldsymbol{\beta} = \boldsymbol{y} \quad \text{where} \quad H_{ik} = h\left(\boldsymbol{w}_k^T \boldsymbol{x}_i\right). \tag{1}$$

Here $h(\cdot)$ is a non-linear activation function applied elementwise. Training this model is simple, as the optimal output weights β_k can be calculated by ordinary least squares. The method relies on the idea of random projection: mapping the data randomly into a sufficiently high-dimensional space means that a linear model is likely to be relatively accurate. As such, the number of hidden-layer neurons needed for achieving equivalent accuracy is often much higher than in a multilayer perceptron trained by back-propagation, but the computational burden for training the model is still considerably lower.

The optimal weights $\boldsymbol{\beta}$ can be calculated as the least squares solution to Eq. (1), or formulated by using the Moore–Penrose pseudoinvorse as follows:

$$\boldsymbol{\beta} = \mathbf{H}^{+}\boldsymbol{y} \tag{2}$$

A high number of neurons in the hidden layer introduces concerns of overfitting, and regularised versions of the ELM have been developed to remedy this issue. These include the *optimally pruned ELM* (OP-ELM) [13], and its Tikhonov-regularised variant TROP-ELM [14]. In the current case, Tikhonov regularisation is applied when solving the least square problem in Eq. (1). The value of the regularisation parameter is selected by minimising the leave-one-out error (efficiently calculated via the PRESS statistic [14]) by a MATLAB minimisation procedure.[1]

3.2 *Multiple Imputation ELM for Incomplete Data*

To handle the missing value problem, Multiple Imputation ELM (MI ELM) [5] is used. The method is based on the established procedure of multiple imputation [15], which is a principled approach to modelling incomplete data sets while avoiding any additional bias.

For ELM, the multiple imputation procedure is as follows. First generate a set of M imputations of the data $\boldsymbol{X}$, denote these as $\boldsymbol{X}_k$, for $1 \leq k \leq M$. The imputations should be randomly drawn from a distribution representing the data. In this case, we fit a Gaussian distribution to the data set by using the EM algorithm [16, 17]. Having the distribution allows us to generate imputed versions of the data by drawing random samples from the conditional distribution of each missing value.

For each imputed version of the data, calculate the corresponding hidden layer representation $\mathbf{H}_k = h\left(\mathbf{W}^T \boldsymbol{X}_k\right)$, using the same set of hidden layer weights $\mathbf{W}$. Then solve for the output weights $\boldsymbol{\beta}_k = \mathbf{H}_k^{+}\boldsymbol{y}$.

When applying the model to a new set of (testing) data $\boldsymbol{X}_t$, in principle each trained ELM is used to generate a separate prediction $\hat{y}_k = \boldsymbol{\beta}_k \mathbf{H}_t$, where $\mathbf{H}_t = h\left(\mathbf{W}^T \boldsymbol{X}_t\right)$, and these are then averaged to produce the final result:

[1] `fminsearch`: https://www.mathworks.com/help/matlab/ref/fminsearch.html.

$$\hat{y} = \frac{1}{M} \sum_k \hat{y}_k \tag{3}$$

However, the equivalent result can be obtained more efficiently by first averaging the weights as

$$\boldsymbol{\beta} = \frac{1}{M} \sum_k \boldsymbol{\beta}_k \tag{4}$$

and then applying the model

$$\hat{y} = \boldsymbol{\beta} \mathbf{H}_t \tag{5}$$

In particular, the average weight in Eq. (4) can be calculated during the training phase, without having access to the testing data. The trained model consists only of the random weight matrix $\mathbf{W}$ and the (average) output layer weight vector $\boldsymbol{\beta}$, just as in the conventional ELM. The multiple imputation approach is only used in the training phase to more accurately find $\boldsymbol{\beta}$ in the presence of missing values. The number of multiple imputations can be dynamically chosen in accordance with available resources, the guiding principle being that a larger number of imputations leads to a more accurate model.

If the data in the test set also have missing values, as in the current case, these can also be handled by multiple imputation. That is, generate several imputed copies, calculate predictions for each copy, and average the results. Note that the multiple imputation procedure for training and testing can be conducted entirely separately from each other, and the number of imputations need not be the same.

3.3 Variable Selection

As the data is high-dimensional, and contains redundant information, a variable selection procedure is applied to condense the problem. First, variables which are highly correlated (correlation coefficient with other variables above 0.99) are discarded. However, this only reduces the dimensionality from 561 to 483, and further reductions are needed.

While many methods for variable selection have been developed, only a few of them can be applied when the data contains missing values. One which is applicable is the Delta test [18, 19], which only requires identifying the nearest neighbor of each sample. This can be accomplished by first estimating distances with another method, filling in missing values and accounting for the uncertainty [2]. By again applying the previously calculated Gaussian distribution of the data, the conditional mean and variance can be calculated for each missing value. Replacing each missing value by its conditional mean produces an imputed version of the data, denoted by $\tilde{X}$. Let the conditional variance for each missing sample be notated as $\sigma_{i,d}^2 = \mathrm{Var}(x_{i,d})$, with $\sigma_{i,d}^2 = 0$ if $x_{i,d}$ is known. Then the expected (squared) distance between two samples

$\boldsymbol{x}_i$ and $\boldsymbol{x}_j$ is

$$\mathrm{E}\left[\|\boldsymbol{x}_i - \boldsymbol{x}_j\|^2\right] = \|\tilde{\boldsymbol{x}}_i - \tilde{\boldsymbol{x}}_j\|^2 + \sum_d \sigma_{i,d}^2 + \sum_d \sigma_{j,d}^2 \tag{6}$$

These distances can then be used to identify nearest neighbours and calculate the Delta test. As the candidate space is too large for an exhaustive search, optimising the Delta test is done by applying genetic algorithms [20]. The end result is a set of 29 selected variables, and these are used for the remainder of the experiments.

3.4 Entire Procedure

The complete training procedure can be summarised as follows:

1. Pre-processing
   (a) Standardise input variables to zero mean and unit variance
   (b) Discard too highly correlated variables
   (c) Construct outputs $\boldsymbol{y}_i$ for each sample (prediction trajectory)
2. Fit Gaussian distribution to the incomplete data set using the EM algorithm
3. Variable selection:
   (a) Generate imputed data with uncertainties, and calculate distances
   (b) Use Genetic Algorithm to select variables which minimise the Delta test
4. Multiple imputation ELM
   (a) Generate weights using a fixed value of 1000 neurons
   (b) Generate multiple imputed copies by drawing from the conditional Gaussian distribution
   (c) Select regularisation parameter by minimising the leave-one-out error
   (d) For each copy, train ELM using selected variables
   (e) Average the weights to get one model.

4 Experiments

Five methods are compared in the accuracy of predicting the diagnosis 0–10 years ahead:

- ELM with multiple imputation
- ELM with missing values imputed with the conditional mean (using the Gaussian distribution)
- ELM using only samples for which all variables are known
- Support Vectors Machines (SVM) [21] using only samples for which all variables are known

- Nearest neighbor classifier (1-NN) with imputation

Each method evaluated on a test set of 30% of the patients, while the remaining data is used for training the models. The experiment is repeated 250 times to obtain more reliable measures of expected performance. Since the output also contains missing values, the training and testing for each prediction horizon is conducted only on those samples where the output is known.

4.1 Results

Overall classification accuracy for 0–10 years from the date of the session is presented in Fig. 2. The imputation-based ELM variants both consistently achieve accuracies above 90%, whereas the other models have poorer performance. However, with unbalanced classes and different costs for false positives and false negatives, it is crucial to study precision and recall separately, and these are shown in Fig. 3. Alternatively, models can be compared by their F-score, which gives a more balanced assessment of the performance than the overall classification accuracy [22]. The F-score, or F_1 measure, can be defined through the precision and recall as

$$F_1 = 2 \cdot \frac{\text{precision} \cdot \text{recall}}{\text{precision} + \text{recall}} \tag{7}$$

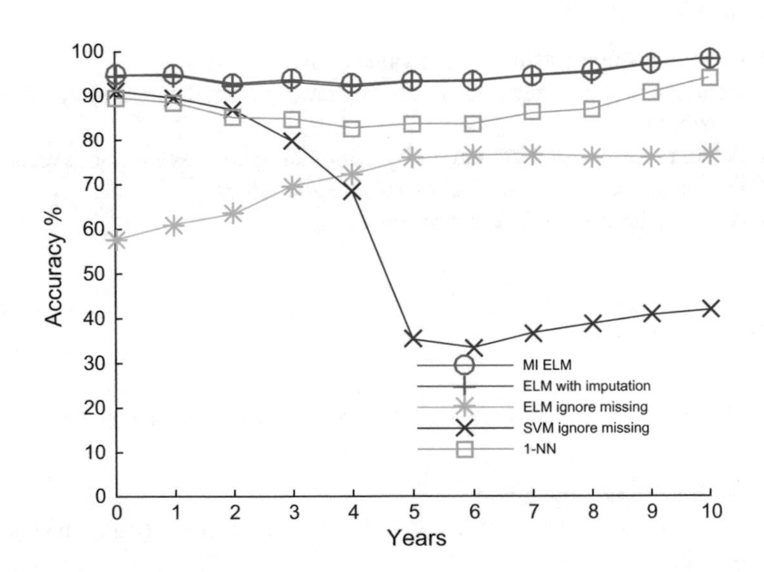

Fig. 2 Results in terms of average classification accuracy for each method for 0–10 years ahead

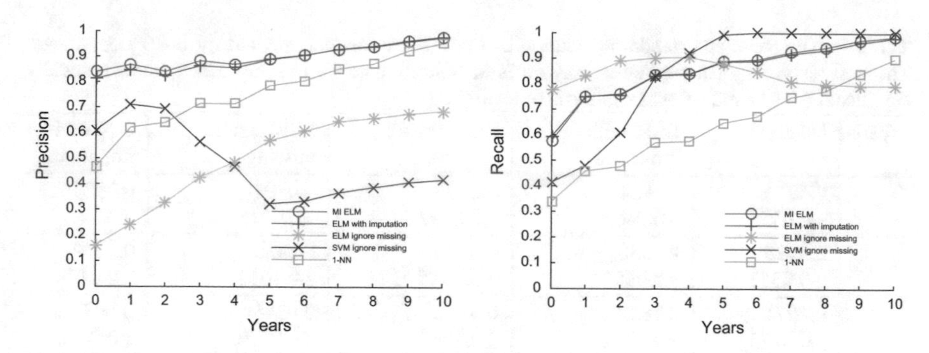

Fig. 3 Results in terms of average precision and recall for each method for 0–10 years ahead

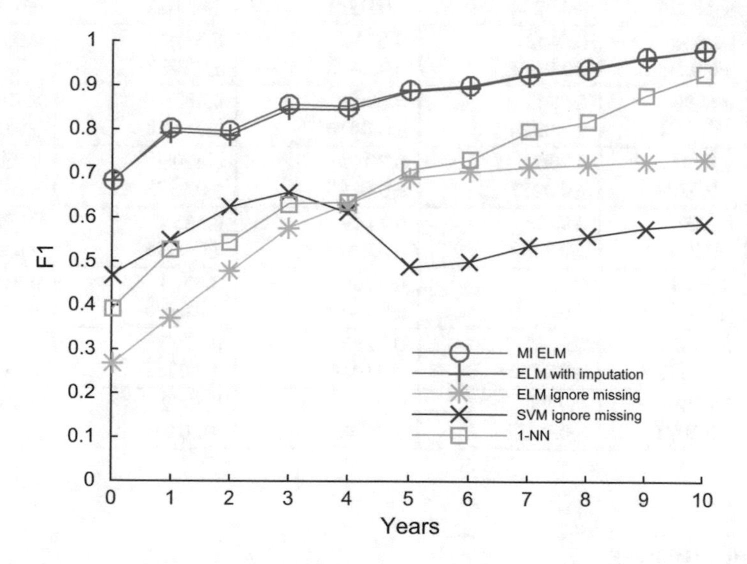

Fig. 4 Results in terms of average F-score for each method for 0–10 years ahead

The results as measured by the F-score are shown in Fig. 4. The same values are presented in Table 1, along with standard deviations. A statistical significance analysis is also done to determine which differences in accuracy can be considered significant.

It can be seen that the Multiple Imputation ELM procedure gives the best results for 1–9 years ahead, and notably is significantly better than ELM with (single) imputation. For 0 and 10 years ahead, the accuracies between the two methods are not statistically distinguishable. In all cases, the two methods perform clearly better than the other three methods (1-NN, SVM, and ELM when ignoring samples with missing values).

Table 1 The mean and standard deviation of the F-score for each method for 0–10 years ahead. The best result for each prediction horizon is in bold font, as well as any values not statistically significant (in a paired t-test at significance level 0.05)

Years	MI ELM	ELM with imputation	ELM ignore missing	SVM ignore missing	1-NN with imputation
0	**0.6814** ±0.0455	**0.6834** ±0.0462	0.2656 ±0.0297	0.4701 ±0.0815	0.3909 ±0.0474
1	**0.8020** ±0.0352	0.7938 ±0.0354	0.3728 ±0.0365	0.5483 ±0.1012	0.5249 ±0.0478
2	**0.7971** ±0.0275	0.7882 ±0.0262	0.4762 ±0.0363	0.6245 ±0.0770	0.5435 ±0.0420
3	**0.8565** ±0.0226	0.8459 ±0.0241	0.5778 ±0.0387	0.6577 ±0.0618	0.6323 ±0.0398
4	**0.8536** ±0.0243	0.8452 ±0.0237	0.6312 ±0.0408	0.6112 ±0.0727	0.6349 ±0.0405
5	**0.8896** ±0.0231	0.8868 ±0.0225	0.6902 ±0.0416	0.4875 ±0.0438	0.7087 ±0.0354
6	**0.8997** ±0.0209	0.8960 ±0.0217	0.7044 ±0.0390	0.4998 ±0.0372	0.7310 ±0.0340
7	**0.9256** ±0.0179	0.9228 ±0.0182	0.7149 ±0.0479	0.5358 ±0.0385	0.7951 ±0.0289
8	**0.9404** ±0.0160	0.9362 ±0.0160	0.7206 ±0.0489	0.5573 ±0.0396	0.8189 ±0.0272
9	**0.9647** ±0.0110	0.9616 ±0.0122	0.7282 ±0.0503	0.5772 ±0.0363	0.8783 ±0.0233
10	**0.9801** ±0.0075	**0.9796** ±0.0085	0.7349 ±0.0493	0.5882 ±0.0405	0.9246 ±0.0163

5 Conclusions

In this paper we study how well variants of the Extreme Learning Machine can be used to predict the diagnosis of Huntington's disease from early MRI scans. The results clearly show that informative predictions are possible with satisfactory accuracy, and predicting onset of symptoms 10 years in advance is realistic.

The Extreme Learning Machine is able to model the scenario accurately. The best results are achieved by applying the principled multiple imputation procedure. Indeed, properly accounting for the missing values is crucial for the machine learning task to perform reliably.

Further investigation is still required to more precisely analyse which variables (or combinations of variables) are the most informative in enabling the early prediction, and whether further refinements to the modelling procedure could lead to even more accurate predictions.

References

1. Little, R.J.A., Rubin, D.B.: Statistical Analysis with Missing Data, 2nd edn. Wiley-Interscience (2002). doi:10.1002/9781119013563
2. Eirola, E., Doquire, G., Verleysen, M., Lendasse, A.: Distance estimation in numerical data sets with missing values. Inf. Sci. **240**, 115–128 (2013). doi:10.1016/j.ins.2013.03.043
3. Huang, G.B., Zhu, Q.Y., Siew, C.K.: Extreme learning machine: theory and applications. Neurocomputing **70**(13), 489–501 (2006). doi:10.1016/j.neucom.2005.12.126
4. Yu, Q., Miche, Y., Eirola, E., van Heeswijk, M., Séverin, E., Lendasse, A.: Regularized extreme learning machine for regression with missing data. Neurocomputing **102**, 45–51 (2013). doi:10.1016/j.neucom.2012.02.040
5. Sovilj, D., Eirola, E., Miche, Y., Björk, K., Nian, R., Akusok, A., Lendasse, A.: Extreme learning machine for missing data using multiple imputations. Neurocomputing **174, Part A**, 220–231 (2016). doi:10.1016/j.neucom.2015.03.108
6. Gao, H., Liu, X.W., Peng, Y.X., Jian, S.L.: Sample-based extreme learning machine with missing data. Math. Prob. Eng. **2015** (2015). doi:10.1155/2015/145156
7. Xie, P., Liu, X., Yin, J., Wang, Y.: Absent extreme learning machine algorithm with application to packed executable identification. Neural Comput. Appl. **27**(1), 93–100 (2016). doi:10.1007/s00521-014-1558-4
8. Yan, Y.T., Zhang, Y.P., Chen, J., Zhang, Y.W.: Incomplete data classification with voting based extreme learning machine. Neurocomputing **193**, 167–175 (2016). doi:10.1016/j.neucom.2016.01.068
9. Paulsen, J.S., Langbehn, D.R., Stout, J.C., Aylward, E., Ross, C.A., Nance, M., Guttman, M., Johnson, S., MacDonald, M., Beglinger, L.J., Duff, K., Kayson, E., Biglan, K., Shoulson, I., Oakes, D., Hayden, M.: Detection of Huntington's disease decades before diagnosis: the predict-HD study. J. Neurol. Neurosurg. Psychiatry **79**(8), 874–880 (2008). doi:10.1136/jnnp.2007.128728
10. Paulsen, J.S., Long, J.D., Ross, C.A., Harrington, D.L., Erwin, C.J., Williams, J.K., Westervelt, H.J., Johnson, H.J., Aylward, E.H., Zhang, Y., et al.: Prediction of manifest Huntington's disease with clinical and imaging measures: a prospective observational study. Lancet Neurol. **13**(12), 1193–1201 (2014). doi:10.1016/S1474-4422(14)70238-8
11. Matsui, J.T., Vaidya, J.G., Wassermann, D., Kim, R.E., Magnotta, V.A., Johnson, H.J., Paulsen, J.S.: Prefrontal cortex white matter tracts in prodromal Huntington disease. Hum. Brain Mapp. **36**(10), 3717–3732 (2015). doi:10.1002/hbm.22835
12. Sturrock, A., Laule, C., Wyper, K., Milner, R.A., Decolongon, J., Santos, R.D., Coleman, A.J., Carter, K., Creighton, S., Bechtel, N., et al.: A longitudinal study of magnetic resonance spectroscopy Huntington's disease biomarkers. Mov. Disord. **30**(3), 393–401 (2015). doi:10.1002/mds.26118
13. Miche, Y., Sorjamaa, A., Bas, P., Simula, O., Jutten, C., Lendasse, A.: OP-ELM: optimally-pruned extreme learning machine. IEEE Trans. Neural Netw. **21**(1), 158–162 (2010). doi:10.1109/TNN.2009.2036259
14. Miche, Y., van Heeswijk, M., Bas, P., Simula, O., Lendasse, A.: TROP-ELM: a double-regularized ELM using LARS and Tikhonov regularization. Neurocomputing **74**(16), 2413–2421 (2011). doi:10.1016/j.neucom.2010.12.042
15. Rubin, D.B.: Multiple Imputation for Nonresponse in Surveys. Wiley (1987)
16. Dempster, A.P., Laird, N.M., Rubin, D.B.: Maximum likelihood from incomplete data via the EM algorithm. J. R. Stat. Soc. Ser. B **39**(1), 1–38 (1977)
17. Eirola, E., Lendasse, A., Vandewalle, V., Biernacki, C.: Mixture of gaussians for distance estimation with missing data. Neurocomputing **131**, 32–42 (2014). doi:10.1016/j.neucom.2013.07.050
18. Eirola, E., Liitiäinen, E., Lendasse, A., Corona, F., Verleysen, M.: Using the Delta test for variable selection. In: Proceedings of ESANN 2008, European Symposium on Artificial Neural Networks, Bruges (Belgium), pp. 25–30 (2008)

19. Eirola, E., Lendasse, A., Corona, F., Verleysen, M.: The Delta test: The 1-NN estimator as a feature selection criterion. In: 2014 International Joint Conference on Neural Networks (IJCNN), pp. 4214–4222. IEEE (2014). doi:10.1109/IJCNN.2014.6889560
20. Sovilj, D.: Multistart strategy using delta test for variable selection. In: International Conference on Artificial Neural Networks, Springer Berlin Heidelberg, pp. 413–420 (2011). doi:10.1007/978-3-642-21738-8_53
21. Cortes, C., Vapnik, V.: Support-vector networks. Mach. Learn. **20**, 273–297 (1995). doi:10.1023/A:1022627411411
22. Rijsbergen, C.J.V.: Information Retrieval, 2nd edn. Butterworth-Heinemann (1979)

Deep-Learned and Hand-Crafted Features Fusion Network for Pedestrian Gender Recognition

Lei Cai, Jianqing Zhu, Huanqiang Zeng, Jing Chen and Canhui Cai

Abstract In this paper, we propose an effective deep-learned and hand-crafted features fusion network (DHFFN) for pedestrian gender recognition. In the proposed DHFFN, the deep-learned and hand-crafted (i.e., HOG) features are extracted for the input image, followed by the feature fusion process that is to combine these two features together for fully exploring the merits from both deep-learned and HOG features. Extensive experiments on multiple public datasets have demonstrated that the proposed DHFFN method is superior to the state-of-the-art pedestrian gender recognition methods.

Keywords Pedestrian gender recognition · Convolutional neural network · Deep-learned feature · Hand-crafted feature

1 Introduction

In recent years, digital video surveillance systems have been widely deployed in various areas for public safety, such as shopping mall, train station, airport, and so on. For the increasing huge amount of video data, video analytic tools, such as face recognition [1], pedestrian re-identification [2], etc., have been developed as effective and essential solutions for identifying the various attributes of pedestrian in quick and accurate manner. Among them, gender is an important attribute of pedestrian in many applications, for example, human-computer interaction, identity recognition, video surveillance, population statistics and multimedia retrieval system [3]. Since the appearance of a pedestrian often changes a lot with the viewpoint, lighting, dress, occlusion etc., pedestrian gender recognition is a challenging task in the computer vision field.

L. Cai · H. Zeng (✉) · J. Chen
School of Information Science and Engineering, Huaqiao University,
Xiamen 361021, China
e-mail: zeng0043@hqu.edu.cn

J. Zhu · C. Cai
School of Engineering, Huaqiao University, Quanzhou 362021, China

J. Cao et al. (eds.), *Proceedings of ELM-2016*, Proceedings in Adaptation, Learning and Optimization 9, DOI 10.1007/978-3-319-57421-9_17

To address the pedestrian gender recognition problem, there are some methods that can be found in the literature. Usually, these methods performed gender recognition task by using the silhouette of the pedestrian, since it is not easy to obtain the clear pedestrian appearance (e.g., face) under the condition of long distance or across views. However, the silhouette of the pedestrian will inevitably be disturbed due to the interference of illumination, image blur, occlusion, and so on. In order to enhance the description ability of pedestrian silhouette, some hand-crafted features, which are original proposed for other object recognition problems, are applied on pedestrian gender recognition. For example, many authors have found that the *histogram of oriented gradient* (HOG) feature [4] is suitable for gender recognition. Cao et al. [5] made the first attempt to employ the HOG feature and Adaboost classifier for exploiting silhouette information and obtained 76% average pedestrian gender recognition rate on MIT dataset. Furthermore, Collins et al. [6] presented an improved HOG feature called PixelHOG, and achieved a higher recognition rate (i.e., 80%) on the VIPeR dataset.

Moreover, with the rapid development of machine learning in recent years, supervised learning regards the deep *convolutional neural network* (CNN) as the first-choice method for image classification problems [1, 7, 8]. CNN [9] has been widely applied and demonstrated its superiority in many computer vision domains [10–14]. However, only a few works adopt deep learning on pedestrian gender recognition problem. Ng et al. [15] trained a CNN for gender recognition on the MIT dataset and achieved 80% recognition rate. Antipov et al. [16] trained a CNN model called Mini-CNN and obtained 80% average precision rate. Moreover, they also fine-tuned a pre-trained CNN designed by Krizhevsky et al. [8] and improved the average precision rate to be 85% on the PETA collection of dataset. In addition, there is another popular machine learning algorithm, named Extreme Learning Machine (ELM) [17, 18]. Due to its fast learning speed and good generalization ability, ELM and its variants are widely applied in various areas, such as image classification, landmark recognition, vehicle detection [19–24]. To our best knowledge, there is no existing method to apply ELM for pedestrian gender recognition.

In this paper, we propose an effective pedestrian gender recognition method based on a specially designed deep-learned and hand-crafted features fusion network, called DHFFN. The superior performance of the proposed method is due to that it can make full use of the merits from both hand-crafted and deep-learned features. More specifically, the proposed DHFFN simultaneously extracts deep-learned and hand-crafted features on the input images, which are fused together to train a two-class classifier by using the specifically designed features fusion network. Extensive experiments on multiple public datasets have shown that the proposed DHFFN method outperforms the state-of-the-art pedestrian gender recognition methods.

The rest of this paper is organized as follows. Section 2 introduces the proposed pedestrian gender recognition method, DHFFN, in detail. Section 3 presents the experimental results and comparisons. Section 4 concludes this paper.

2 Proposed Deep-Learned and Hand-Crafted Features Fusion Network (DHFFN) for Pedestrian Gender Recognition

2.1 Architecture of Proposed DHFFN

Figure 1 shows the architecture of proposed deep-learned and hand-crafted features Fusion Network (DHFFN) for Pedestrian Gender Recognition. The proposed DHFFN consists of two feature extraction parts: (1) deep-learned feature extraction: this part mainly copes with the processing of the convolution, max-pooling and activation for the input images; (2) hand-crated feature extraction: this part extracts the HOG feature and performs the HOG feature dimension reduction. These two feature extraction parts are finally connected together to produce a more discriminative fused feature in Concatenate layer (i.e., Fusion layer in Fig. 1). The details of proposed DHFFN will be introduced in the following sub-sections.

2.2 Feature Extraction

Deep-Learned Feature The deep-learned feature is extracted as shown in the upper part of Fig. 1. The input images used in our experiments are with the size of 48×128 and three input channels (i.e., RGB) for color images. In the designed CNN, layer C1 contains 32 filters with the size of 5×5, and the learned feature maps are 44×124

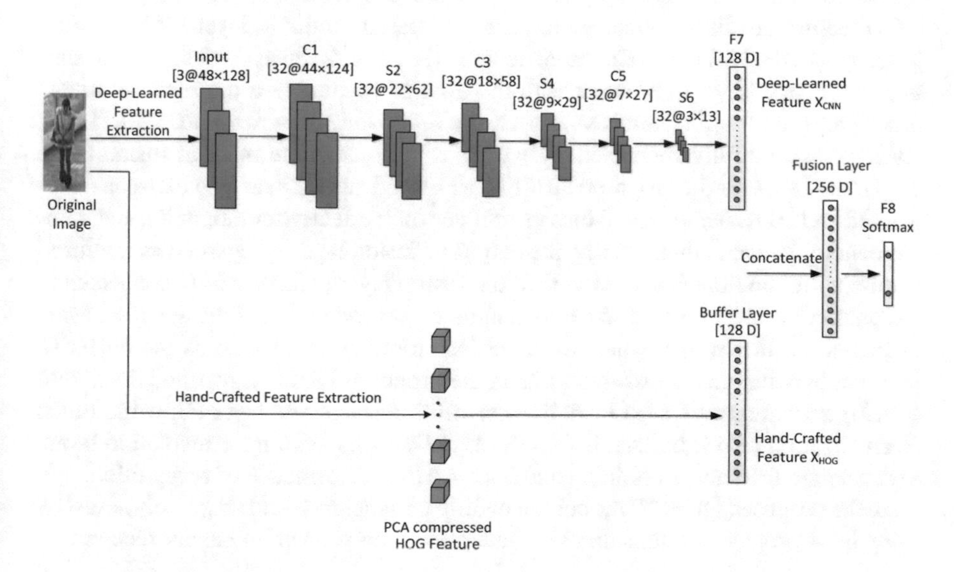

Fig. 1 The architecture of proposed DHFFN

after the first convolution. Maximum-pooling is used behind the first convolution layer C1, resulting in 22×62 feature maps in layer S2. Layer C3 contains 32 feature maps with the size of 18×58 resulted from 5×5 filters. After the operation of downsample using 2×2 max pooling, each feature map in layer S4 is 9×29. Layer C5 also has 32 feature maps. Unlike the Layers C1 and C3, the size of filters are 3×3. Hence, the size of resulted feature maps in layer C5 are 7×27. Meanwhile, different from layers S2 and S4, layer S6 uses the filters with the size of 3×3 and produces 3×13 feature maps.

Hand-Crafted Feature The HOG has been demonstrated to be successful for gender recognition [5, 6]. For simplicity, we directly exploit HOG [4] as the hand-crafted feature in this work. The HOG feature is extracted as shown in the lower part of Fig. 1. Firstly, the basic processing unit of HOG is 16×16 block, which is further divided into 4 square cell 8×8 to compute the histogram of gradient with 9 bins. Hence, the feature dimension for each 16×16 block is 36. For the input image (48×128), with the stride 8 pixels, we can obtain the HOG feature with the dimension $36 \times 5 \times 15 = 2700$. To be consistent to the dimension of deep-learned feature, the traditional Principal Component Analysis [25] is further performed to reduce the HOG feature dimension as 128.

2.3 *Feature Fusion*

In this work, the feature fusion aims to make the deep-learned feature be complementary to HOG feature for further improving the classification accuracy. To conduct the feature fusion, the obtained deep-learned feature (i.e., Layer S6) and 128-dimension HOG feature are fully connected to each 128 neuron units in layer F7 in the first stage, respectively. Then, a Concatenate layer (i.e., Fusion layer) is designed to combine these two kinds of features with the goal of producing a more robust image descriptor (i.e., fused feature). Moreover, the *Local Response Normalization* (LRN) layer behind the fully connected layer F7 is to normalize the input of fusion layer. This is essential for the proposed DHFFN, as LRN deals with the gap between deep-learned and HOG features, and thus guarantees their effective combination and complementary to each other. Finally, the output of fusion layer is regarded as the fused feature with 256 dimension. After that, the fusion layer is further fully connected to the prediction layer F8 with 2 neurons units corresponding to 2 classes (i.e., Male or Female in this work), where Softmax loss function is adopted as the objective function to compute the loss. Note that in the proposed DHFFN, rectified linear unit (ReLU) activations are used in all the convolution layers and layer F7, while Batch Normalization [26] is utilized before ReLU activations in all the convolution layers to accelerate the network training and improve the performance of recognition.

In the proposed DHFFN, the corresponding parameters could adaptively adjust by using back propagation to achieve self-adaptation on pedestrian gender recognition problem. Let the input of Fusion layer be x_{CNN} and x_{HOG}. Then, the output of the fusion layer (i.e., fused feature) can be written as:

$$x_{Fusion} = [x_{CNN}, x_{HOG}]. \tag{1}$$

According to the forward propagation algorithm, the prediction score output by the layer F8 can be computed as:

$$S(x_{Fusion}) = W^T x_{Fusion} + b, \tag{2}$$

where W and b denote the weights and bias term, respectively. As shown in Eq. (2), W and b are used to project the fused feature x_{Fusion} into the prediction score. With the proposed method, the x_{HOG} part of x_{Fusion} is fixed in the optimization of the proposed network, the x_{CNN} part is automatically adjusted to work with x_{HOG} to improve the performance.

3 Experimental Results and Discussions

3.1 Dataset and Evaluation Criteria

To evaluate the performance, the proposed DHFFN method is tested and then compared with the state-of-the-art gender recognition method on multiple widely-used and challenging datasets, including CUHK, PRID, GRID, MIT, and VIPeR. Figure 2 shows some samples drawn from these datasets [27]. One can see that appearances of pedestrian greatly change due to the different camera angles and environments. Following [16], we also filter out some images that consist of the same person or unidentified target or are with very low resolution. Finally, there are 8404 images used in our experiments. Table 1 shows the training and testing images from each dataset.

In our experiments, the proposed DHFFN and DFN are implemented using caffe deep learning framework [28]. Moreover, a preprocessing is performed to resize all

Fig. 2 Samples from each dataset

Table 1 Training and testing images from each dataset

Dataset	Training size (♂ + ♀)	Testing size (♂ + ♀)
CUHK	3844 = (2715 + 1129)	379 = (190 + 189)
PRID	947 = (458 + 489)	101 = (50 + 51)
GRID	928 = (531 + 397)	100 = (50 + 50)
MIT	788 = (538 + 250)	84 = (42 + 42)
VIPeR	1113 = (546 + 567)	120 = (60 + 60)
Total	7620	784

the images to 48×128 and subtract their mean values. And the mirrored copies of the training images are also used to augment the training data. Furthermore, the commonly-used criteria, i.e., Mean Average Precision (MAP), Area Under ROC Curve (AUC) [29], are used to evaluate the performance.

3.2 Results and Discussions

The performance evaluation of the proposed DHFFN and the state-of-the-art method (i.e., [16]) is compared. Moreover, the proposed method is also compared with the Hierarchical ELM directly applied in pedestrian gender recognition [18]. And the results are shown in the Table 2. Note that the proposed DHFFN method can be viewed as an enhanced version of the *deep-learned feature network* (DFN) by further fusing the HOG feature. Therefore, the performance resulted from DFN is also evaluated to show how much of the contribution coming from this part, besides the performance evaluation of the proposed DHFFN method (i.e., the enhanced DFN version with HOG feature incorporated).

It can be clearly observed from Table 2 that the proposed DFN can achieve 0.93 MAP and 0.93 AUC while the proposed DHFFN yields 0.95 MAP and 0.95 AUC. Hence, compared with DFN, the proposed DHFFN is able to yield better performance. The same observation can also be obtained from the receiver operating characteristic (ROC) curves of the proposed DFN and DHFFN as shown in Fig. 3. This

Table 2 Performance comparison

Features	MAP	AUC
HOG [16]	0.72	0.84
Mini-CNN [16]	0.80	0.88
AlexNet-CNN [16]	0.85	0.91
Hierarchical ELM [18]	0.92	0.92
Proposed DFN	**0.93**	**0.93**
Proposed DHFFN	**0.95**	**0.95**

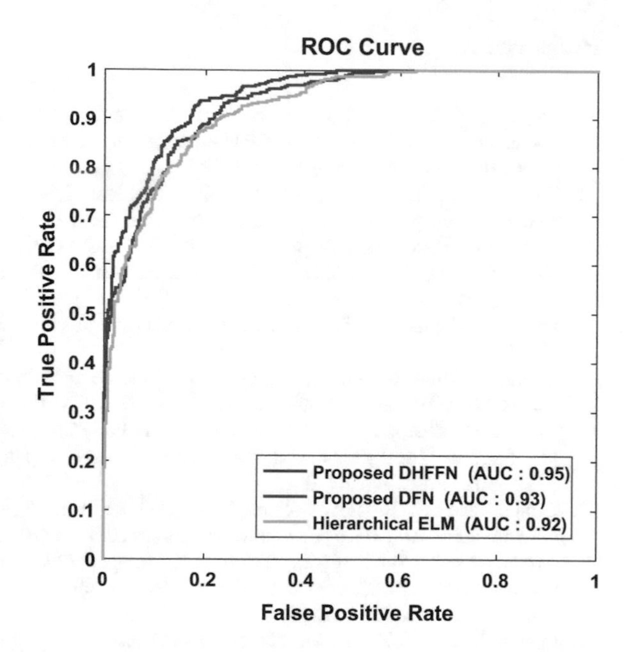

Fig. 3 ROC curves

study shows that the hand-crafted feature (i.e., HOG) plays an effective complementary role to the DFN. Furthermore, the proposed DHFFN method consistently outperforms the state-of-the art method [16], and hierarchial ELM [18] in terms of both MAP and AUC. This result further demonstrates that the fused feature is more representative and discriminative in the task of pedestrian gender recognition.

4 Conclusion

In this paper, a novel deep-learned and hand-crafted features fusion network (DHFFN) method is proposed for pedestrian gender recognition. By designing a special features fusion network, the deep-learned and HOG features are extracted for the input image. Then, these two kinds of features are combined and fused together to improve the recognition rate. Moreover, we investigate and demonstrate the complementary behaviors of the HOG feature to the deep-learned feature. Experiments on multiple challenging datasets show that the proposed DHFFN method outperforms the state-of-the-art pedestrian gender recognition method.

Acknowledgements This work was supported in part by the National Natural Science Foundation of China under the Grants 61401167, 61372107 and 61602191, in part by the Natural Science Foundation of Fujian Province under the Grant 2016J01308, in part by the Opening Project of State Key Laboratory of Digital Publishing Technology under the grant FZDP2015-B-001, in part by the Zhejiang Open Foundation of the Most Important Subjects, in part by the High-Level Talent Project Foundation of Huaqiao University under the Grants 14BS201, 14BS204 and 16BS108, and in part by the Graduate Student Scientific Research Innovation Project Foundation of Huaqiao University.

References

1. Taigman, Y., Yang, M., Ranzato, M., Wolf, L.: Deepface: closing the gap to human-level performance in face verification. In: IEEE Conference on Computer Vision and Pattern Recognition, pp. 1701–1708. IEEE Press, USA (2014)
2. Wu, S., Chen, Y.C., Li, X., Wu, A.C., You, J.J., Zheng, W.S.: An enhanced deep feature representation for person re-identification. In: 2016 IEEE Winter Conference on Applications of Computer Vision (WACV), pp. 1–8. IEEE Press, USA (2016)
3. Ng, C.B., Tay, Y.H., Goi, B.M.: Vision-based human gender recognition: A survey (2012). arXiv:1204.1611
4. Dalal, N., Triggs, B.: Histograms of Oriented Gradients for Human Detection. In CVPR, USA (2005)
5. Cao, L., Dikmen, M., Fu, Y., Huang, T.S.: Gender recognition from body. In: 16th ACM International Conference on Multimedia, pp. 725–728. ACM Press, USA (2008)
6. Collins, M., Zhang, J., Miller, P., Wang, H.: Full body image feature representations for gender profiling. In: 12th IEEE International Conference on Computer Vision Workshops, pp. 1235–1242. IEEE Press, Japan (2009)
7. Goodfellow, I.K., Bulatov, Y., Ibarz, J., Arnoud, S.: Multi-digit number recognition from street view imagery using deep convolutional neural networks (2013). arXiv:1312.6082
8. Krizhevsky, A., Sutskever, I., Hinton, G.E.: Imagenet classification with deep convolutional neural networks. In: 25th NIPS Advances in Neural Information Processing Systems, pp. 1097–1105. NIPS Press, USA (2012)
9. LeCun, Y., Bengio, Y.: Convolutional networks for images, speech, and time series. Handbook Brain Theor. Neural Netw. 3361(10) (1995)
10. LeCun, Y., Bottou, L., Bengio, Y., Haffner, P.: Gradient-based learning applied to document recognition. Proc. IEEE **86**(11), 2278C–2324 (1998)
11. Lawrence, S., Giles, C.L., Tsoi, A.C., Back, A.D.: Face recognition: a convolutional neural-network approach. IEEE Trans. Neural Netw./ A Publ. IEEE Neural Net. Council **8**(1), 98C–113 (1997)
12. Osadchy, M., Cun, Y., Miller, M.: Synergistic face detection and pose estimation with energy-based models. J. Mach. Learn. Res. **8**, 1197C–1215 (2007)
13. Ciresan, D., Meier, U., Masci, J., Schmidhuber, J.: A committee of neural networks for traffic sign classification. In: The 2011 International Joint Conference on Neural Networks (IJCNN), vol. 1(1), pp. 1918C–1921. IEEE press, USA (2011)
14. Ji, S., Xu, W., Yang, M., Yu, K.: 3D convolutional neural networks for human action recognition. IEEE Trans. Pattern Anal. Mach. Intell. **35**(1), 221C–231 (2013)
15. Ng, C.B., Tay, Y.H., Goi, B.M.: A convolutional neural network for pedestrian gender recognition. In: 10th International Symposium on Neural Networks, pp. 558–564. Springer, Heidelberg press, China (2013)
16. Antipov, G., Berrani, S.A., Ruchaud, N., Dugelay, J.L.: Learned vs. hand-crafted features for pedestrian gender recognition. In: 23th ACM International conference on Multimedia, pp. 1263–1266. ACM Press, USA (2015)
17. Huang, G.B., Zhu, Q.Y., Siew, C.K.: Extreme learning machine: a new learning scheme of feed-forward neural networks. In: 2004 IEEE International Joint Conference on Neural Networks, pp.985–990. IEEE Press, Hungary (2004)
18. Zhu, W., Miao, J., Qing, L., Huang, G.B.: Hierarchical extreme learning machine for unsupervised representation learning. In: 2015 International Joint Conference on Neural Networks, pp. 1–8. IEEE Press, Ireland (2015)
19. Cao, J., Wang, W., Wang, J., Wang, R.: Excavation equipment recognition based on novel acoustic statistical features. IEEE Trans. Cybern. (2016)
20. Cao, J., Hao, J., Lai, X., Vong, C.M., Luo, M.: Ensemble extreme learning machine and sparse representation classification. J. Franklin Inst. (2016)
21. Cao, J., Zhang, K., Luo, M., Yin, C., Lai, X.: Extreme learning machine and adaptive sparse representation for image classification. Neural Netw. **81**, 91–102 (2016)

22. Cao, J., Chen, T., Fan, J.: Landmark recognition with compact BoW histogram and ensemble ELM. Multimed. Tools Appl. **75**(5), 2839–2857 (2016)
23. Tang, J., Deng, C., Huang, G.B.: Extreme learning machine for multilayer perceptron. IEEE Trans. Neural Netw. Learn. Syst. **27**(4), 809–821 (2016)
24. Zhu, W., Miao, J., Hu, J., Qing, L.: Vehicle detection in driving simulation using extreme learning machine. Neurocomputing **128**, 160–165 (2014)
25. Abdi, H., Williams, L.J.: Principal component analysis. Wiley Interdiscip. Rev. Comput. Stat. **2**(4), 433–459 (2010)
26. Ioffe, S., Szegedy, C.: Batch normalization: Accelerating deep network training by reducing internal covariate shift (2015). arXiv:1502.03167
27. Deng, Y., Luo, P., Loy, C.C., Tang, X.: Pedestrian attribute recognition at far distance. In: 22th ACM International Conference on Multimedia, pp. 789–792. ACM Press, USA (2014)
28. Jia, Y., Shelhamer, E., Donahue, J., Karayev, S., Long, J., Girshick, R., et al.: Caffe: convolutional architecture for fast feature embedding. In: 22th ACM International Conference on Multimedia, pp. 675–678. ACM Press, USA (2014)
29. Hanley, J.A., McNeil, B.J.: The meaning and use of the area under a receiver operating characteristic (ROC) curve. Radiology **143**, 29–36 (1982)

Facial Landmark Detection via ELM Feature Selection and Improved SDM

Peng Bian, Yi Jin and Jiuwen Cao

Abstract Model initialization and feature extraction are crucial in supervised landmark detection. Mismatching caused by detector error and discrepant initialization is very common in these existing methods. To solve this problem, we have proposed a new method based on ELM feature selection and Improved Supervised Descent Method (ELMFS-iSDM), which also includes an automatic initialization model, for the robust facial landmark localization. In our new method, firstly, a fast detection will be processed to locate the eyes and mouth, and the initialization model will adapt to the real location according to fast facial points detection. Secondly, ELM based feature selection is adopted on our Improved Supervised Descent Method model to achieve a better performance. For each task, multiple features will be jointly learned by ELM feature selection and their weights will be calculated during training process. Experiments on four benchmark databases show that our method achieves state-of-the-art performance.

Keywords Facial landmark detection · Self-adapted model · Extreme learning machines (ELM) · Feature selection · Supervised descent method

P. Bian
College of Mechanical and Material Engineering, North China University of Technology, Beijing 100144, People's Republic of China
e-mail: bianpeng@ncut.edu.cn

Y. Jin (✉)
School of Computer and Information Technology, Beijing Jiaotong University, Beijing 100044, People's Republic of China
e-mail: yjin@bjtu.edu.cn

Y. Jin
Beijing Key Lab of Traffic Data Analysis and Mining, Beijing 100044, People's Republic of China

J. Cao
Institute of Information and Control, Hangzhou Dianzi University, Zhejiang 310018, China
e-mail: jwcao@hdu.edu.cn

J. Cao et al. (eds.), *Proceedings of ELM-2016*, Proceedings in Adaptation, Learning and Optimization 9, DOI 10.1007/978-3-319-57421-9_18

1 Introduction

Facial landmark detection, which is an essential prerequisite step for most face understanding tasks, has drawn more and more attention in human face recognition area for its convenience in location alignment [1]. Locating fiducial face landmarks such as eyes corners, nose tip, and mouth corners even face contours can make face understanding tasks, such as, face recognition, facial expression recognition, face tracking et al., much simpler and more accurate. Existing Facial landmark detection methods have exploited the structure information, which benefits from the strong joint constrain of all the landmark points, and built their own models to address the problem. However, it is still a challenging problem due to the face appearance variations caused by occlusion, illumination, pose rotation and facial expressions [1, 2].

Models that describe facial appearance and face contours are quite critical in facial landmark detection methods. Thus, how to reduce the face appearance variations and extract the most discriminative features become an essential issue in facial landmark detection or face alignment tasks. Generally speaking, facial landmark detection methods can be divided into two categories: global feature methods and local feature methods. Global methods aim to extract and model appearance from the entire face, and the most typical methods include Active Appearance Models (AAMs) [3]. The models are parametric for appearance and shape, and the parameters are optimized by efficient gradient descent. Local methods, on the contrary, model local appearance information like color and texture from local patches around each landmark. Active Shape Models (ASM) [4] and Constrained Local Methods (CLM) [5] are examples of the second category methods which use local methods.

Since the face appearance and face shape can be modeled, how to find the relationship between them was the key to facial landmark annotation. Although SDM is an efficient and accurate approach for facial landmark detection, regression performance decreases when the initialization model is poor or the appearance difference of the training set is too large. Thus, an intuitive way to promote the efficiency of SDM is to improve the feature representation and the initialization model. In this paper, we focus on the learning-based feature representation and propose the ELM feature selection and Improved Supervised Descent Method (ELMFS-iSDM) for Facial Landmark Detection. In the new proposed method, a Self-Adaption Model is firstly proposed for the initialization of our algorithm. Then, we exploit the discriminant and correlative image patterns using the ELM feature selection, by which the discriminant ability is strengthened by learning feature representation from two different feature descriptors, such as HOG and SIFT. Experimental results on four different facial landmark detection applications demonstrate the effectiveness of our proposed method.

The remainder of this paper is organized as follows. Section 2 is the related work and Sect. 3 describes the Self Adapted Model and the ELM Feature selection based Improved SDM. Experimental results and discussion on three face annotation databases are presented in Sect. 4. Section 5 draws the conclusion of this paper.

2 Related Work

2.1 Cascade Regression Methods

Previous work on facial landmark detection can be grouped to two categories: (1) holistic feature based methods and (2) local feature methods. Recently local models with regression methods make great performance on face alignment, especially SDM [6]. Other local methods like ASMs and CLMs aim to model face appearance in a complicated way to fit the final face shape. Unlike these methods, regression methods extract simple features instead of models from local face appearance and intend to figure out the mapping function between features and face shape. But a single regression is always too week to get the final location due to linear and non-linear changes on face. To address this, more than one stage of regression procedure is cascaded to approach the ground-truth step by step, and we dont need to learn a complex function to get the final location by one step.

The idea was firstly proposed by Piotr Dollar [7] in 2010 and the Cascade Pose Regression (CPR) model got good results at that time. The existing regression methods are mainly base on this idea. To address the occlusion problem, Xavier P. Burgos-Artizzu proposed Robust Cascaded Pose Regression (RCPR) based on CPR. Another remarkable work when DRMF published was Supervised Descent Method (SDM) [6] proposed by Xuehan Xiong from Carnegie Mellon University. SDM regard the regression as a non-linear function and intended to figure out the function. But because of the huge matrix and complicated procedure figuring out the exact function was almost impossible. So SDM simplified the regression and consider it as a linear regression according to Newton's method. By doing so, figuring out the descent gradient matrix and bias vector s can simulate the mapping function. Unlike DRMF, SDM uses SIFT feature to describe face appearance. Regressing Local Binary Features (LBF) also uses linear regression model like SDM, but LBF proposed a more efficient feature called local sparse binary feature which can be learnt from random forest [8]. The computing cost is greatly reduced by the sparse binary feature so the method is more efficient than SDM and DRMF. A novel work [9] explores the use of context on regression-based methods for facial landmarking and it analyse the key methodological aspects of the performance of SDM. The work also made an effective extension of the SDM by using an optimal amount of context within the input features.

2.2 ELM Feature Selection

Many computer vision problems can be solved by applying Extreme Learning Machines (ELM). ELM is recently proposed for efficiently training single-hidden-layer feed forward neural networks (SLFNs). And ELM perform more consistently with a much faster training speed [10]. The essence of ELM is that ELM performs classification by projecting original data to a high dimensional vector and changes the classification task into a multi-output functional regression problem [11, 12].

With its high learning efficiency, ELM [13] has attracted increasing attention on a widespread type of applications, e.g. pattern classification, object recognition and data analysis. ELM based approaches are also proposed in FR tasks, such as, Zong and Huang [14] propose a ELM based method in multi-label FR applications. Zong et al. [15] later propose a kernelized ELM method in FR. Long et al. [16] propose a graph regularized discriminative non-negative matrix factorization (GDNMF), where the projection matrix is learned jointly by both the graph Laplacian and supervised label information.

Feature selection is a crucial pre-processing in computer vision applications. Nowadays, ELM has been attracting the attentions from more and more researchers and was originally developed for feature selection and feature mapping [10, 17]. For instance, Cao et al. [18] propose a SPK-BoW approach, which is first employed ELM based FNN combined with the SRC to extract features and construct an over complete dictionary for landmark image feature learning. Rajendra Kumar Roul et al. [19] propose a new clustering based feature selection using ELM, which both chooses the most relevant features and reduces the feature size. Mangy Zhai et al. [20] propose a ELM based feature selection algorithm which uses a feature ranking criterion to measure the significance of each feature, which also improves the speed of the algorithm. Motivated by this, we thus introduce the ELM feature selection framework into our facial landmark detection method, and prove the effectiveness and efficiency on four different datasets.

3 ELM Feature Selection Based Improved Supervised Descent Method

In this section, we present the new method called ELM Feature Selection based Improved Supervised Descent Method for Facial Landmark Detection. Section 3.1, we introduce the original Supervised Descent Method (SDM). Then, we explain how to use Self-Adaption Model to improve the performance of SDM in Sect. 3.2. Finally, the ELM Feature Selection based Improved Supervised Descent Method is proposed in the Sect. 3.3.

3.1 Original SDM Models

To address the nonlinear optimization method problem, Xuehuan et al. proposed a Supervised Descent Method [6]. The descent gradient is computed based on the Newtons method. However, the Hessian and Jacobian matrix during the Newtons process are 2nd ordered and the optimization can be impractical to carry on. A sequence of update created by Newtons method can be formulated as,

$$\mathrm{x}_{k+1} = \mathrm{x}_k - \mathrm{H}^{-1}\left(\mathrm{x}_k\right) \mathrm{J}_f\left(\mathrm{x}_k\right) \tag{1}$$

where H(x) and $J_f(\mathrm{x}_k)$ are Hessian and Jacobian matrix which is quadratic to convergence evaluated at x_k. So generating the H(x) and $J_f(\mathrm{x}_k)$ directly can be very complicated. To simplify the computation, [6] turned (1) into

$$\mathrm{x}_{k+1} = \mathrm{x}_k + R_k \varphi_k + b_k \tag{2}$$

R_k is genetic descent direction, b_k is bias term and is feature vector extracted around previous landmark locations x_k. In this way, we only need to learn the sequence of descent direction and bias instead of Hessian and Jacobian matrix. According to this, SDM uses l_2-loss function for the learning of descent gradient and the bias vector:

$$\arg\min_{\mathrm{R}_0,\mathrm{b}_0} \sum_{d^t} \int p(\mathrm{x}_0^i) \left\| \Delta \mathrm{x}^i - \mathrm{R}_0 \varphi_0^i - \mathrm{b}_0 \right\|^2 d\mathrm{x}_0^i \tag{3}$$

and

$$\arg\min_{\mathrm{R}_k,\mathrm{b}_k} \sum_{d^t} \sum_{\mathrm{x}_k^i} \left\| \Delta \mathrm{x}_k^i - \mathrm{R}_k \varphi_k^i - \mathrm{b}_k \right\|^2 \tag{4}$$

3.2 The Self-adaption Initialization Model

Supervised learning is strict about initial model since the first stage of regression can be determinative. SDM as well as CFAN both use a Haar-like feature based detector [21] for face detection and a mean shape model for initialization. So the detection rate is limited to the capability of the detection method. The detector will locate face regions of an image, then the mean shape model will adapt to the face region. But the mean shape only adapts the size of the region, that is, it is only a scaling procession. The initial landmarks are placed far away from the patches they should be in, some times in the wrong side. This can cause difficulty in the following update computation, even cause wrong direction during the whole regression.

We proposed an improved initial model according to the mean shape model. For the original mean shape model, the landmarks were located manually, and only the size will be change for different initialization of different faces. Although it is convenient and the robustness of SDM can confirm final results, there are still exceptions that this mean shape model cant provide SDM an efficient initialization and result for error alignment. We will use the inter-ocular distance to normalize the locations of each parts of human face. We also used a detector to locate face region of the image. After face region detection, an eye detect procedure base on Gabor Kernel method [22] is worked to compute the distance between the two eyeballs. Then the distance is applied to the mean shape to relocate the spread of the landmarks of parts of face. We can formulate the whole procedure as:

$$x_0 = y_0 + M(G(I)) \tag{5}$$

where $G(x)$ is Gabor eye detector, $M(x)$ is a points moving function, I is an original face image, y_0 is the coordinates of mean shape and x_0 is the adapted coordinates.

As a result, the eye region, nose region and the mouth region will be placed in normalized locations that close to the ground-truth.

3.3 ELM Feature Selection

In the formulation of SDM, φ_{k} is the feature vector extract from the patches around landmarks. We suppose $\mathrm{h}(x_k, \varphi_k)$ is the original SDM method that calculates landmarks for each stage. Experiments in [6] use SIFT [23] features during learning and testing. But during testing we found that in some cases the SIFT feature can't get a good result but other features can. So a feature selection is needed.

To tackle the issue of facial landmark detection method and improve the general representation of different features, an ELM feature selection is proposed in our model for feature representation. In our proposed ELMFS-iSDM, ELM is firstly trained with the a fixed number of hidden nodes and the same activation function. The hidden layer parameters of ELM are randomly initialized independently. Actually, the ELM projects the facial data from the high-dimensional input space to the low-dimensional hidden-layer feature space (ELM feature space), in which the improved SDM is performed. Then, the predicted label is determined by an iterative k-means cluster method. The ELM feature selection method utilized in the proposed recognition approach can be described as follows.

Assuming that the available training feature dataset is $\left\{\left(\boldsymbol{x}_i, t_i\right)\right\}_{i=1}^{\mathbf{N}}$, where $\boldsymbol{x}_i$, t_i, and $\mathbf{N}$ represent the feature vector of the i-th face image, its corresponding category index and the number of images, respectively, the SLFN with K nodes in the hidden layer can be expressed as

$$S(\boldsymbol{x})_i = \sum_{j=1}^{K} \theta_j g(\boldsymbol{a}_j, \boldsymbol{b}_j, \boldsymbol{x}_i), \quad i = 1, 2, \ldots, \mathbf{N} \tag{6}$$

where $S(\boldsymbol{x})_i$ is the output obtained by the SLFN associated with the i-th input data, $\boldsymbol{a}_j \in \mathbb{R}^d$ and $\boldsymbol{b}_j \in \mathbb{R}$ $(j = 1, 2, \ldots, K,)$ are parameters of the jth hidden node, respectively. The variable $\boldsymbol{\theta}_j \in \mathbb{R}^m$ is the link connecting the jth hidden node to the output layer and $\boldsymbol{h}(\cdot)$ is the hidden node activation function. With all training samples, (6) can be expressed in the compact form as

$$S(\boldsymbol{x}) = \boldsymbol{h}\boldsymbol{\theta} \tag{7}$$

where $\boldsymbol{\theta} = (\boldsymbol{\theta}_1, \boldsymbol{\theta}_2, \ldots, \boldsymbol{\theta}_K)$ and $\boldsymbol{H}(\boldsymbol{x})$ are the output weight matrix and the network outputs, respectively. The variable $\boldsymbol{h}$ denotes the hidden layer output matrix with the entry $\boldsymbol{h}_{ij} = g(\boldsymbol{a}_j, \boldsymbol{b}_j, \boldsymbol{x}_i)$. And $\boldsymbol{h}$ can be considered as feature mapping, by which the data from the L-dimensional input space to the K-dimensional hidden-layer ELM feature space. The goal of ELM is to reach not only the smallest training error but also the smallest norm of output weights [13]. Thus, the ELM classification problem can be formulated as,

$$\min_{\text{s.t.}: \boldsymbol{h}(x)\boldsymbol{\theta}=t_i-\boldsymbol{e}_i,\, i=1,\ldots,N} \frac{1}{2}\|\boldsymbol{\theta}\|^2 + \mu\frac{1}{2}\sum_{i=1}^{N} \boldsymbol{e}_i^2 \tag{8}$$

Equation (8) can be solved based on the KKT conditions and the output functions of ELM classifier is

$$S(\boldsymbol{x}) = h(x)\mathbf{h}^T(\frac{1}{\mu} + \mathbf{h}\mathbf{h}^T)^{-1}T \tag{9}$$

where $\boldsymbol{T} = [t_1, \ldots, t_N]^T, \mathbf{h} = [h(x_1), \ldots, h(x_N)]^T$.

4 Experiments

This section reports experimental results on our approach and some baselines. The first experiment compares the SDM with our self-adapted model and SDM without self-adapted model. In the second experiment, we are gonging to test the performance of our ELM feature selection based iSDM approach.

4.1 Analyses on Self-adaption Model

This experiment is aim to compare the performance of the model we proposed against the mean shape model, not only in accuracy but also in speed. The experi-

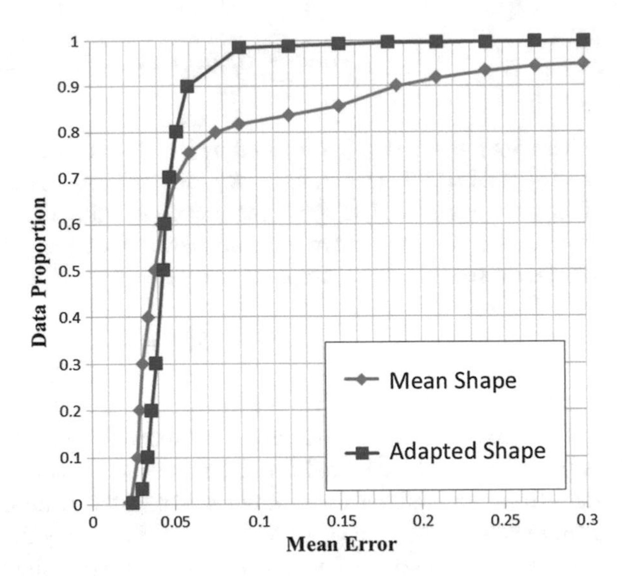

Fig. 1 Comparison on different initial models on AFW dataset

ment is carried out on two famous in the wild databases. They are the I-bug database [8] and AFW database [2]. We found that the faces in these databases have a wide range of ages, including babies, teenagers, adults as well as the olds. Moreover, the poses, expressions and lights are in different conditions. So, initial experiments on these dataset can generate an obvious result.

There is also an export of the mean time we do the detection. The mean detection time for each face is 0.51 s for original SDM and 0.82 s for our model.

From Fig. 1 we can see that by using our model, the performance of SDM has a significant improve at about 0.05 mean shape error.

4.2 ELM Feature Selection Based iSDM

In this part we are going to compare the SDM with feature selection against the original SDM and other baselines. To evaluate the efficiency of our approach, we have used the following public databases for our experiment, i.e., LFPW [24], HELEN [25], AFW [2] and IBUG [26]. The images of those datasets contain different conditions, including illumination, rotation, gestures, expressions changes, etc. We set our method for experiments on these datasets, and the result was shown on Fig. 2. A four-stage regression and two features are utilized for task learning. We used each dataset for both learning and testing. The ground truth annotations include 68 points, but we only extract 49 points among them for our experiment.

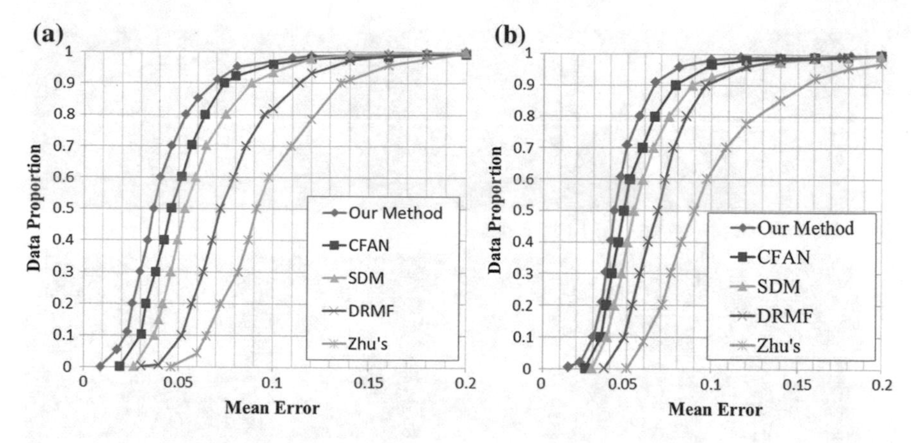

Fig. 2 Performance comparison with state-of-the-art methods, where **a** shows the comparison of our method with SDM, DRMF, CFAN on LFPW database and **b** shows the comparison of different methods on HELEN database

Table 1 The normalized mean error of each stage on AFW

Methods	Datasets			
	AFW	IBUG	LFPW	HELEN
SDM [6]	6.55	15.40	5.67	5.50
DRMF [27]	6.94	19.75	6.57	6.70
Zhu's [2]	7.36	18.33	8.16	8.29
CFAN [28]	–	–	5.44	5.53
Our method	5.35	12.75	5.18	5.21

From Fig. 2a, b, we can see that our method is more effective than the other compared methods and it achieves good performance in the compared two datasets. It also shows that both our method and CFAN have a better performance than SDM and DRMF. However, our method is more efficient when the mean error is less than 0.1. Additionally, we can see that our method has achieved an effective and stable performance on much bigger datasets, such as HELEN and LFPW. The reason may be that the faces in these datasets were more likely to be detected in the first step.

The cumulated mean errors of our method and the compared state-of-the-art methods are given on Table 1. Table 1 displays the details of the cumulated mean errors on the LFPW [10], HELEN [11], AFW [9] and IBUG [8] datasets. We can see that our method outperform those previous methods on these datasets. Some of the annotation results from four different databases are shown on Fig. 3.

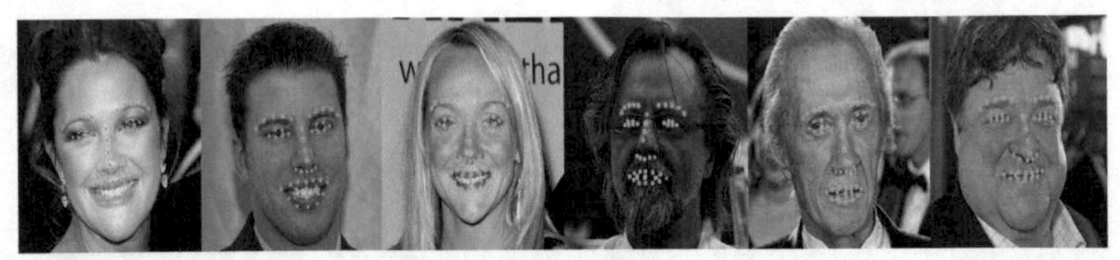

(a). The annotation results on LFPW dataset

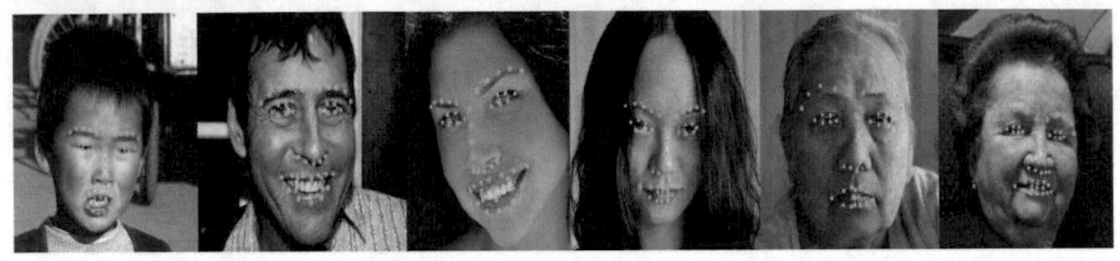

(b).The annotation results on HELEN dataset

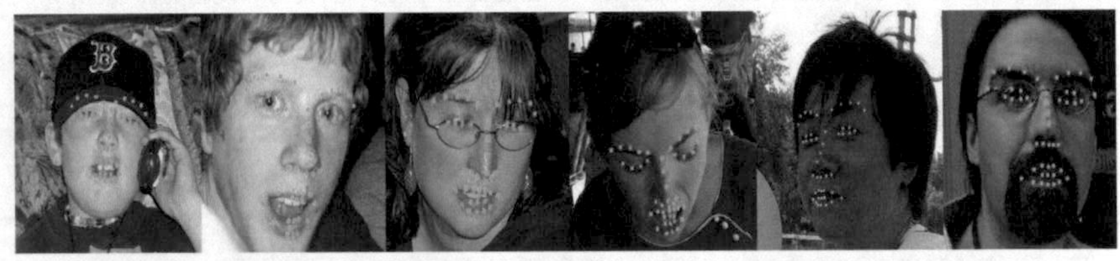

(c).The annotation results on AFW dataset

(d) .The annotation results on IBUG dataset

Fig. 3 Performance comparison with state-of-the-art methods, where **a** shows the comparison of our method with SDM, DRMF, CFAN on LFPW database and **b** shows the comparison of different methods on HELEN database

5 Conclusion

This paper presents novel method for facial landmark detection by using ELM based feature selection and improved SDM. In our new method, a self-adapted model based face detection is proposed to improve the accuracy of the initialization. Then, an ELM feature selection based Supervised Descent Method (SDM) is applied to improve the accuracy of the regression stage. Experimental results show that the effectiveness and efficiency of our new model on four different face databases.

Acknowledgements This work was supported by the National Natural Science Foundation of China (No. 51505004, 61403024, 61471032), the National Key Basic Research Program of China (2012CB316304) and the Beijing Natural Science Foundation (No.4163075, 4162048).

References

1. Gupta, O.P., et al.: Robust facial landmark detection using a mixture of synthetic and real images with dynamic weighting: a survey. Sci. Eng. Tech. **25** (2016)
2. Zhu, X., Ramanan, D.: Face detection, pose estimation, and landmark localization in the wild. In: 2012 IEEE Conference on Computer Vision and Pattern Recognition (CVPR), pp. 2879–2886 (2012)
3. Matthews, I., Baker, S.: Active appearance models revisited. Int. J. Comput. Vis. **60**(2), 135–164 (2004)
4. Blake, A., Isard M.: Active shape models. In: Active Contours, pp. 25–37. Springer (1998)
5. Cristinacce, D., Cootes, T.F.: Feature detection and tracking with constrained local models. In: BMVC, vol.1, p. 3 (2006)
6. Xiong, X., De la Torre, F.: Supervised descent method and its applications to face alignment. In: Proceedings of the IEEE Conference on Computer Vision and Pattern Recognition, pp. 532–539 (2013)
7. Dollár, P., Welinder, P., Perona, P.: Cascaded pose regression. In: 2010 IEEE Conference on Computer Vision and Pattern Recognition (CVPR), pp. 1078–1085 (2012)
8. Ren, S., Cao, X., Wei, Y., Sun, J.: Face alignment at 3000 fps via regressing local binary features. In: Proceedings of the IEEE Conference on Computer Vision and Pattern Recognition, pp. 1685–1692 (2014)
9. Worth, C.L., Preissner, R., Blundell, T.L.: Sdma server for predicting effects of mutations on protein stability and malfunction. Nucleic Acids Res. **39**(suppl 2), W215–W222 (2011)
10. Huang, G.-B., Wang, D.H., Lan, Y.: Extreme learning machines: a survey. Int. J. Mach. Learn. Cybern. **2**(2), 107–122 (2011)
11. Cao, J., Lin, Z.: Extreme learning machines on high dimensional and large data applications: a survey. Math. Probl. Eng. **501**, 103796 (2015)
12. Cao, J., Zhang, K., Luo, M., Yin, C., Lai, X.: Extreme learning machine and adaptive sparse representation for image classification. Neural Netw. **81**, 91–102 (2016)
13. Huang, G.-B., Zhou, H., Ding, X., Zhang, R.: Extreme learning machine for regression and multiclass classification. IEEE Trans. Sys. Man Cybern. Part B: Cybern. **42**(2), 513–529 (2012)
14. Zong, W., Huang, G.-B.: Face recognition based on extreme learning machine. Neurocomputing **74**(16), 2541–2551 (2011)
15. Zong, W., Zhou, H., Huang, G.-B., Lin, Z.: Face recognition based on kernelized extreme learning machine. In: Autonomous and Intelligent Systems. Lecture Notes in Computer Science, vol. 6752, pp. 263–272 (2011)
16. Long, X., Lu, H., Peng, Y., Li, W.: Graph regularized discriminative non-negative matrix factorization for face recognition. Multimed. Tools Appl. **72**(3), 2679–2699 (2014)
17. Cao, J., Chen, T., Fan, J.: Landmark recognition with compact BoW histogram and ensemble ELM. Multimed. Tools Appl. **75**(5), 2839–2857 (2016)
18. Cao, J., Zhao, Y., Lai, X., Ong, M.E.H., Yin, C., Koh, Z.X., Liu, N.: Landmark recognition with sparse representation classification and extreme learning machine. J. Frankl. Inst. **352**(10), 4528–4545 (2015)
19. Roul, R.K., Gugnani, S., Kalpeshbhai, S.M.: Clustering based feature selection using extreme learning machines for text classification. In: 2015 Annual IEEE India Conference (INDICON) pp. 1–6 (2015)
20. Zhai, M.-Y., Yu, R.-H., Zhang, S.-F., Zhai, J.-H.: Feature selection based on extreme learning machine. In: 2012 International Conference on Machine Learning and Cybernetics, vol. 1, pp. 157–162 (2012)

21. Viola, P., Jones, M.J.: Robust real-time face detection. Inter. J. Comput. Vis. **57**(2), 137–154 (2004)
22. Liu, C.: Gabor-based kernel PCA with fractional power polynomial models for face recognition. IEEE Trans. Pattern Anal. Mach. Intell. **26**(5), 572–581 (2004)
23. Lowe, D.G.: Distinctive image features from scale-invariant keypoints. Inter. J. Comput. Vis. **60**(2), 91–110 (2004)
24. Belhumeur, P.N., Jacobs, D.W., Kriegman, D.J., Kumar, N.: Localizing parts of faces using a consensus of exemplars. IEEE Trans. Pattern Anal. Mach. Intell. **35**(12), 2930–2940 (2013)
25. Le, V., Brandt, J., Lin, Z., Bourdev, L., Huang, T.S.: Interactive facial feature localization. In: European Conference on Computer Vision, pp. 679–692. Springer (2012)
26. Sagonas, C., Tzimiropoulos, G., Zafeiriou, S., Pantic, M.: 300 faces in-the-wild challenge: the first facial landmark localization challenge. In: Proceedings of the IEEE International Conference on Computer Vision Workshops, pp. 397–403 (2013)
27. Asthana, A., Zafeiriou, S., Cheng, S. Pantic, M.: Robust discriminative response map fitting with constrained local models. In: Proceedings of the IEEE Conference on Computer Vision and Pattern Recognition, pp. 3444–3451 (2013)
28. Zhang, J., Shan, S., Kan, M., Chen, X.: Coarse-to-fine auto-encoder networks (cfan) for real-time face alignment. In: European Conference on Computer Vision, pp. 1–16. Springer (2014)

Online Sequential Extreme Learning Machine with Under-Sampling and Over-Sampling for Imbalanced Big Data Classification

Jie Du, Chi-Man Vong, Yajie Chang and Yang Jiao

Abstract In this paper, a novel method called *online sequential extreme learning machine with under-sampling and over-sampling* (OSELM-UO) for imbalanced Big data classification is proposed which combines the structures of under-sampling and over-sampling and applies online sequential extreme learning machine as its base model. The novel structure enables OSELM-UO performs well on both minority and majority classes and simultaneously overcomes the issues of information loss and overfitting. Moreover, when the dataset keeps growing, OSELM-UO can be applied without retraining all previous data. Experiments have been conducted for OSELM-UO and several imbalance learning methods over real-world datasets respectively under high imbalance ratio (IR) and large amount of samples and features. Through the analysis of the experimental results, OSELM-UO is shown to give the best results in various aspects.

Keywords Big data · Imbalance learning · OS-ELM · Under-sampling · Over-sampling

1 Introduction

Big data problem [1] has attracted growing attentions in recent years. However, in very large or complex data, traditional data processing methods become inadequate for the learning and/or extraction of useful information [2]. If the data are also imbalanced, the problem becomes even more challenging. In imbalanced data, the critical and highly interested class (called minority) is with significantly less amount of samples than the other one (called majority) so that the critical class is easily misclassified or even ignored. Actually, in most Big data applications, the data are always imbalanced [3]. Therefore, Big data learning always occur together with class imbalance in practical applications.

J. Du · C.-M. Vong (✉) · Y. Chang · Y. Jiao
Department of Computer of Information Science, University of Macau, Macau, China
e-mail: cmvong@umac.mo

J. Cao et al. (eds.), *Proceedings of ELM-2016*, Proceedings in Adaptation, Learning and Optimization 9, DOI 10.1007/978-3-319-57421-9_19

In the literature, imbalance learning methods can be divided into two categories: cost sensitive methods [4] at algorithm level and resampling methods [5] at data level. In fact, cost sensitive methods are unable to handle big data, because a class weight matrix $\mathbf{W}_{class}$ is employed to control the attentions on minority and majority samples [6]. Usually, the weights associated with minority samples are larger than the ones associated with majority samples [7], and hence more attention is focused on minority samples. However, the size of $\mathbf{W}_{class}$ is determined by the training data size N, i.e., $\|\mathbf{W}_{class}\|$ is N × N [8]. It can be easily seen that the memory storage for $\mathbf{W}_{class}$ is prohibitively expensive with large N, resulting to intractable computation.

Resampling methods aim to preprocess and balance the training data. These methods include random under-sampling [5], oversampling [5] and synthetic minority over-sampling technique (SMOTE) [9]. Random under-sampling is to randomly remove some samples from the majority class until the data becomes balanced. Obviously, significant information loss occurs, leading to low accuracy on majority class [10]. On the other hand, over-sampling is to directly duplicate the minority samples to balance the data. However, over-sampling may cause overfitting problem [11] because the classifier is constructed / learned under the same rule for multiple copies of the same sample [11]. SMOTE is another over-sampling method which generates some synthetic minority samples along the k lines between a minority sample and its selected k nearest neighbors rather than duplication. However, the artificially synthetic minority samples may be noises rather than the true minority ones [12], and hence deteriorated accuracy. Moreover, over-sampling and SMOTE sharply increase the training data size so that big dataset becomes even bigger. As a result, random under-sampling without information loss is the best choice for preprocessing Big data with class imbalance problem.

In this work, we propose a novel method which employs online sequential extreme learning machine (OS-ELM) [13] and under-sampling to deal with the imbalanced Big data problem. OS-ELM is an online sequential version of extreme learning machine (ELM) [14], which includes initial and sequential learning phase. (Remark: ELM is a very popular single layer feedforward neural network which has high efficiency and accuracy [15]. Its input weights are randomly generated while the output weights are to be analytically determined by least square solution [16]) In initial learning phase, the learning procedure of OS-ELM is same as ELM where a basic model is constructed. In sequential learning phase, the basic model is updated/learned sequentially by a recursive algorithm [13] using the new incoming data rather than all previously arrived data. Therefore, OS-ELM can learn a model for Big data by sequentially training multiple small clusters sampled from the original Big data.

In this work, random under-sampling is employed in every learning phase to sequentially sample multiple small and balanced clusters. The training procedure is as follows:

(i) In initial learning phase, the original training data is divided into one balanced cluster $\mathbf{D}_0$ generated by under-sampling and another cluster that includes the remaining majority samples for future learning. Then $\mathbf{D}_0$ is trained to learn a basic model.

(ii) In $(k+1)$ th sequential learning phase, under-sampling is re-employed to generate another balanced cluster $\mathbf{D}_{k+1}$ that includes all learned minority samples and the similar size of untrained majority ones. Then the model is updated by training $\mathbf{D}_{k+1}$

(iii) After all majority samples in the original Big data have been trained, the whole training procedure is completed.

Although under-sampling is employed in every learning phase, the whole training procedure actually is over-sampling because minority samples are retrained in every phase. Therefore, we name this novel method as OSELM-UO (OSELM with under-sampling and over-sampling). The main contributions of OSELM-UO are enumerated as follows:

(i) Compared to cost sensitive method, OSELM-UO can be directly applied in processing Big dataset because the training data used in each learning phase is only one small set of the whole data.

(ii) Compared to random under-sampling method, there is no information loss in OSELM-UO because all majority samples are learnt.

(iii) Compared to over-sampling method, there is no overfitting in OSELM-UO because in each learning phase, the objective function (or learned rule) is different and sequentially updated by recursive algorithm. In other words, the learned model is constructed/learned under different rules for multiple copies of the same sample.

(vi) If the dataset keeps growing, OSELM-UO also can be applied in such case without retraining all data again while other discussed methods cannot.

The organization of this paper is as follows. A short review of OS-ELM is presented in Sect. 2, followed by the details of the training procedure of proposed OSELM-UO in Sect. 3. Section 4 shows the experimental results compared with random under-sampling, over-sampling, SMOTE and cost sensitive method. Finally, a conclusion is drawn in Sect. 5.

2 Related Works

In this section, we briefly review *online sequential extreme learning machine* OS-ELM [13], because OS-ELM is the base model of our proposed OSELM-UO.

OS-ELM is a variant of ELM that can sequentially update a learned model with data in chunk by chunk and one by one (a special case of chunk by chunk). Two steps are consisted in OS-ELM.

Step 1: **Initial learning phase**

In this step, a chunk of training data $n_0 = \{x_i, t_i\}$ $i = 1, \ldots, N_0$ is necessary to learn an initial model, where N_0 is the size of initial chunk of data, x_i represents the ith sample and t_i is the label for the ith sample in the initial chunk of data. Identical to batch ELM, the input weights and bias of the basic model are randomly generated and the initial output weight β_0 is calculated based on the least square solution as follows

$$\beta_0 = \mathbf{P}_0 \mathbf{H}_0^{\mathrm{T}} \mathbf{T}_0 \tag{1}$$

and

$$\mathbf{P}_0 = \left(\mathbf{H}_0^{\mathrm{T}} \mathbf{H}_0\right)^{-1}$$

where $\mathbf{H}_0$ is the initial hidden layer output matrix, and $\mathbf{T}_0$ is the label matrix of the initial chunk of data.

In order to make the learning performance of OS-ELM identical to the batch ELM, N_0 should be more than the number of hidden nodes [13].

Step 2: **Sequential learning phase**

As the new set of training data arrives, the $(k + 1)$ th chunk of data $n_{k+1} = \{x_i, t_i\}$ where $k \geq 0$ and N_{k+1} denotes the size of $(k + 1)$ th chunk, the partial hidden layer output matrix $\mathbf{H}_{k+1}$ is calculated firstly. Then the output weight matrix β_{k+1} with $\mathbf{T}_{k+1}$ and β_k is computed as follows

$$\beta_{k+1} = \beta_k + \mathbf{P}_{k+1} \mathbf{H}_{k+1}^T \left(\mathbf{T}_{k+1} - \mathbf{H}_{k+1} \beta_k\right) \tag{2}$$

and

$$\mathbf{P}_{k+1} = \mathbf{P}_k - \mathbf{P}_k \mathbf{H}_{k+1}^T \left(I + \mathbf{H}_{k+1} \mathbf{P}_k \mathbf{H}_{k+1}^T\right)^{-1} \mathbf{H}_{k+1} \mathbf{P}_k$$

The above equation is similar to the recursive least squares algorithm [17] and when there is new chunk of data arriving, the output weight is updated following Eq. (2).

3 Proposed Method

In this section, we will detail the training procedure of proposed OSELM-UO. There are three steps in OSELM-UO.

Step 1: **Preprocess the training data**

The small and balanced training clusters used in every learning phase are firstly generated. Given a set of original training data $\mathbf{D} = \{\mathbf{x}_i, t_i\}$, $i = 1, \ldots, N$ N is the data size, $t_i = \{-1, +1\}$, $t_i = +1$ indicates that ith sample belongs to minority class. Similarly, $t_i = -1$ indicates majority one. We firstly divide $\mathbf{D}$ into minority class (called **Mi**) and majority class (called **Ma**), i.e.,$\mathbf{D} = \mathbf{Mi} \cup \mathbf{Ma}$. Then **Ma** is randomly divided into different subsets of equal size $\mathbf{Ma}_l$, $l = 0, \ldots, last$. Every $\mathbf{Ma}_l$ has the same or similar data size with **Mi**. For instance, if the size of **Mi** is 76 and the size of **Ma** is 450, $last = \lceil 450/76 \rceil - 1 = 5$. Then, **Ma** is divided into 6 subsets and each subset has 450 / 6 = 75 samples. Finally, the small and balanced training clusters are $\mathbf{D}_l = \mathbf{Mi} \cup \mathbf{Ma}_l, l = 0, ..., last$.

Step 2: **Initial learning phase**

In this step, cluster $\mathbf{D}_0$ is chosen as initial chunk of data to learn an initial model. The training procedure is the same with the initial learning phase in OS-ELM and the initial output weight β_0 is calculated using Eq. (1). Therefore, the initial output (or initial decision) of the output layer is given by

$$f_0 = \mathbf{H}_0 \beta_0 \tag{3}$$

Actually, Eq. (3) is identical to the result obtained by ELM combined with random under-sampling because only $\mathbf{Ma}_0$ is trained.

Step 3: **Sequential learning phase**

In this step, cluster $\mathbf{D}_{k+1}$ is chosen as $(k+1)$th sequential chunk of data to be trained and the output weight matrix β_{k+1} is computed using Eq. (2). Up to now, $(k+1)$ subsets of majority class are trained and β_{k+1} has preserved the information of all trained data. If $(k+1) == last$, the whole training procedure is completed and all majority samples are trained. Therefore, the final decision is as follows:

$$f_{last} = \mathbf{H}_{last} \beta_{last} \tag{4}$$

The workflow of OSELM-UO is illustrated in Fig. 1.

4 Experiments

In this section, OSELM-UO is compared with random under-sampling, over-sampling, SMOTE and weighted ELM (W-ELM) [6]. Under-sampling and over-sampling are chosen to verify the properties of proposed OSELM-UO. SMOTE is chosen as a popular and effective resampling method nowadays [9]. W-ELM is a cost sensitive version of ELM, which can effectively solve the class imbalance problem [6]. Moreover, after preprocessed by under-sampling, over-sampling and SMOTE, the preprocessed data are trained by ELM for a fair comparison.

The comparison is conducted over 8 real-word datasets including UCI machine learning repository [18] and KEEL [19], as shown in Tables 1 and 2. Among these 8 datasets, 5 datasets are with high imbalance ratio (IR). IR is calculated by majority class size/minority class size. High IR indicates the data are with highly imbalanced class distribution. For instance, if the IR is 99, it means the dataset includes only 1% of minority samples but 99% of majority samples. Table 1 shows the 5 datasets with high IR but small amount of samples. Table 2 shows the 3 datasets that have large volume of samples or features but with low IR. (Remark: the datasets in Tables 1 and 2 are chosen from KEEL and UCI respectively.) This section mainly presents: (i) experimental setup; (ii) the results comparison of OSELM-UO with random

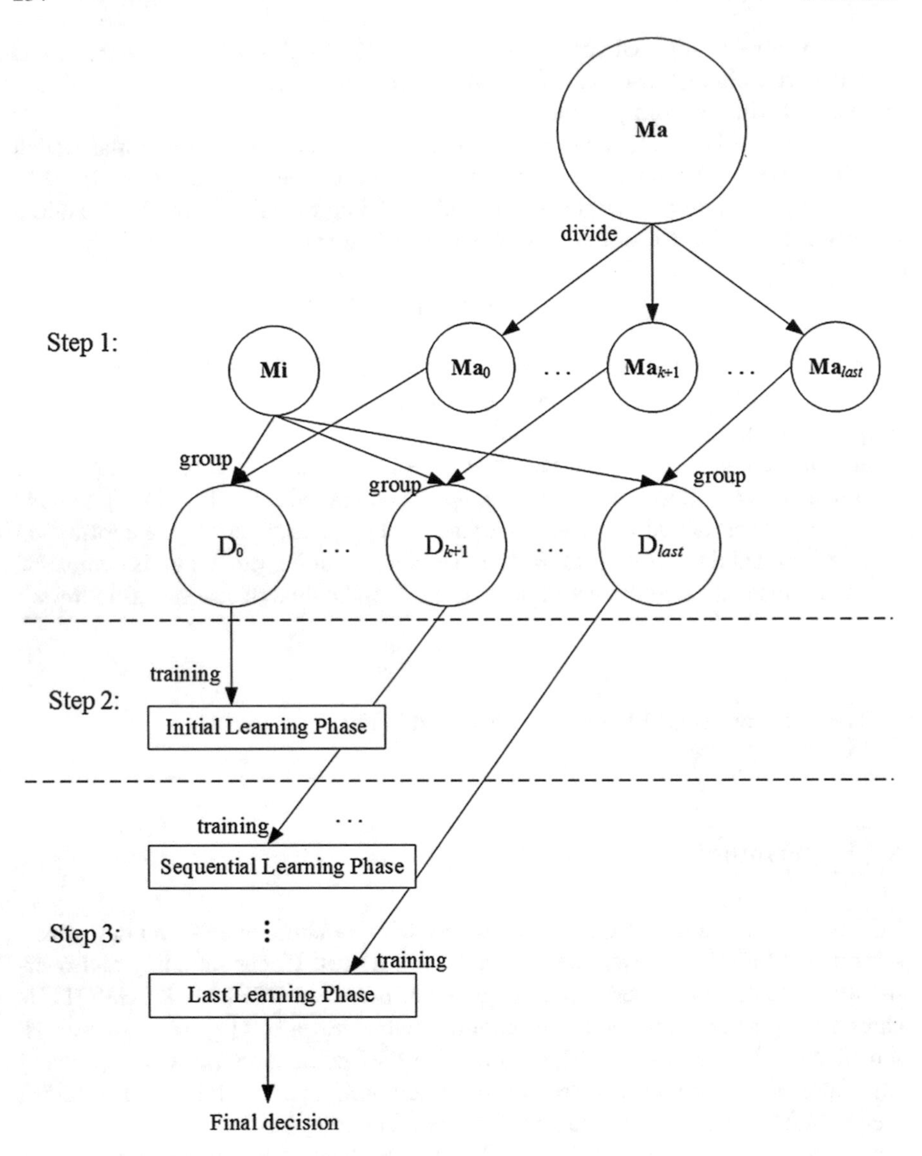

Fig. 1 Workflow of OSELM-UO

under-sampling, over-sampling, SMOTE and W-ELM to verify the effectiveness in terms of G-mean [11], minority and majority accuracy; (iii) the analysis of the properties of OSELM-UO. For all experiments, G-mean is employed as evaluation metric because it is very popular in imbalance learning [20].

Table 1 Properties of datasets with high IR

Datasets	# Features	# Instances	IR
abalone19	8	4174	129.44
kddcup-rootkit-imap_vs_back	41	2225	100.14
poker-8-9_vs_5	10	2075	82
kr-vs-k-one_vs_fifteen	6	2193	80.22
abalone-20_vs_8-9-10	8	1916	72.69

Table 2 Properties of datasets with large amount of samples or attributes

Datasets	# Features	# Instances	IR
handwritten	649	2000	9
ijcnn1	13	91701	9.53
skin_nonskin	3	201280	27.43

4.1 Experimental Setup

In order to have a fair comparison, we run 50 times of training for every compared method. Every time, the training data is randomly drawn. The parameters (e.g., number of hidden nodes) for each method are also optimized in a user-specified range (e.g., $\{10, 20, \ldots, 300\}$). The result comparison over all methods is conducted in terms of G-mean. The detailed results such as the means of minority and majority accuracy are also illustrated. All the experiments were conducted on MATLAB over a PC of 3.60 GHz with 16 GB RAM. Some notes for the experiments are as follows:

(1) All the features of each dataset are linearly scaled into $[-1, 1]$
(2) The datasets *handwritten* is originally balanced and have multiple classes. In order to test OSELM-UO, we select one class as minority and the remaining classes as majority so that these datasets become imbalanced.
(3) The training data account for 80% of total data, while testing data account for 20%.

4.2 G-Mean

We firstly compare and evaluate the overall performance of the proposed OSELM-UO with random under-sampling, over-sampling, SMOTE and W-ELM on all 8 datasets. For the evaluation, testing results of G-mean are reported in Table 3 and the best results are labeled in italic and underline. OSELM-UO gives the best G-means in all datasets, even though the data are with high IR and large size. Compared to

Table 3 Mean and standard deviation of G-mean (%) in testing results on all compared datasets

Datasets	Under-sampling	Over-sampling	SMOTE	W-ELM	OSELM-UO
abalone19	72.26 ± 11.66	75.15 ± 10.51	75.30 ± 12.65	75.99 ± 10.54	*78.56 ± 8.74*
kddcup-rootkit-imap_vs_back	99.95 ± 0.07	95.98 ± 6.47	*100.0 ± 0.0*	97.32 ± 5.64	*100.0 ± 0.0*
poker-8-9_vs_5	65.01 ± 11.53	74.42 ± 12.84	74.98 ± 7.64	73.87 ± 15.57	*81.47 ± 11.01*
kr-vs-k-one_vs_fifteen	99.89 ± 0.14	*100.0 ± 0.0*	*100.0 ± 0.0*	*100.0 ± 0.0*	*100.0 ± 0.0*
abalone-20_vs_8-9-10	86.62 ± 7.18	89.94 ± 4.69	89.88 ± 5.06	90.49 ± 8.69	*93.13 ± 5.36*
handwritten	95.69 ± 1.44	97.35 ± 1.63	97.62 ± 0.37	98.52 ± 1.68	*99.02 ± 0.59*
ijcnn1	97.58 ± 0.07	97.81 ± 0.10	97.82 ± 0.13	*null*	*97.84 ± 0.12*
skin_nonskin	*99.81 ± 0.04*	99.79 ± 0.10	99.75 ± 0.13	*null*	*99.81 ± 0.11*

Table 4 Mean of minority accuracy (%) in testing results on all compared datasets

Datasets	Under-sampling	Over-sampling	SMOTE	W-ELM	OSELM-UO
abalone19	71.66	71.66	76.66	76.66	*78.33*
kddcup-rootkit-imap_vs_back	*100.0*	92.50	*100.0*	95.00	*100.0*
poker-8-9_vs_5	74.00	66.00	66.00	66.00	*78.00*
kr-vs-k-one_vs_fifteen	*100.0*	*100.0*	*100.0*	*100.0*	*100.0*
abalone-20_vs_8-9-10	88.00	86.00	88.00	88.00	*92.00*
handwritten	97.75	97.75	98.75	97.50	*100.0*
ijcnn1	98.72	98.62	98.56	*null*	*98.73*
skin_nonskin	*100.0*	*100.0*	*100.0*	*null*	*100.0*
Average	91.27	89.07	90.99	*null*	*93.38*

Table 5 Mean of majority accuracy (%) in testing results on all compared datasets

Datasets	Under-Sampling	Over-Sampling	SMOTE	W-ELM	OSELM-UO
abalone19	72.72	*80.50*	75.91	77.17	79.80
kddcup-rootkit-imap_vs_back	99.90	*100.0*	*100.0*	*100.0*	*100.0*
poker-8-9_vs_5	59.07	86.12	86.14	86.09	*86.58*
kr-vs-k-one_vs_fifteen	99.79	*100.0*	*100.0*	*100.0*	*100.0*
abalone-20_vs_8-9-10	85.89	94.36	92.11	94.02	*94.55*
handwritten	93.72	96.97	96.52	*99.58*	98.06
ijcnn1	96.45	97.02	*97.09*	*null*	96.94
skin_nonskin	99.63	99.58	99.48	*null*	*99.64*
Average	88.39	94.31	92.81	*null*	*94.45*

random under-sampling and over-sampling, OSELM-UO can improve the performance up to 16 and 7% (e.g., *poker-8-9_vs_5*), respectively. In Table 3, OSELM-UO gets up to 7 and 8% more of G-mean (e.g., *poker-8-9_vs_5*) than SMOTE and W-ELM, respectively. Due to the large data size in *ijcnn1* and *skin_nonskin*, W-ELM cannot run and hence no results are obtained (labeled by *null* in Tables 3, 4 and 5).

4.3 Minority and Majority Accuracy

Subsequently, minority and majority accuracies on all datasets are shown in Tables 4 and 5. In Table 4, OSELM-UO also gets best minority accuracy. In Table 5, compared with other methods, random under-sampling gets the worst majority accuracy in almost all datasets, because significant majority information loss occurs. In Table 4, compared with other methods, over-sampling gives poor minority accuracy, because overfitting happens in training. From the observation of average accuracy shown in Tables 4 and 5, OSELM-UO gives the best performance on both minority and majority classes. In addition, random under-sampling gets good minority accuracy but poor majority accuracy, while over-sampling does the opposite, i.e., good majority accuracy but poor minority accuracy. These results verify that OSELM-UO successfully integrates the advantages of under-sampling (good performance on minority class) and over-sampling (good performance on majority class), and also eliminates their disadvantages (i.e., information loss and overfitting) simultaneously.

5 Conclusion

In this paper, we present a novel method called OSELM-UO that aims to effectively tackle imbalanced Big data problem. In OSELM-UO, multiple small and balanced clusters are sequentially generated by random under-sampling. Although under-sampling is employed in every learning phase, the whole training procedure is actually over-sampling the minority samples to be retrained in every phase. Therefore, the structure of OSELM-UO is a combination of under-sampling and over-sampling.

The performance of OSELM-UO is assessed through testing some typical datasets with high IR (e.g., the IR of *abalone19* is 129.44), large amount of samples (e.g., *skin_nonskin* has 201280 samples) and features (e.g., *handwritten* has 649 features), respectively. The results are compared with popular imbalance learning methods including random under-sampling, over-sampling, SMOTE and W-ELM. Experimental results verify the following conclusions:

(i) OSELM-UO gives the best G-mean results in all compared datasets, which can improve the performance up to 16%. Moreover, it also gets the best average accuracy on both minority and majority classes.
(ii) OSELM-UO is with the advantages of both under-sampling and over-sampling but without their disadvantages. In other words, OSELM-UO obtains good performance on both minority and majority class, and does not suffer from information loss and overfitting.

In a nutshell, OSELM-UO can efficiently produce an accurate classification model for imbalance Big data. Therefore, OSELM-UO is suitable to many practical applications with imbalanced Big data problem.

Acknowledgements The work is financially supported by funding from University of Macau, project number MYRG2014-00083-FST, and from FDCT Macau, project number 050/2015/A.

References

1. Katal, A., Wazid, M., Goudar, R.: Big data: issues, challenges, tools and good practices. In: 2013 Sixth International Conference on Contemporary Computing (IC3), pp. 404–409. IEEE (2013)
2. Chen, C.P., Zhang, C.-Y.: Data-intensive applications, challenges, techniques and technologies: a survey on big data. Inf. Sci. **275**, 314–347 (2014)
3. del Río, S., López, V., Benítez, J.M., Herrera, F.: On the use of mapreduce for imbalanced big data using random forest. Inf. Sci. **285**, 112–137 (2014)
4. Ling, C.X., Sheng, V.S.: Cost-sensitive learning. In: Encyclopedia of Machine Learning, pp. 231–235. Springer (2011)
5. Gershunskaya, J., Jiang, J., Lahiri, P.: Resampling methods in surveys. Handb. Stat. **29**, 121–151 (2009)
6. Zong, W., Huang, G.-B., Chen, Y.: Weighted extreme learning machine for imbalance learning. Neurocomputing **101**, 229–242 (2013)
7. Gao, X., Chen, Z., Tang, S., Zhang, Y., Li, J.: Adaptive weighted imbalance learning with application to abnormal activity recognition. Neurocomputing **173**, 1927–1935 (2016)
8. Sharma, R., Bist, A.S.: Genetic algorithm based weighted extreme learning machine for binary imbalance learning. In: 2015 International Conference on Cognitive Computing and Information Processing (CCIP), pp. 1–6. IEEE (2015)
9. Chawla, N.V., Bowyer, K.W., Hall, L.O., Kegelmeyer, W.P.: Smote: synthetic minority oversampling technique. J. Artif. Intell. Res. **16**, 321–357 (2002)
10. Tang, Y., Zhang, Y.-Q.: Granular svm with repetitive undersampling for highly imbalanced protein homology prediction. In: 2006 IEEE International Conference on Granular Computing, pp. 457–460. IEEE (2006)
11. He, H., Garcia, E.A.: Learning from imbalanced data. IEEE Trans. knowl. data Eng. **21**(9), 1263–1284 (2009)
12. He, H., Bai, Y., Garcia, E.A., Li, S.: Adasyn: Adaptive synthetic sampling approach for imbalanced learning. In: 2008 IEEE International Joint Conference on Neural Networks (IEEE World Congress on Computational Intelligence), pp. 1322–1328. IEEE (2008)
13. Liang, N.-Y., Huang, G.-B., Saratchandran, P., Sundararajan, N.: A fast and accurate online sequential learning algorithm for feedforward networks. IEEE Trans. Neural Netw. **17**(6), 1411–1423 (2006)
14. Huang, G.-B., Zhu, Q.-Y., Siew, C.-K.: Extreme learning machine: theory and applications. Neurocomputing **70**(1), 489–501 (2006)
15. Huang, G.-B., Zhou, H., Ding, X., Zhang, R.: "Extreme learning machine for regression and multiclass classification. IEEE Trans. Syst. Man Cybern. Part B (Cybernetics) **42**(2), 513–529 (2012)
16. Huang, G.-B., Wang, D.H., Lan, Y.: Extreme learning machines: a survey. Int. J. Mach. Learn. Cybern. **2**(2), 107–122 (2011)
17. Benesty, J., Paleologu, C., Gänsler, T., Ciochină, S.: Recursive least-squares algorithms. In: A Perspective on Stereophonic Acoustic Echo Cancellation, pp. 63–69. Springer (2011)
18. Frank, A., Asuncion, A.: Uci machine learning repository [http://archive.ics.uci.edu/ml]. irvine, ca: University of california, School of Information and Computer Science, vol. 213 (2010)
19. Alcalá, J., Fernández, A., Luengo, J., Derrac, J., García, S., Sánchez, L., Herrera, F.: Keel data-mining software tool: data set repository, integration of algorithms and experimental analysis framework. J. Mult.-Valued Log. Soft Comput. **17**(2–3), 255–287 (2010)
20. Gu, Q., Zhu, L., Cai, Z.: Evaluation measures of the classification performance of imbalanced data sets. In: International Symposium on Intelligence Computation and Applications, pp. 461–471. Springer (2009)

An Automatic Identification System (AIS) Database for Maritime Trajectory Prediction and Data Mining

Shangbo Mao, Enmei Tu, Guanghao Zhang, Lily Rachmawati, Eshan Rajabally and Guang-Bin Huang

Abstract In recent years, maritime safety and efficiency become very important across the world. Automatic Identification System (AIS) tracks vessel movement by onboard transceiver and terrestrial and/or satellite base stations. The data collected by AIS contain broadcast kinematic information and static information. Both of them are useful for maritime anomaly detection and vessel route prediction which are key techniques in maritime intelligence. This paper is devoted to construct a standard AIS database for maritime trajectory learning, prediction and data mining. A path prediction method based on Extreme Learning Machine (ELM) is tested on this AIS database and the testing results show this database can be used as a standardized training resource for different trajectory prediction algorithms and other AIS data based mining applications.

S. Mao · E. Tu (✉)
Rolls-Royce@NTU Corporate Lab, Nanyang Technological University, Singapore, Singapore
e-mail: hellotem@hotmail.com; emtu@ntu.edu.sg

S. Mao
e-mail: SBMAO@ntu.edu.sg

G. Zhang · G.-B. Huang
School of Electrical & Electronic Engineering, Nanyang Technological University, Singapore, Singapore
e-mail: GZHANG009@e.ntu.edu.sg

G.-B. Huang
e-mail: EGBHuang@ntu.edu.sg

L. Rachmawati
Computational Engineering Team, Advanced Technology Centre, Rolls-Royce Singapore Pte Ltd, Singapore, Singapore
e-mail: lily.rachmawati@rolls-royce.com

E. Rajabally
Strategic Research Center, Rolls-Royce Plc, London, UK
e-mail: Eshan.Rajabally@Rolls-Royce.com

J. Cao et al. (eds.), *Proceedings of ELM-2016*, Proceedings in Adaptation, Learning and Optimization 9, DOI 10.1007/978-3-319-57421-9_20

Keywords Automatic identification system (AIS) database · Maritime trajectory learning · Data mining

1 Introduction

In the modern globalized economy, ocean shipping becomes the most efficient method for transporting commodities over long distance. The persistent growth of the world economy leads to increasing demand of maritime transportation with larger ship capacity and higher sailing speed [1]. In this circumstance, safety and security become key issues in maritime transportation. Intelligent maritime navigation system using Automatic Identification System (AIS) data improves the maritime safety with less cost compared with conventional maritime navigation system using human navigators. The AIS is a maritime safety and vessel traffic system imposed by the International Maritime Organization (IMO). Autonomously broadcasted AIS messages contain kinematic information (including ship location, speed, heading, rate of turn, destination and estimated arrival time) and static information (including ship name, ship MMSI ID, ship type, ship size and current time), which can be transformed into useful information for intelligent maritime traffic manipulations, e.g. vessel path prediction and collision avoidance, and thus plays a central role in the future autonomous maritime navigation system. Over the last several years, receiving AIS messages from vessels and coastal stations has become increasingly ordinary.

Although sufficient AIS data can be obtained from many data providers, e.g. Marinecadastre (MarineC.) [2] and Sailwx [3], to the best of our knowledge, there is no existing standard AIS benchmark database in maritime research area, which makes it quite inconvenient for researchers and practitioners in the field, since collecting a usable dataset will cost a lot of time and effort. Furthermore, as the intelligent maritime system develops rapidly, many researchers proposed anomaly detection and motion prediction algorithms and it is quite important to have a database that could be served as a benchmark for comparing the performance of different methods and algorithms. For example, in 2008, Ristic et al. [4] proposed an anomaly detection and motion prediction algorithm based on statistical analysis of motion pattern in AIS data. In 2013, Premalatha Sampath generated vessel trajectory from raw AIS data and analyzed the trajectory to identify the kinematic pattern of vessel in New Zealand waterways [5]. So in this paper, a ready-to-use standard AIS database is constructed for maritime path learning, prediction and data mining.

The remaining parts of the paper are organized as follows: Sect. 2 describes the AIS data type and data source. Section 3 describes the detailed process of constructing the AIS database. Then the structure and static information of our AIS database are summarized and described in Sect. 4. Finally, we conduct an experiment based Extreme Learning Machine (ELM) on the AIS database to show the usefulness of it in Sect. 5.

2 Properties of AIS Database

This section describes the attributes of AIS data and introduces some popular AIS data providers. And AIS data have some special attributes which would lead to the difference between maritime trajectory prediction and route prediction in other fields.

2.1 AIS Data Attributes

AIS technology broadcast ship information and voyage information at regular time interval. The information can be received by onboard transceiver and terrestrial and/or satellite base station. There are some important attributes of AIS data: longitude, latitude, speed over ground (SOG), course over ground (COG), vessel's maritime mobile service identity (MMSI), base date time, vessel type, vessel dimension, rate of turn (ROT), navigation status and heading. In this paper, the standard AIS database contains longitude, latitude, SOG, COG, MMSI and base date time, which are the most useful attributes for maritime trajectory learning and prediction.

2.2 AIS Data Providers

There are many existing AIS data providers e.g. Marine Traffic (Marine T) [6], VT explorer (VT E.) [7], FleetMon [8], Marinecadastre (MarineC.) [2] and Aprs [9]. Among these providers, MarineC can be downloaded for free and have good data quality according to data completeness and position precision. So in this paper, MarineC is selected to collect AIS data online. MarineC contains historical records from 2009 to 2014 in America at a minute interval. We can choose and download AIS data files in specific month and specific interest area. We downloaded February 2009 AIS data in UTM zone ten. However, the AIS data downloaded from MarineC contains some data missing. To solve these problem, we use the linear interpolation and the detail will he introduced in this paper later.

3 AIS Database Construction

This section describes the data processing tool and the detail of constructing the standard AIS database we proposed. The whole process contains four parts: raw data pre-processing, raw data selecting, candidate data cleaning and missing data interpolating.

3.1 Raw Data Pre-processing

The first step of constructing an AIS database is to download the raw database file with dbf format from http://www.marinecadastre.gov/ais/. Prior to download raw data online, the selection of interest area is necessary. As shown in Fig. 1, zone ten is on west coast of the United States and it contains considerable amount of ships. And these AIS data are open-sourced. In this paper, zone ten is chosen as interested area because it contains sufficient amount of AIS data. The shaded part in Fig. 1 is the chosen interested area whose longitude is from −120 to −126° and latitude is from 30 to 50°.

In order to pre-process the AIS data and pick out the useful data, an application which can transfer raw database file with dbf format to csv (comma-separated values) format is required since csv format file is constituted by lines of tabular data records and is much easier to be handled by researchers. Arcmap is the most frequent cited Geographic Information System (GIS) software and is mainly used to view, edit, create and analyze geospatial data. Since Arcmap is selected as our data transformation software, a tool of it named "export feature attribute to ASCII" was used for exporting feature class coordinates and attribute values to a space, comma, or semicolon-delimited ASCII text file. The exporting result is presented in Fig. 12 in Sect. 4.

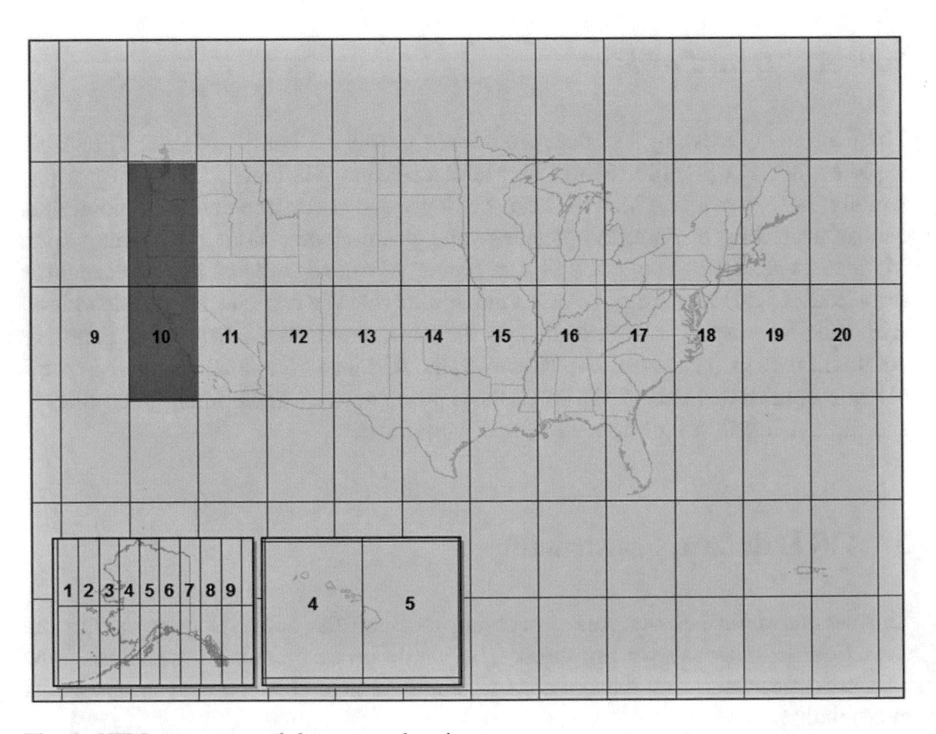

Fig. 1 UTM zone map and data source location

3.2 Raw Data Selecting

After raw data pre-processing, it is necessary to select the candidate data from the raw data with Excel format. Data selection contains two steps. First, for manipulation convenience in the following operations, the whole raw data are sorted in increasing time order and then sorted again by MMSI. One MMSI represents one single vessel. Thus in this way, the track of each single ship can be displayed in chronological order and easier to process. The second step is to calculate route complexity and longest duration of navigation.

- Longest duration of navigation

If the SOG value of a vessel meets the following inequality, we call this vessel is in the navigation condition.

$$SOG \neq 0 \tag{1}$$

Therefore, the longest duration of navigation is defined as the longest continuous nonzero SOG sequence of the AIS messages in the navigation condition. As a standard AIS database for maritime trajectory prediction and data mining, one single route should contain substantial information. Thus, the route with short duration which contains not enough data for training and testing on the route prediction and data mining algorithms. Based on our experience, the selection requirement of this property is that the trajectory data contain more than 500 AIS messages.

- Route complexity

For each single route, the $\cos\theta$ of each ship position is calculated and the definition of route complexity is the mean value of $\cos\theta$ (Fig. 2), which can be calculated by the following equation:

$$\cos\theta_i = \frac{\overrightarrow{P_iP_{i-1}} \bullet \overrightarrow{P_{i+1}P_i}}{\left|\overrightarrow{P_iP_{i-1}}\right|\left|\overrightarrow{P_{i+1}P_i}\right|} \tag{2}$$

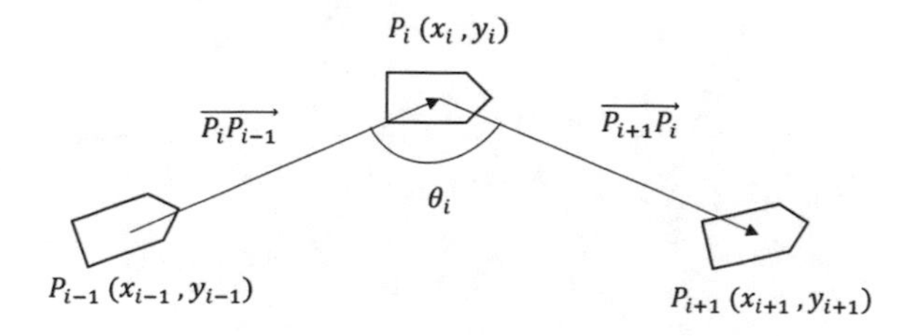

Fig. 2 Sample of route complexity

Where $P_i(x_i, y_i)$ is vessel position at T_i; x_i is the longitude of vessel at T_i; y_i is the latitude of vessel at T_i and $\overrightarrow{P_iP_{i-1}}$ is the vector $(x_i - x_{i-1}, y_i - y_{i-1})$. The route complexity should be larger than 0.8 in our database since the trajectory whose complexity is lower than 0.8 is tangled.

3.3 Candidate Data Cleaning

After candidate data were obtained, further selection based on trajectories is required. All trajectories of candidate data are plotted by MATLAB. Among all the trajectories, we defined three noisy trajectory types as follows. (showed from Figs. 3 to 6 and the horizontal axis is longitude and the vertical axis is latitude), we then removed them all:

- The discontinuous trajectory is showed in Figs. 3 and 4.
- The loose trajectory is showed in Fig. 5.
- The tangled trajectory is shown in Fig. 6.

Because these noisy trajectories have some inherent drawbacks. Routes prediction and data mining algorithms cannot learn the patterns of routes. And the shapes of noisy routes are not typical. Once the noisy trajectories are identified, they should be removed. Finally, 200 useful trajectories which contain 403599 AIS records are saved and used to construct the standard AIS database. Figure 7 shows some typical trajectories in our database (The horizontal axis is longitude and the vertical axis is latitude).

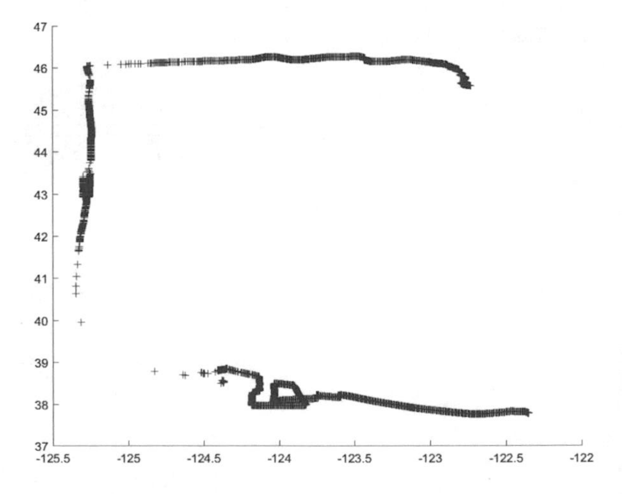

Fig. 3 Discontinuous trajectory sample-1

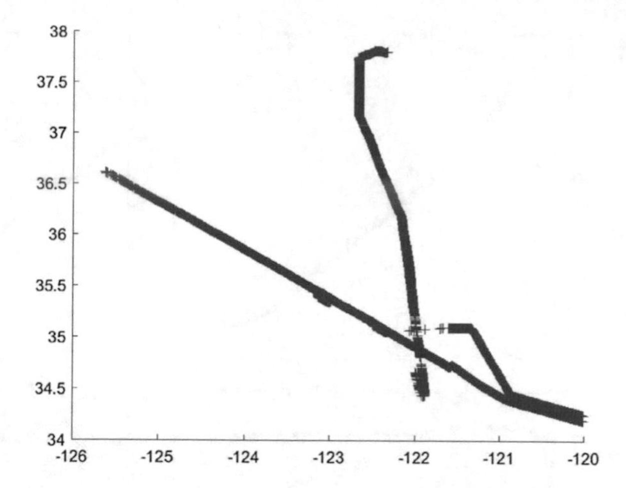

Fig. 4 Discontinuous trajectory sample-2

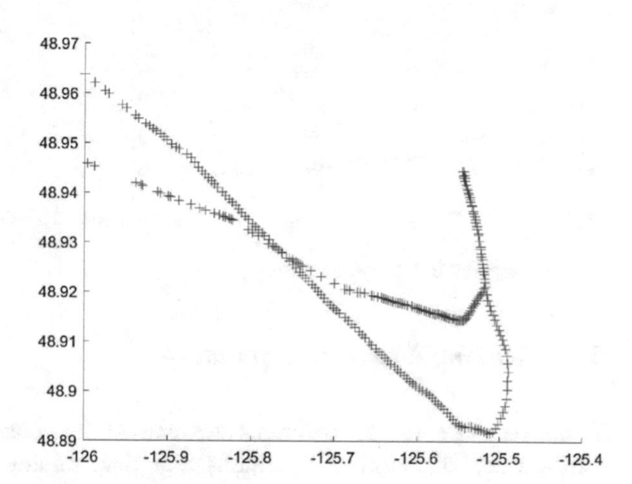

Fig. 5 Loose trajectory sample

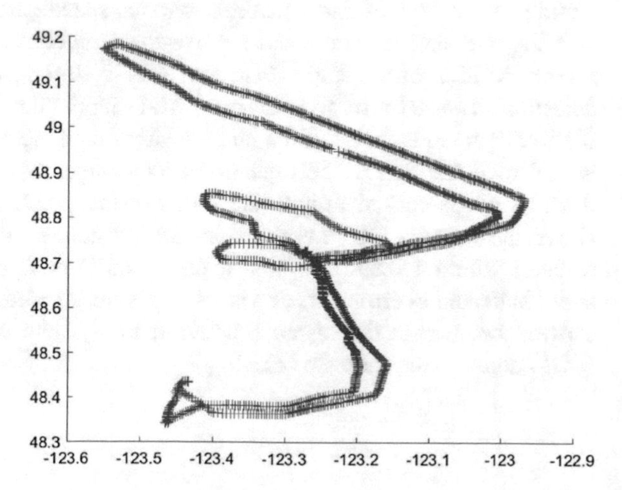

Fig. 6 Tangled trajectory sample

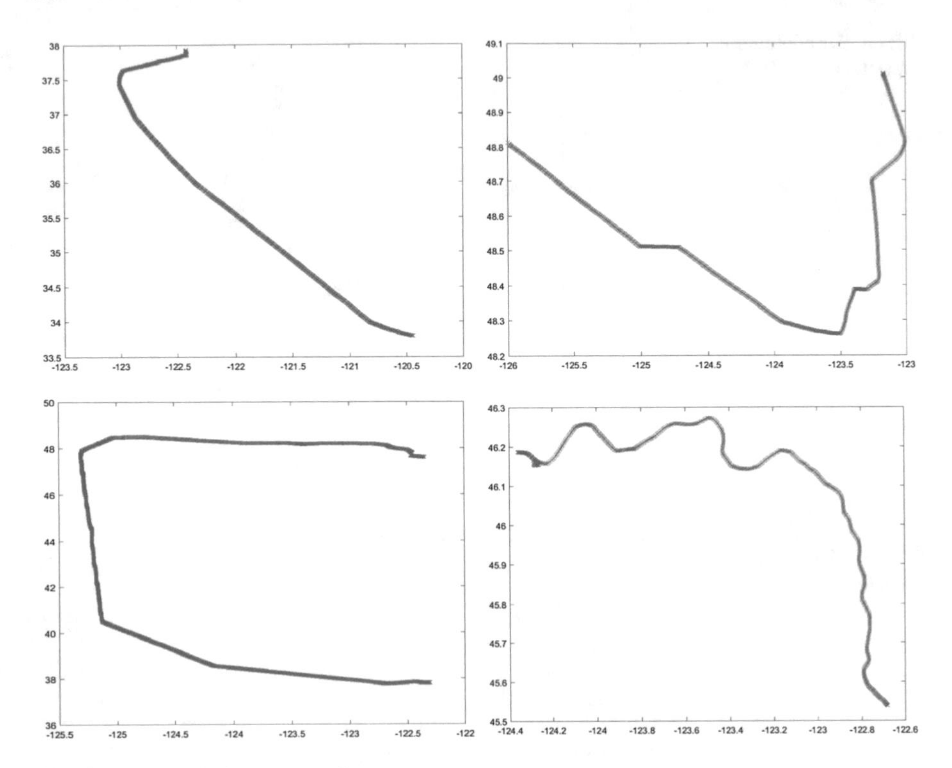

Fig. 7 Reserved trajectory sample

3.4 Missing Value Interpolating

In our database, the discontinuousness caused data missing values may affect the performance of learning algorithms and data mining quality of the database. Besides, the raw data also contain erroneous speed data. Before performing interpolation, we have to detect and remove the erroneous data. Each AIS record represents position of the ship. There are 403599 ship positions in the database. The detection of speed errors is based on SOG jump (the difference between current SOG and previous SOG). If the jump is larger than the threshold we set in advance, we calculate the distance between the two messages using the latest speed and test if this distance is consistent with the actual distance between the messages given by Haversine formula [10], i.e. the calculated distance should be close to the actual distance within a small threshold if the speed jump is correct. If not so, the latest speed is treated as erroneous and is set to previous speed. The Harversine formula is showed below.d is the distance between two points with longitude and latitude (ψ, ϕ) and r is the radius of Earth.

XCoord	YCoord	SOG	COG	time	MMSI
-121.1481	34.825067	20	330	200901071138	366882000
-121.151967	34.830567	102	360	200901071139	366882000
-121.155453	34.83544	21	329	200901071140	366882000

Fig. 8 Example of erroneous SOG jump

XCoord	YCoord	SOG	COG	ROT	BASEDATETIME	MMSI
-124.9991	43.2833	12	359	0	200902011307	258919000
-124.999217	43.298783	12	0	0	200902011311	258919000

Fig. 9 Exmaple of missing data pair

$$d = 2r\sin^{-1}\left(\sqrt{\sin^2\left(\frac{\phi_2 - \phi_1}{2}\right) + \cos(\phi_1)\cos(\phi_2)\sin^2\left(\frac{\psi_2 - \psi_1}{2}\right)}\right) \quad (3)$$

The second row in Fig. 8 is an example of incorrect SOG jump. In order to interpolate the missing values efficiently, all of the AIS records with speed errors should be corrected in advance.

For path interpolation, there are three steps: detecting data missing, judging if it needs interpolation and making linear interpolation. Data missing occurs when the time interval period between two consecutive messages is larger than one chosen interval. We choose 1 min as the threshold interval in this paper. Once detected, these two row data are defined as the missing data pair. A sample of missing data pair is shown in Fig. 9 in which there is a 5 min interval between the two consecutive messages. Then the missing time period is defined as the time range between the missing data pair and the great-circle distance between missing data pair calculated by Haversine formula [10]. The computed distance divides the SOG (km/min) of the earlier position. The division result, that is larger than two, requires linear interpolation. Figures 10 and 11 show a trajectory example before and after the interpolation (From Figs. 10 to 11, the horizontal axis is longitude and the vertical axis is latitude).

The principle of linear interpolation is that we presume that the ship is in uniform linear motion during the missing time period and the speed is considered as the SOG of the earlier position. The calculation and interpolation of the missing data are based on these two assumptions.

4 Description of AIS Database

In this section, we introduce the standard AIS database we constructed in two parts: the structure and statistical information of the database.

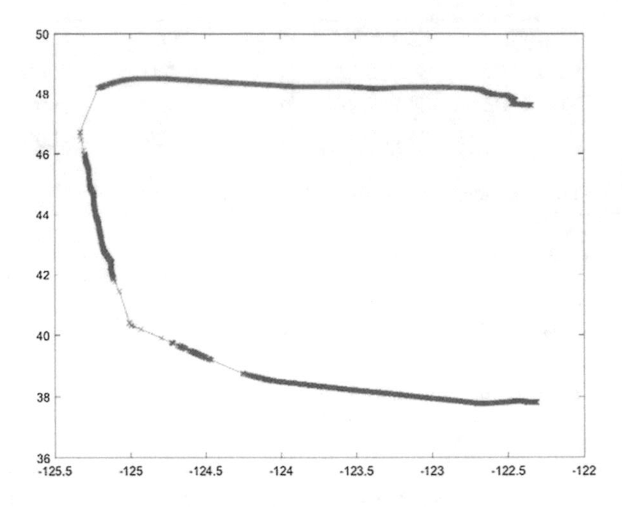

Fig. 10 Before interpolation

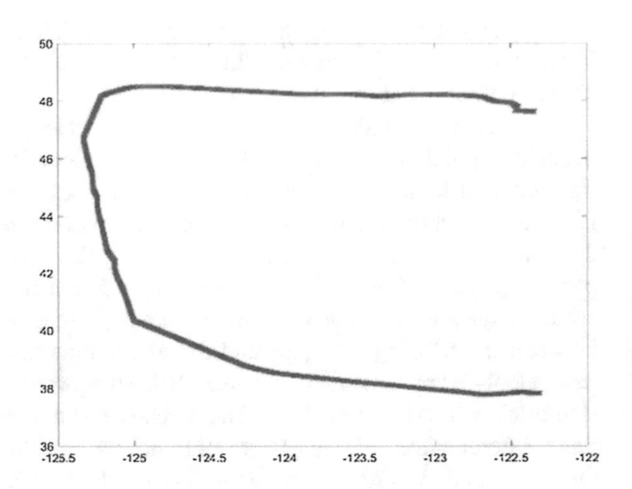

Fig. 11 After interpolation

4.1 Structure

The whole AIS database contains 200 clean trajectories stored in 200 csv file.[1] Each file is named by the MMSI and sorted in increasing time order. Each csv file contains latitude, longitude, SOG, COG, ROT, time and MMSI of a single ship. Figure 12 presents part of one xlsx file as an example.

[1]The files will be uploaded to UCI Machine Learning repository (http://archive.ics.uci.edu/ml/).

XCoord	YCoord	SOG	COG	ROT	BASEDATETIME	MMSI
-120.003717	34.242683	20	285	0	200902012013	235844000
-120.0102	34.244133	20	285	0	200902012014	235844000
-120.016783	34.245633	20	285	0	200902012015	235844000

Fig. 12 Example of a csv file

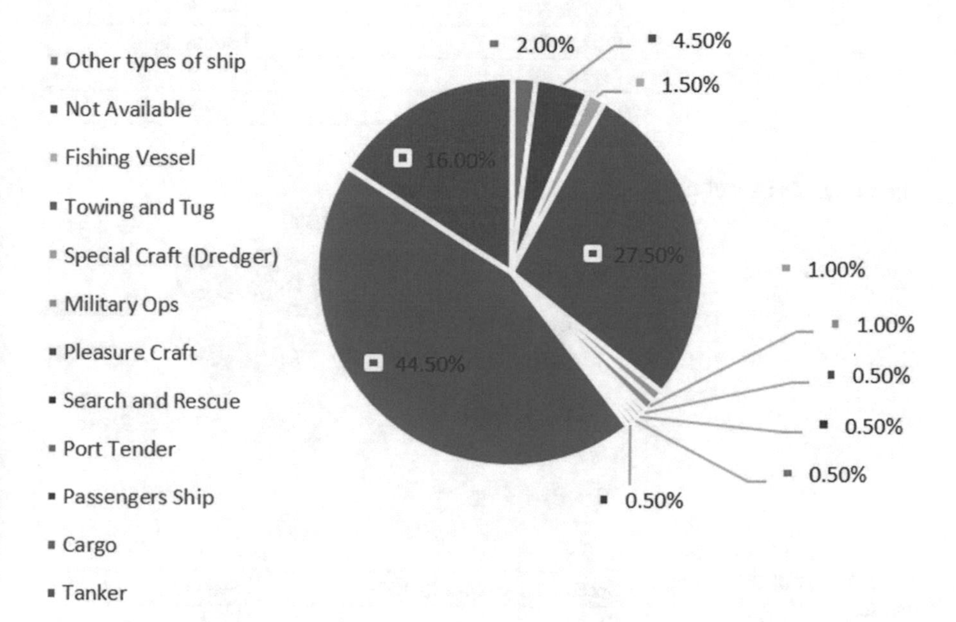

Fig. 13 Distribution of vessel types

4.2 *Statistical Information*

The raw data was chosen from limited area and time periods. The range of longitude is from −120 to −126° and the latitude is from 30 to 50° as we showed in Fig. 1. The complexities of routes are larger than 0.87.

There are many ship types within this database including cargo ships, tankers, tugs, ships engaged in military operations, etc. The detail distribution of vessel types is shown in the Fig. 13. COG is the actual direction of a vessel and is often affected by the weather over sea. All AIS data are divided into eight statuses according to the COG values as we showed in Table 1 [11]. Figure 14 shows the distribution of vessels' direction. SOG represents the speed of vessel and is an important parameter for us to analyze the AIS data. Five statues of SOG are defined [11] according to AIS dynamic information and listed in Table 2. The percentage of each SOG status is showed in Fig. 15. The length of each route is another important property of AIS data and could help us analyze the route length. According to the data amount of each ship, all two hundred ship routes are separated into four types:

Table 1 List of COG statuses

Course over ground (COG)	Statuses
[337.5, 360] ∪ [0, 22.5)	North
[22.5, 67.5)	Northeast
[67,5, 112.5)	East
[112.5, 157.5)	Southeast
[157.5, 202.5)	South
[202.5, 247.5)	Southwest
[247.5, 292.5)	West
[292.5, 337.5)	Northwest

Fig. 14 Distribution of COG

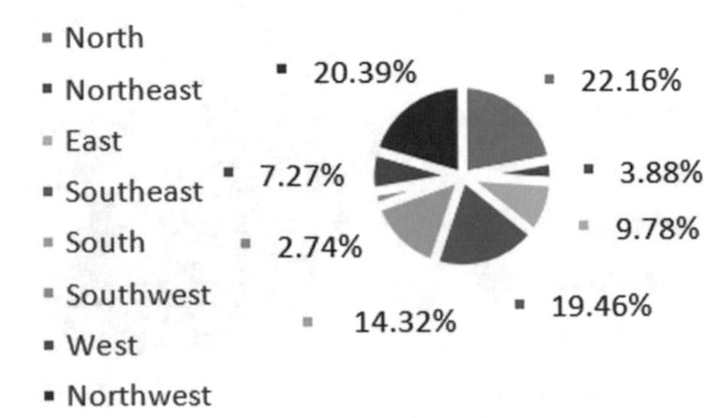

Table 2 List of SOG statuses

Speed over ground (SOG)	Statuses
[0, 3)	Slow
[3, 14)	Medium
[14, 23)	High
[23, 99)	Very high
Over 99	Exception

Fig. 15 Distribution of SOG

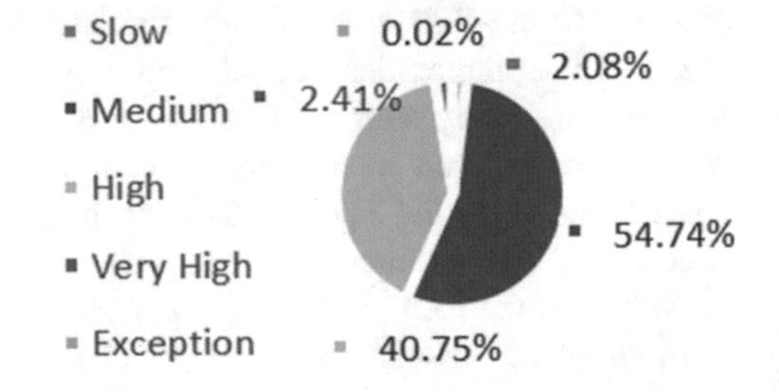

short route, medium route, long route and exception route as we listed in Table 3. Figure 16 shows the proportion of each route type with different trajectory length. After linear interpolation, the length of each trajectory has changed so we summarize the distribution of interpolated trajectories in Fig. 17, from which we can see

Table 3 List of route types

Data quantity range	Route types
[530, 1000)	Short
[1000, 2000)	Medium
[2000, 10000)	Long
Over 10000	Exception

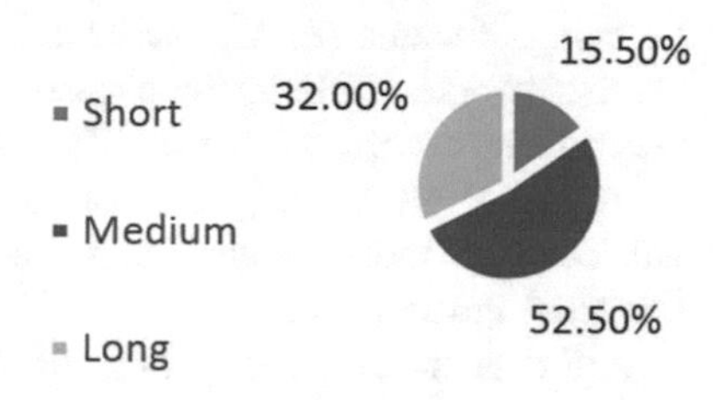

Fig. 16 Length of original route distribution

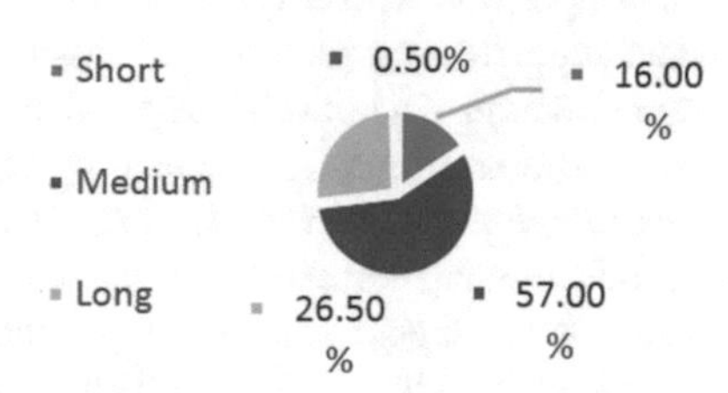

Fig. 17 Length of interpolated route distribution

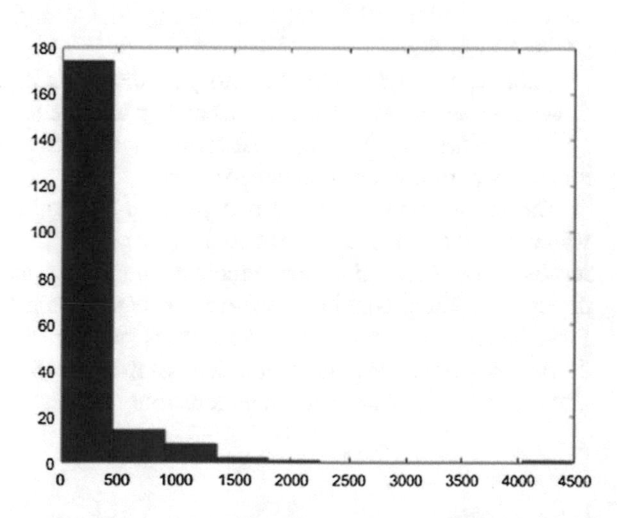

Fig. 18 Histogram of interpolated length

that most of the trajectories in the processed database belong to medium and long categories. Figure 18 is the histogram of how many positions are interpolated. (The horizontal axis is interpolated length and the vertical axis is number of trajectories for each interpolated length).

5 Experiments

In this section, we give an experimental demonstration of the usefulness of this database and summarize the experimental results. The experiments are conducted for maritime trajectory learning and prediction that are aimed to predict future motion of ship based on the current position and historic movement. Extreme Learning Machine (ELM) is a learning algorithm for single hidden layer feedforward networks (SLFNs) with random hidden nodes [12, 13]. ELM have good generalization performance and fast learning speed. As a preliminary experiment, we run ELM on our AIS database to predict the vessel trajectory. In the future, we will compare more different algorithms on our AIS database e.g. Kalman Filter, Gaussian Process, etc.

AIS data are time series data and new data of each feature comes continuously with potentially uneven time interval. It is a question how these machine learning algorithms could make the utmost use of historical data to predict different future vessel position with a dynamic prediction time. In this project, we use the following data segmentation method, as illustrated in Fig. 19. Suppose the training set contains s samples and each sample feature has length l. The prediction time is t_p. We start at current time t_c, indicated by red line. The first training sample is cut at time tick t_c-t_p-l with time length l and its target value is vessel position at time t_c. The second training sample is cut at time tick t_c-t_p-l-1 with time length l and its target value is vessel position at time t_c-1 and so on for the rest of all training samples. The testing sample is cut backwards at t_c with feature length l. This makes sure that the latest data can be utilized for both training and testing, without any dependence to future information.

ELM is used to make the trajectory prediction in these experiments. We analyze the performance of the algorithm according to these testing results. In order to make a comprehensive evaluation of ELM, predicting the same trajectory in 20 min and 40 min is performed in this experiment.

The testing results contain two parts: prediction results and error distribution which are described and discussed accordingly in this section. The prediction results show the original and predicted trajectory and algorithm performance in direct way. The prediction results of 20 and 40 min interval are presented in the Figs. 20 and 21. From these two figures, we can find the ELM performance of 20 min experiment is much better than 40 min one. Since the vessel motion is often affected by the dynamic and unpredictable sea weather, the task of predicting the

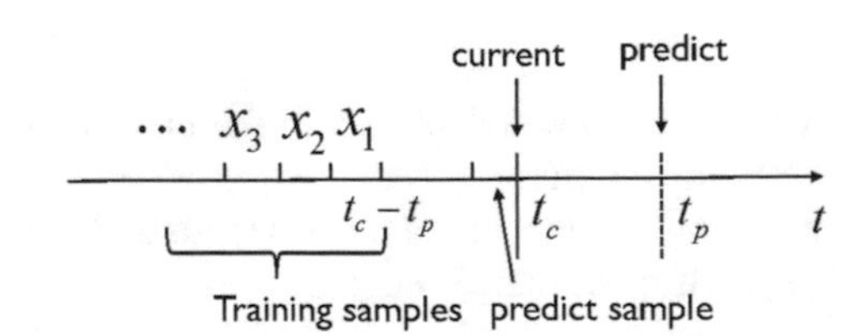

Fig. 19 Sample segmentation of trajectory prediction

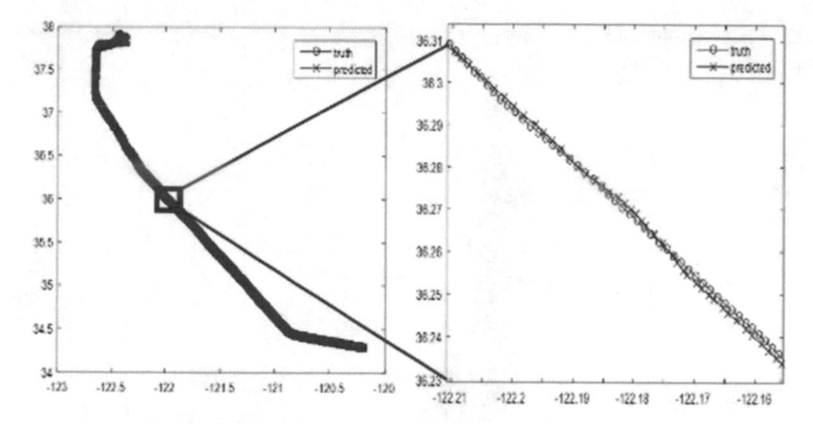

Fig. 20 Prediction results: 20 min

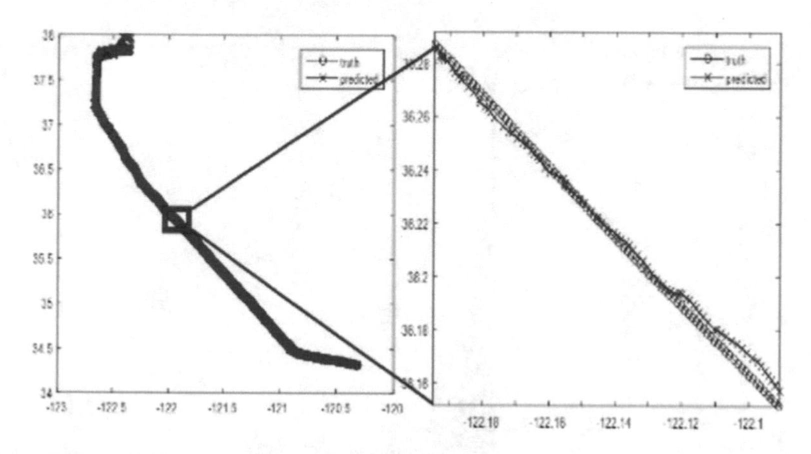

Fig. 21 Prediction results: 40 min

route in 40 min is more challenging and complex. (From Figs. 20 to 21, the horizontal axis is longitude and the vertical axis is latitude)

The error distribution helps us to analyze the algorithm performance in a quantitative way. Error is defined as the earth surface distance between real position and predicted position. And this distance can be calculated by Harversine formula which has been showed in Sect. 3.4. As we showed in Fig. 22, the error is from 0 to 2.5 when the experiment is to predict the trajectory in 20 min. Most of them are from 0 to 0.5. Figure 23 shows the error of 40 min distributing from 0 to 6 and most of them are in the range between 0 and 1 (From Figs. 22 to 23, the horizontal axis is average error (Nautical mile) and the vertical axis is number of predictions for each error).

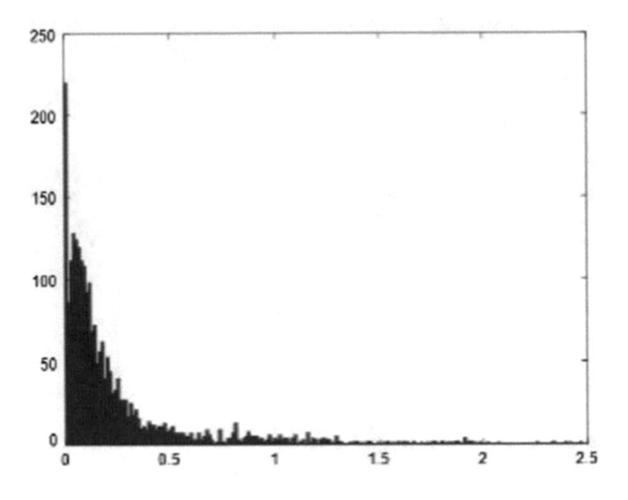

Fig. 22 Error distribution: 20 min prediction results

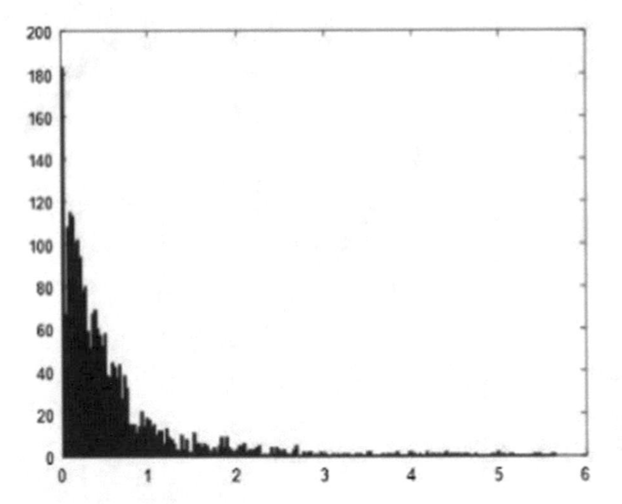

Fig. 23 Error distribution: 40 min prediction results

6 Conclusions

In conclusion, in this paper we present a standard AIS database and described the details of constructing it. The testing results on our AIS database demonstrate its potential value of serving as a benchmark for maritime trajectory learning, prediction and data mining. Our following work will be focused on conducting more experiments on the database by including more machine learning algorithms, such as manifold clustering algorithm [14] and semi-supervised learning algorithms [15]. In the future, this database can also be used as a benchmark database to verify the efficiency of other novel AIS data mining algorithms and compare their performances.

Acknowledgements This work was conducted within the Rolls-Royce@NTU Corporate Lab with support from the National Research Foundation (NRF) Singapore under the Corp Lab@University Scheme.

References

1. Kaluza, P., Kölzsch, A., Gastner, M.T., Blasius, B.: The complex network of global cargo ship movements. J. R. Soc. Interface **7**(48), 1093–1103 (2010)
2. Marinecadastre website. http://marinecadastre.gov/ais/
3. The Sailwx website. http://sailwx.info/
4. Ristic, B., La Scala, B., Morelande, M., Gordon, N.: Statistical analysis of motion patterns in AIS data: anomaly detection and motion prediction. In: 11th International Conference on Information Fusion, pp. 1–7 (2008)
5. Sampath, P., Parry, D.: Trajectory analysis using automatic identification system in New Zealand Waters. In: International Journal of Computer and Information Technology, vol. 2, pp. 132–136 (2013)
6. Marine Traffic website. http://www.marinetraffic.com/
7. The VT Explorer website. http://www.vtexplorer.com/
8. The FleetMon website. https://www.fleetmon.com/
9. The Aprs website. http://aprs.fi/page/api
10. Chopde, Nitin R., Nichat, M.: Landmark based shortest path detection by using A* and Haversine formula. Int. J. Innov. Res. Comput. Commun. Eng. **1**(2), 298–302 (2013)
11. Deng, F., Guo, S., Deng, Y., Chu, H., Zhu, Q., Sun, F.: Vessel track information mining using AIS data. In: 2014 International Conference on Multisensor Fusion and Information Integration for Intelligent Systems (MFI), pp. 1–6. IEEE (2014)
12. Huang, G.B., Zhu, Q.Y., Siew, C.K.: Extreme learning machine: Theory and applications. Neurocomputing **70**(1–3), 489–501 (2006)
13. Huang, G.-B., Zhou, H., Ding, X., Zhang, R.: Extreme learning machine for regression and multiclass classification. In: IEEE Transactions on Systems, Man, and Cybernetics, Part B (Cybernetics), vol. 42, no. 2, pp. 513–529 (2012)
14. Tu, E., Cao, L., Yang, J., Kasabov, N.: A novel graph-based k-means for nonlinear manifold clustering and representative selection. In: Neurocomputing, vol. 143 pp. 109–122 (2014)
15. Tu, E., Yaqian, Z., Lin, Z., Yang, J., Kasabov, N.: A Graph-Based Semi-Supervised k Nearest-Neighbor Method for Nonlinear Manifold Distributed Data Classification (2016). arXiv preprint arXiv: 1606.00985

Back Propagation Convex Extreme Learning Machine

Weidong Zou, Fenxi Yao, Baihai Zhang and Zixiao Guan

Abstract Recently, extreme learning machine has greatly improved in training speed and learning effectiveness of feedforward neural network which includes one hidden layer. However, the random initialization of ELM model parameters can bring randomness and affect generalization ability. The paper proposed back propagation convex extreme learning machine (BP-CELM), in which the hidden layer parameters $(\mathbf{a}, \mathbf{b})$ can be calculated by formulas. The convergence of BP-CELM is proved in the paper. Simulation results show that BP-CELM has higher training speed and better generalization performance than other randomized neural network algorithms.

Keywords Back propagation convex extreme learning machine · Learning effectiveness · Generalization performance · Hidden layer parameter

1 Introduction

Owing to its universal approximation performance, ELM [1, 2] has been successfully applied in plenty of practical application, such as clustering learning [3], data analysis [4], automation control [5, 6] and semi-supervised learning [7]. An ELM is composed of an input layer, a hidden layer and an output layer. For N arbitrary distinct samples $\{(\mathbf{x}_i, \mathbf{t}_i)\}_{i=1}^{N}$, where $\mathbf{x}_i \in \mathbf{R}^n$ and $\mathbf{t}_i \in \mathbf{R}^m$, ELM model with n additive nodes can be indicated by

$$f_n(\mathbf{x}) = \sum_{i=1}^{L} \beta_i h(\mathbf{a}_i \cdot \mathbf{x}_j + b_i) = \sum_{i=1}^{L} \mathbf{H}_i \cdot \boldsymbol{\beta}_i, j = 1, \ldots, N, \mathbf{a}_i \in \mathbf{R}^n, \beta_i, b_i \in R \quad (1)$$

W. Zou · F. Yao (✉) · B. Zhang · Z. Guan
School of Automation, Beijing Institute of Technology, Beijing, China
e-mail: yanfenxi1964@163.com

J. Cao et al. (eds.), *Proceedings of ELM-2016*, Proceedings in Adaptation, Learning and Optimization 9, DOI 10.1007/978-3-319-57421-9_21

where $\mathbf{a}_i$, b_i and β_i are the parameters of ELM, and h is the mapping functions of hidden node, $\mathbf{a}_i \cdot \mathbf{x}$ indicates the inner product of vectors.

During the past ten years, various improvements have been made to primordial ELM algorithm, such as bidirectional ELM [8], constrained ELM [9], and hessian semi-supervised ELM [10], to obtain better learning effectiveness and faster training speed.

However, randomly generating hidden nodes parameters **(a, b)** result in certain randomness [11] and inefficient use of hidden nodes [12]. The generalization ability of ELM is under expectation by randomly generated many hidden nodes. This will spend much time in training process and testing process, which is useless in practical applications.

In order to solve these problems, we proposed back propagation convex extreme learning machine (BP-CELM) algorithm that can quickly reduce the residual error of neural network by finding more appropriate hidden nodes parameters **(a, b)**. The contributions and novelty of this paper reside in the following two aspects.

(1) The learning speed of BP-CELM has higher training speed than other neural network algorithms, such as Support Vector Machine (SVM), back propagation neural network (BP), I-ELM, EM-ELM and so on.
(2) Different from other ELM methods, all the parameters $(\mathbf{a}, \mathbf{b}, \boldsymbol{\beta})$ of ELM can be calculated by some formulas. It can provide better generalization capability than other methods.

2 Preliminaries

2.1 CI-ELM

According to Barron's convex optimization method [13], Huang et al. [14] proposed convex I-ELM which denotes the residual error function of f_n as $e_n = f - f_n$ where $f \in L^2(\mathbf{X})$ is the target function. For the additive hidden nodes, the mathematical form of CI-ELM is

$$f_n(\mathbf{x}) = (1 - \beta_{n-1}) f_{n-1}(\mathbf{x}) + \beta_n h(\mathbf{a}_n \cdot \mathbf{x} + b_n) \tag{2}$$

The following lemma has been proved by Huang et al. [14].

Lemma 2.1 ([14]) *For any continuous target function f, a SLFN with additive hidden nodes* $\mathbf{H}(\mathbf{a} \cdot \mathbf{x} + \mathrm{b})$ *which are nonconstant piecewise continuous, then any randomly generated sequence of function* $\mathbf{H}_n^r(\mathbf{x}) = \mathbf{H}(\mathbf{a}_n \cdot \mathbf{x} + \mathrm{b}_n)$ *based on any continuous sampling distribution,* $\lim_{n \to \infty} \left\| f - ((1 - \beta_{n-1}) f_{n-1} + \beta_n \mathbf{H}_n^r) \right\| = 0$ *holds with probability 1 if*

$$\beta_n = \frac{\langle e_{n-1}, \mathbf{H}_n^r - f_{n-1} \rangle}{\left\| \mathbf{H}_n^r - f_{n-1} \right\|^2} \quad (3)$$

where $e_n = f - f_n$ *presents the residual error of ELM* f_n *with n additive hidden nodes.*

2.2 ELM with Subnetwork Hidden Layer Nodes

For improving the learning speed and learning effectiveness of ELM, Yang et al. [15] proposed ELM with subnetwork hidden layer nodes which can produce subnetwork hidden layer nodes by drawing back neural network residual error to hidden layer. All the parameters $(\mathbf{a}, \mathbf{b}, \boldsymbol{\beta})$ of ELM-based learning algorithm are got by Lemma 2.2.

Lemma 2.2 ([15]) *For N arbitrary distinct samples* $\{(\mathbf{x}_i, \mathbf{t}_i)\}_{i=1}^N$, *where* $\mathbf{x}_i \in \mathbf{R}^n$ *and* $\mathbf{t}_i \in \mathbf{R}^m$, *if the mapping function of hidden layer node h is sigmoidal, given a normalized function* $u: \mathbf{R} \to (0, 1]$, *if the mapping function of hidden layer node h is cosine or sine, given a normalized function* $u: \mathbf{R} \to [0, 1]$, *and then any continuous target outputs t, we have* $\lim_{n \to \infty} \left\| t - \left(u^{-1}(h(\hat{\mathbf{a}}_1 \cdot \mathbf{x} + \hat{b}_1)) \cdot \boldsymbol{\beta}_1 + \ldots + u^{-1}(h(\hat{\mathbf{a}}_n \cdot \mathbf{x} + \hat{b}_n)) \cdot \boldsymbol{\beta}_n \right) \right\| = 0$ *holds with probability 1 if*

$$\hat{\mathbf{a}}_n = h^{-1}(u(e_{n-1})) \cdot \mathbf{x}^T \left(\frac{C}{\mathbf{I}} + \mathbf{x}\mathbf{x}^T \right)^{-1}, \hat{\mathbf{a}}_n \in \mathbf{R}^{n \times m} \quad (4)$$

$$\hat{b}_n = sum(\hat{\mathbf{a}}_n \cdot \mathbf{x} - h^{-1}(u(e_{n-1})))/N, \hat{b}_n \in \mathbf{R} \quad (5)$$

$$\boldsymbol{\beta}_n = \frac{\langle e_{n-1}, u^{-1}(h(\hat{\mathbf{a}}_n \cdot \mathbf{x} + \hat{b}_n)) \rangle}{\left\| u^{-1}(h(\hat{\mathbf{a}}_n \cdot \mathbf{x} + \hat{b}_n)) \right\|^2} \quad (6)$$

where $\mathbf{x}^T (\frac{C}{\mathbf{I}} + \mathbf{x}\mathbf{x}^T)^{-1} = \mathbf{x}^{-1}$ *is the Moore-Penrose generalized inverse of training samples, and* h^{-1} *and* u^{-1} *indicate its reverse function, if h is sigmoidal function,* $h^{-1}(\cdot) = -\ln(\frac{1}{(\cdot)} - 1)$, *if h is sine function,* $h^{-1}(\cdot) = \arcsin(\cdot)$.

3 Back Propagation Convex Extreme Learning Machine

The structure of BP-CELM will be indicated in Sect. 3.1. In Sect. 3.2, we prove that BP-CELM with additive hidden layer nodes can approximate any continuous target functions. The code of BP-CELM is given in Sect. 3.3.

3.1 Structure of the Proposed BP-CELM

BP-CELM can quickly reduce the residual error of ELM by finding better hidden layer nodes parameters $(\mathbf{a}, b)$ and the corrected output weights $\hat{\beta}$. The parameters of hidden layer nodes $(\mathbf{a}, b)$ and the corrected output weights $\hat{\beta}$ are got by Theorem 3.1. Figure 1 illustrates the structure of BP-CELM.

3.2 Proposed BP-CELM

Theorem 3.1 *For any continuous target function f, a SLFN with additive hidden nodes which are nonconstant piecewise continuous,randomly generated output weight* β_n, *and obtained output feedback function sequence* $\mathbf{H}_n$, $n \in \mathbf{Z}$, $\lim_{n \to \infty} \left\| f - \left((1 - \hat{\beta}_n) f_{n-1} + \hat{\beta}_n \hat{\mathbf{H}}_n(\mathbf{a}_n \mathbf{x} + b_n) \right) \right\| = 0$ *holds with probability 1 if*

$$\mathbf{H}_n = \mathbf{e}_{n-1} (\beta_n)^{-1} \tag{7}$$

$$\mathbf{a}_n = h^{-1}(u(\mathbf{H}_n)) \cdot \mathbf{x}^T (\frac{C}{\mathbf{I}} + \mathbf{x}\mathbf{x}^T)^{-1}, \mathbf{a}_n \in \mathbf{R}^{n \times m} \tag{8}$$

$$b_n = sum(\mathbf{a}_n \cdot \mathbf{x} \; - h^{-1}(u(\mathbf{H}_n)))/N, b_n \in \mathbf{R} \tag{9}$$

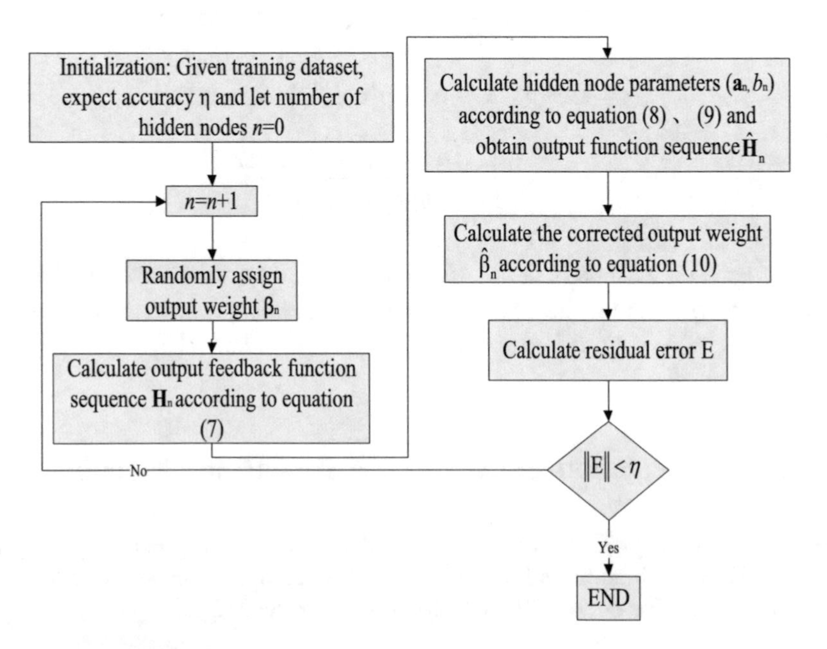

Fig. 1 Structure of BP-CELM

$$\hat{\beta}_n = \frac{\langle \mathbf{e}_{n-1}, \hat{\mathbf{H}}_n(\mathbf{a}_n \cdot \mathbf{x} + b_n) - f_{n-1} \rangle}{\left\| \hat{\mathbf{H}}_n(\mathbf{a}_n \cdot \mathbf{x} + b_n) - f_{n-1} \right\|^2} \tag{10}$$

where $\hat{\beta}_n$ is the corrected output weight, $\hat{\mathbf{H}}_n$ is output function sequence, C is regularized parameter.

Proof We first prove that the error sequence $\|e_n\|$ is decreasing, and then we further prove $\lim_{n\to\infty} \|e_n\| = 0$.

If we set activation function $h(x)$, we have $\mathbf{H}_n = h(\lambda_n)$, $\lambda_n \in \mathbf{R}$. If $h(x)$ is a sigmoidal mapping function, $\lambda_n = h^{-1}(u(\mathbf{H}_n)) = -\log(\frac{1}{u(\mathbf{H}_n)} - 1)$, if $h(x)$ is a sine mapping function, $\lambda_n = h^{-1}(u(\mathbf{H}_n)) = \arcsin(u(\mathbf{H}_n))$.

Let $\lambda_n = \hat{\mathbf{a}}_n \cdot \mathbf{x}$, if $h(x)$ is a sine mapping function, $\mathbf{a}_n = h^{-1}(u(\mathbf{H}_n)) \cdot \mathbf{x}^{-1} = \arcsin(u(\mathbf{H}_n)) \cdot \mathbf{x}^{-1}$, if $h(x)$ is a sigmoidal mapping function, $\mathbf{a}_n = h^{-1}(u(\mathbf{H}_n)) \cdot \mathbf{x}^{-1} = -\log(\frac{1}{u(\mathbf{H}_n)} - 1) \cdot \mathbf{x}^{-1}$, $\mathbf{x}^{-1}$ is the Moore-Penrose generalized inverse [16] of training samples.

In [2], we get that the least-squares solution of linear system $\hat{\mathbf{a}}_n \cdot \mathbf{x} = \lambda_n = h^{-1}(u(\mathbf{H}_n))$ is $\mathbf{a}_n = h^{-1}(u(\mathbf{H}_n)) \cdot \mathbf{x}^{-1}$, meaning that the residual error is the smallest by this least-squares solution.

$$\|\mathbf{a}_n \cdot \mathbf{x} - \lambda_n\| = \min\left\|\hat{\mathbf{a}}_n \cdot \mathbf{x} - h^{-1}(u(\mathbf{H}_n))\right\|, \mathbf{a}_n \in [-1, 1] \tag{11}$$

We get $b_n = sum(\mathbf{a}_n \cdot \mathbf{x} \ - h^{-1}(u(\mathbf{H}_n)))/N, b_n \in \mathbf{R}$.

So we have

$$\begin{aligned} &\min\left\|u^{-1}(h(\hat{\mathbf{a}}_n \cdot \mathbf{x})) - u^{-1}(\lambda_n)\right\| \\ &= \left\|u^{-1}(h(\mathbf{a}_n \cdot \mathbf{x})) - u^{-1}(\lambda_n)\right\| > \left\|u^{-1}(h(\mathbf{a}_n \cdot \mathbf{x} + b_n)) - u^{-1}(\lambda_n)\right\| = \|\sigma\| \end{aligned} \tag{12}$$

Let $\hat{\mathbf{H}}_n - f_{n-1} = u^{-1}(h(\mathbf{a}_n \cdot \mathbf{x} + b_n))$, the residual error as

$$\begin{aligned} \Delta = \|e_{n-1}\|^2 - \left\|e_{n-1} - (\hat{\mathbf{H}}_n - f_{n-1})\right\|^2 &= 2\langle e_{n-1}, \hat{\mathbf{H}}_n - f_{n-1}\rangle - \left\|\hat{\mathbf{H}}_n - f_{n-1}\right\|^2 \\ &= \left\|\hat{\mathbf{H}}_n - f_{n-1}\right\|^2 \left(\frac{2\langle e_{n-1}, \hat{\mathbf{H}}_n - f_{n-1}\rangle}{\left\|\hat{\mathbf{H}}_n - f_{n-1}\right\|^2} - 1 \right) \end{aligned} \tag{13}$$

Let

$$\hat{\mathbf{H}}_n - f_{n-1} = u^{-1}(h(\mathbf{a}_n \cdot \mathbf{x} + b_n)) = e_{n-1} \pm \sigma = \hat{e}_{n-1} \tag{14}$$

Because $\|\hat{e}_{n-1}\| \geq \|\sigma\|$, we have

$$\Delta = \left\|\hat{\mathbf{H}}_n - f_{n-1}\right\|^2 \left(\frac{2\langle \hat{e}_{n-1} \pm \sigma, \hat{e}_{n-1}\rangle}{\|\hat{e}_{n-1}\|^2} - 1\right)$$
$$= \left\|\hat{\mathbf{H}}_n - f_{n-1}\right\|^2 \left(1 \pm \frac{\left\|\sigma \cdot \hat{e}_{n-1}^T\right\|}{\|\hat{e}_{n-1}\|^2}\right) \geq \left\|\hat{\mathbf{H}}_n - f_{n-1}\right\|^2 \left(1 \pm \frac{\|\sigma\|}{\|\hat{e}_{n-1}\|^2}\right) \geq 0 \quad (15)$$

According to (15), we get $\|e_{n-1}\| \geq \left\|e_n - (\hat{\mathbf{H}}_n - f_{n-1})\right\|$, so we can prove $\|e_{n-1}\| \geq \left\|e_n - (\hat{\mathbf{H}}_n - f_{n-1})\right\| \geq \left\|e_n - \boldsymbol{\beta}_n(\hat{\mathbf{H}}_n - f_{n-1})\right\|$.

Let $\boldsymbol{\beta}_n = \mathbf{I}$, we get

$$\left\|e_{n-1} - (\hat{\mathbf{H}}_n - f_{n-1})\right\| = \left\|e_{n-1} - (\hat{\mathbf{H}}_n - f_{n-1}) \cdot \boldsymbol{\beta}_n\right\| \quad (16)$$

Because the hidden nodes parameters $(\mathbf{a}_n, b_n)$ is fixed, $\hat{\beta}_n = \frac{\langle \mathbf{e}_{n-1}, \hat{\mathbf{H}}_n(\mathbf{a}_n \cdot \mathbf{x} + b_n) - f_{n-1}\rangle}{\left\|\hat{\mathbf{H}}_n(\mathbf{a}_n \cdot \mathbf{x} + b_n) - f_{n-1}\right\|^2}$ is one of the least-square solutions of $\left(\hat{\mathbf{H}}_n(\mathbf{a}_n \cdot \mathbf{x} + b_n) - f_{n-1}\right) \cdot \beta_n = \mathbf{e}_{n-1}$.

We have

$$\left\|\left(\hat{\mathbf{H}}_n(\mathbf{a}_n \cdot \mathbf{x} + b_n) - f_{n-1}\right) \cdot \beta_n - \mathbf{e}_{n-1}\right\| = \min_{\beta}\left\|\left(\hat{\mathbf{H}}_n(\mathbf{a}_n \cdot \mathbf{x} + b_n) - f_{n-1}\right) \cdot \hat{\beta}_n - \mathbf{e}_{n-1}\right\|$$
$$\leq \left\|\left(\hat{\mathbf{H}}_n(\mathbf{a}_n \cdot \mathbf{x} + b_n) - f_{n-1}\right) - \mathbf{e}_{n-1}\right\| \quad (17)$$

Based on (15)–(17), we get $\|e_{n-1}\| \geq \left\|e_n - (\hat{\mathbf{H}}_n(\mathbf{a}_n \cdot \mathbf{x} + b_n) - f_{n-1})\right\| \geq \| e_n - \boldsymbol{\beta}_n(\hat{\mathbf{H}}_n(\mathbf{a}_n \cdot \mathbf{x} + b_n) - f_{n-1})\|$, so the sequence $\|e_n\|$ is decreasing.

In [2], when the following three sufficient conditions are satisfied, we can prove $\lim_{n\to\infty} \|e_n\| = 0$.

(1) $span\{\hat{\mathbf{H}}_n(\mathbf{a}_n \cdot \mathbf{x} + \mathrm{b}_n), (\mathbf{a}_n, \mathrm{b}_n) \in \mathbf{R}^d \times \mathrm{R}\}$ is dense in L^2;
(2) $e_n \perp (e_{n-1} - e_n)$;
(3) $\hat{\mathbf{H}}$ is a nonconstant piecewise continuous function.

The preconditions of Theorem 3.1 have satisfied conditions (1) and (3), in order to prove $\lim_{n\to\infty} \|e_n\| = 0$, we need to prove $e_n \perp (e_{n-1} - e_n)$.

since $\|e_n\| = \left\|e_{n-1} - \hat{\beta}_n\left(\hat{\mathbf{H}}_n(\mathbf{a}_n \cdot \mathbf{x} + b_n) - f_{n-1}\right)\right\|$, we have

$$\langle e_n, \hat{\mathbf{H}}_n(\mathbf{a}_n \cdot \mathbf{x} + b_n) - f_{n-1}\rangle = \langle e_{n-1} - \hat{\beta}_n\left(\hat{\mathbf{H}}_n(\mathbf{a}_n \cdot \mathbf{x} + b_n) - f_{n-1}\right), \hat{\mathbf{H}}_n(\mathbf{a}_n \cdot \mathbf{x} + b_n) - f_{n-1}\rangle$$
$$= \langle e_{n-1}, \hat{\mathbf{H}}_n(\mathbf{a}_n \cdot \mathbf{x} + b_n) - f_{n-1}\rangle - \hat{\beta}_n\left\|\hat{\mathbf{H}}_n(\mathbf{a}_n \cdot \mathbf{x} + b_n) - f_{n-1}\right\|^2 = 0 \quad (18)$$

According to formula (18), we further have

$$\langle e_n, e_{n-1} - e_n \rangle = \langle e_n, e_n + \hat{\beta}_n(\hat{\mathbf{H}}_n(\mathbf{a}_n \cdot \mathbf{x} + b_n) - f_{n-1}) \rangle - \|e_n\|^2$$
$$= \|e_n\|^2 + \hat{\beta}_n \langle e_n, \hat{\mathbf{H}}_n(\mathbf{a}_n \cdot \mathbf{x} + b_n) - f_{n-1} \rangle - \|e_n\|^2 = 0 \quad (19)$$

which means $e_n \perp (e_{n-1} - e_n)$.

This completes the proof of Theorem 3.1.

3.3 *Pseudo-code for BP-CELM*

The pseudo-code for BP-CELM can be summarized as follows:

Algorithm BP-ELM. Given a training set $\{(\mathbf{x}_i, t_i)\}_{i=1}^{N} \subset \mathbf{R}^n \times \mathbf{R}$, activation function $\mathbf{H}$, the continuous target function f and maximum number of hidden nodes $L_{\max}$, the expected learning accuracy ε,

Table 1 Specification of 24 benchmark data sets

Datasets	Type	Attribution	Training datasets	Testing datasets
Air quality	Regression	15	5358	4000
BlogFeedback	Regression	281	30021	30000
Fertility	Regression	10	60	40
Energy efficiency	Regression	8	408	360
NoisyOffice	Regression	216	110	106
SML2010	Regression	24	2100	2037
Solar Flare	Regression	10	1189	1200
UJIIndoorLoc	Regression	529	11000	10048
wiki4HE	Regression	53	513	400
YearPredictionMSD	Regression	90	315000	200345
Student performance	Regression	33	349	300
Servo	Regression	4	87	80
Abalone	Classification	8	2100	2077
Artificial characters	Classification	7	3500	2500
Cardiotocography	Classification	23	1100	1016
CNAE-9	Classification	857	580	500
Covertype	Classification	54	291012	290000
Dow Jones Index	Classification	16	450	300
Echocardiogram	Classification	12	72	60
EEG eye state	Classification	15	7980	7000
Folio	Classification	20	337	300
Gisette	Classification	5000	7500	6000
Libras movement	Classification	91	200	160
DNA	Classification	180	1046	1186

Step Initialization: Let the number of hidden nodes $L = 0$ and the residual error $E = t$, where $t = [t_1, \ldots, t_N]$.

Step Learning step:

While $L < L_{\max}$ and $\|E\| > \varepsilon$

Increase the number of hidden nodes by 1 L: $L = L + 1$.

Randomly assign output weight β_L.

Calculate output feedback function sequence $\mathbf{H}_L$ according to (7).

Calculate hidden node parameters $(\mathbf{a}_L, b_L)$ according to (8), (9), and obtain output function sequence $\hat{\mathbf{H}}_L$.

Calculate the corrected output weight $\hat{\beta}_L$ according to (10).

Calculate the residual error after adding the new hidden node L: $E = E - \hat{\beta}_L \hat{\mathbf{H}}_L$

End while

4 Simulation Verification

For testing the generalization performance of BP-CELM, in this section, we test it on regression applications and classification applications. The simulations are conducted in Matlab 2010a running on Windows 7 with at 4 GB of memory and two Dual-Core E5300 (2.60 GHZ) processors. Learning algorithms are tested with I-ELM, CI-ELM, EI-ELM, BP, SVM, SOM, Elman and BP-CELM.

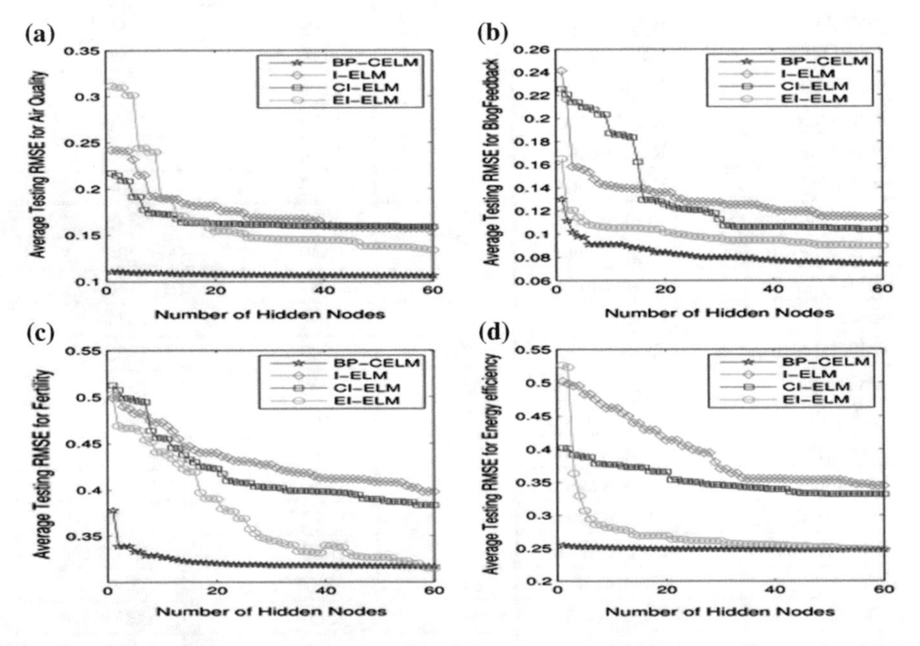

Fig. 2 Testing RMES of different algorithms with sine hidden nodes in air quality, BlogFeedback, fertility and energy efficiency

4.1 Data Description

Twelve regression datasets and twelve classification datasets are selected for the experiments and described in Table 1. The data sets are collected from the University of California at Irvine (UCI) Machine Learning Repository. We pre-process all datasets in the same way.

4.2 Experiments

BP-CELM, I-ELM, CI-ELM and EI-ELM are compared in eight regression applications and eight classification applications. The number of hidden nodes is selected from 1 to 60 by step 1. The performances of these methods with hidden nodes of sine activation function and sigmoid activation function are illustrated in Figs. 2, 3, 4, 5, 6, 7, 8 and 9.

As shown in Fig. 9, BP-CELM can get much better generalization capability than other learning algorithms. More importantly, the testing RMSE of BP-CELM

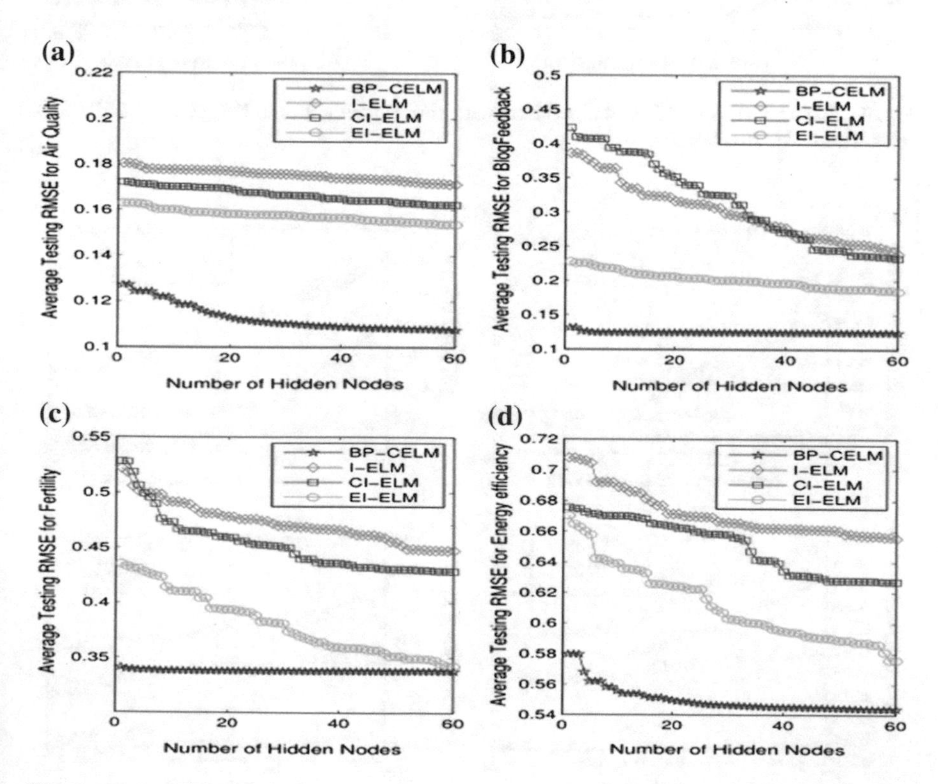

Fig. 3 Testing RMES of different algorithms with sigmoidal hidden nodes in air quality, BlogFeedback, fertility and energy efficiency

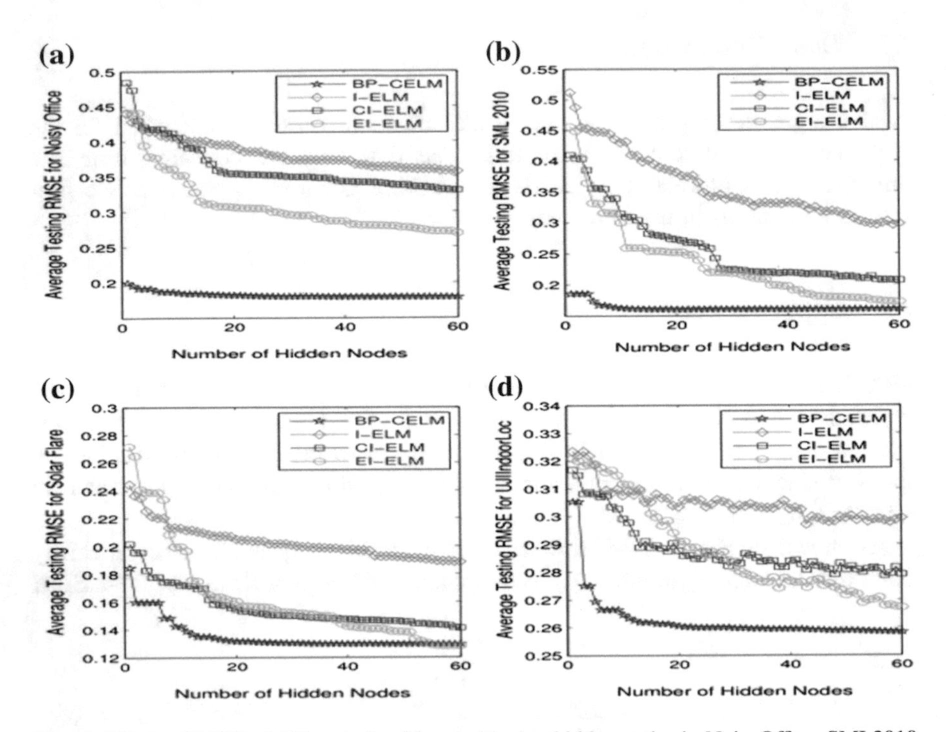

Fig. 4 Testing RMES of different algorithms with sine hidden nodes in NoisyOffice, SML2010, solar flare and UJIIndoorLoc

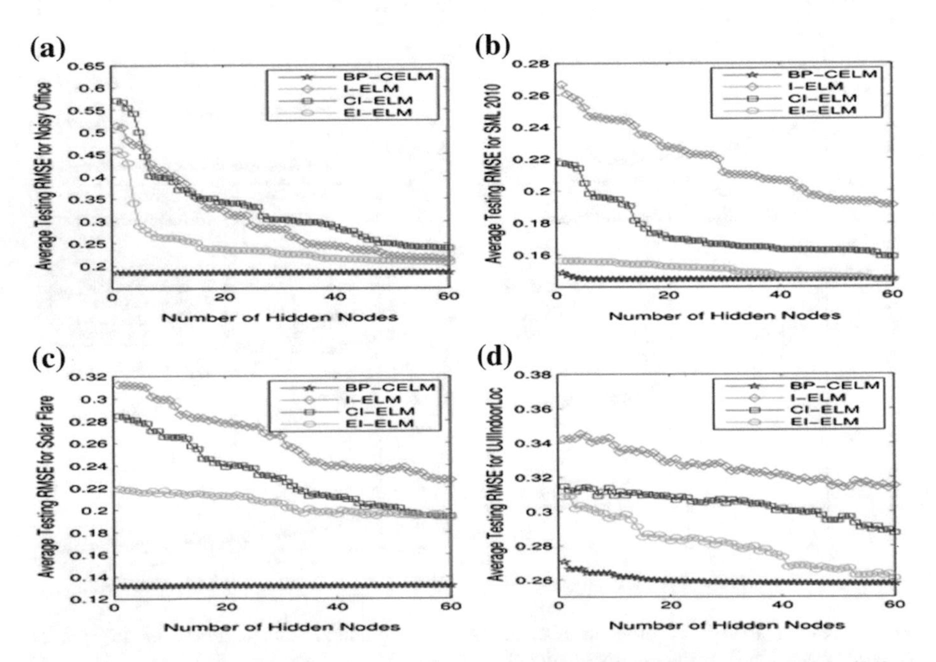

Fig. 5 Testing RMES of different algorithms with sigmoidal hidden nodes in NoisyOffice, SML2010, solar flare and UJIIndoorLoc

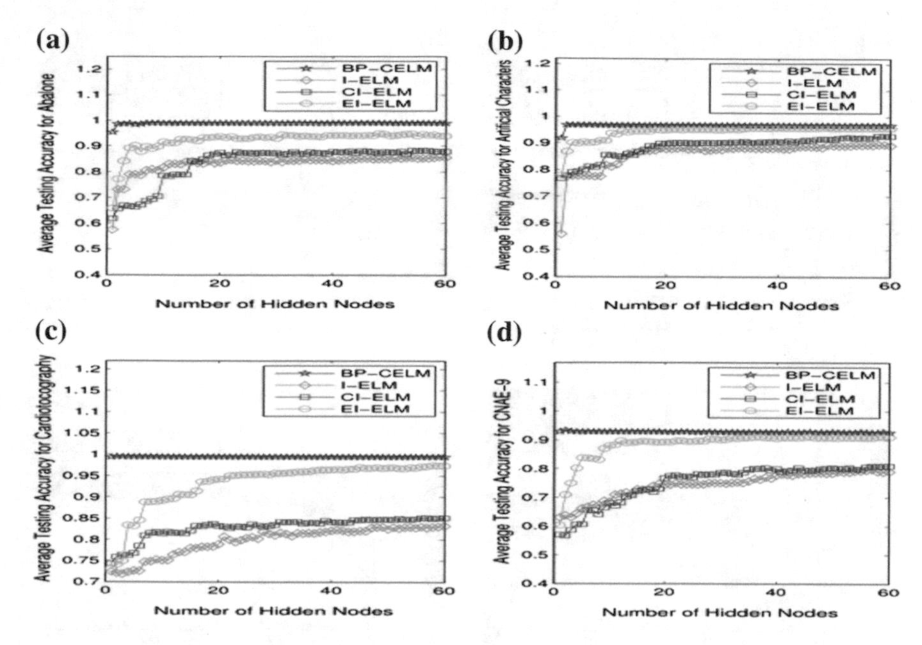

Fig. 6 Testing accuracy of different algorithms with sine hidden nodes in abalone, artificial characters, cardiotocography and CNAE-9

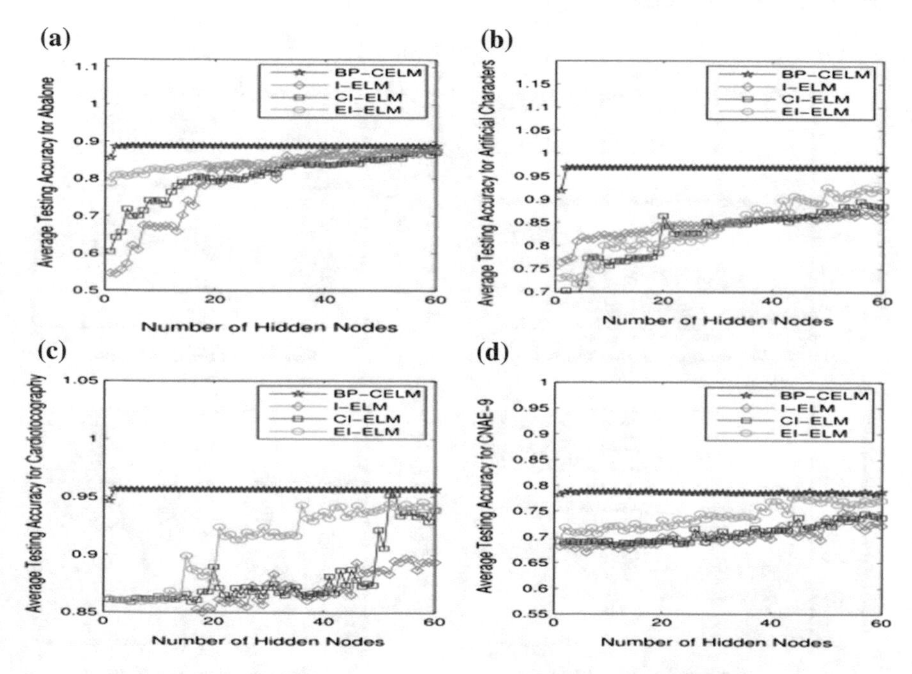

Fig. 7 Testing accuracy of different algorithms with sigmoidal hidden nodes in abalone, artificial characters, cardiotocography and CNAE-9

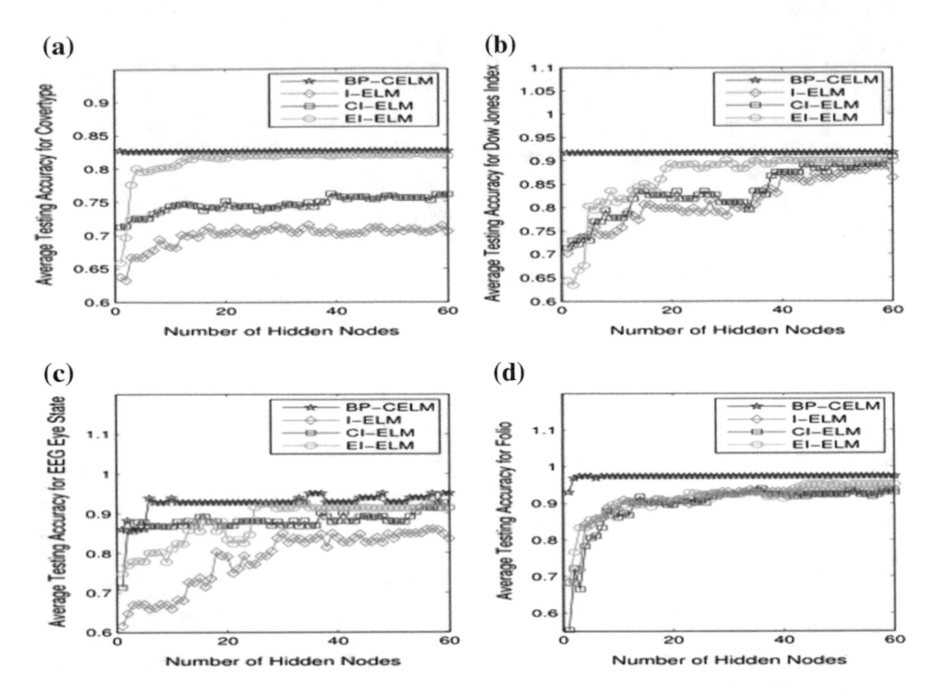

Fig. 8 Testing accuracy of different algorithms with sine hidden nodes in Covertype, Dow Jones index, EEG eye state and folio

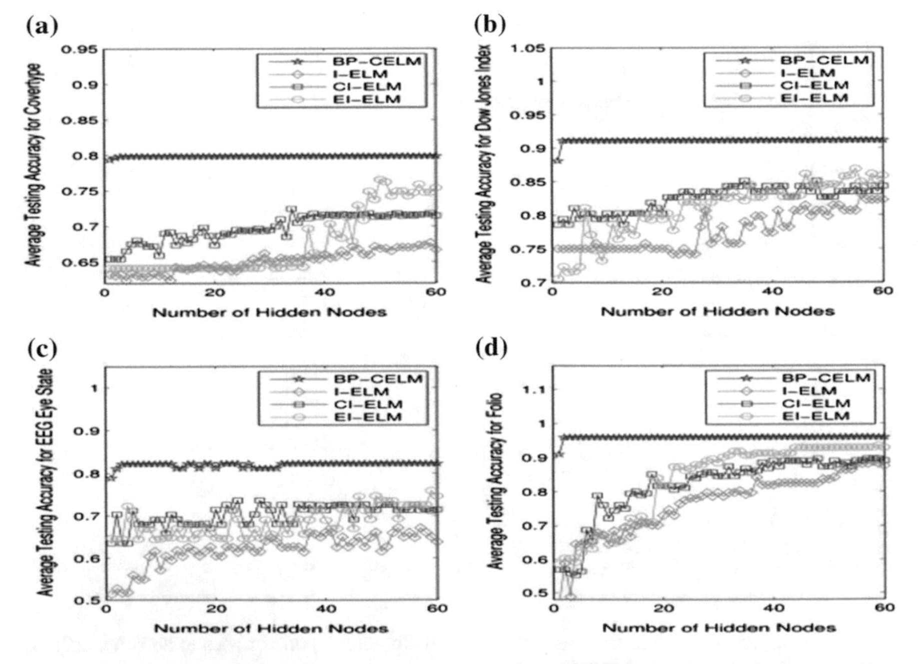

Fig. 9 Testing accuracy of different algorithms with sigmoidal hidden nodes in Covertype, Dow Jones index, EEG eye state and folio

Table 2 Generation performance comparision (average testing RMSE and average training time)

Datasets	BP		Elman		BP-CELM (10nodes)	
	RMSE	Time	RMSE	Time	RMSE	Time
Solar flare	0.3011	1.0563	0.2613	1.6221	**0.1062**	**0.0291**
UJIIndoorLoc	0.5175	0.7042	0.6239	0.8543	**0.3642**	**0.0078**
wiki4HE	0.1691	1.2054	0.1286	1.6847	**0.0854**	**0.0297**
YearPredictionMSD	0.7044	1.5811	0.9208	3.5142	**0.2147**	**0.0401**
Student performance	0.5071	2.3054	0.4505	3.2586	**0.2213**	**0.0501**
Servo	0.0452	0.8421	0.0326	1.1231	**0.0288**	**0.0201**

Table 3 Generation performance comparison (average testing accuracy and average training accuracy)

Datasets	SVM		SOM		BP-CELM	
	Training accuracy (%)	Testing accuracy (%)	Training accuracy (%)	Testing accuracy (%)	Training accuracy (%)	Testing accuracy (%)
Echocardiogram	78.32	77.51	91.42	90.81	**92.48**	**92.33**
EEG eye state	88.67	86.32	90.31	85.43	**97.39**	**89.72**
Folio	90.45	89.86	89.65	85.41	**99.52**	**97.35**
Gisette	99.32	97.34	80.57	78.22	**100**	**98.21**
Libras movement	88.28	86.44	89.13	87.89	**89.52**	**88.67**
DNA	98.16	93.22	84.81	80.11	**99.45**	**94.52**

with 10 hidden nodes can reach a very small value for the regression applications. When the number of hidden nodes is 5, the testing accuracy of BP-CELM is above other incremental ELM algorithms all the time for classification applications. In practical applications, the residual error of neural network reduces very slowly will lead to neural network growing procedure stop, but BP-CELM can reach expected learning accuracy at early learning stage.

Tables 2 and 3 particularly highlight the generation performance comparison between BP-CELM, BP, SVM, SOM and Elman. And among the comparisons of these algorithms, apparently, the similar results are underlined and better test results are given in boldface.

5 Conclusion

In this paper, a new learning algorithm called back propagation convex extreme learning machine (BP-CELM) is presented. Unlike other incremental ELM methods, All the parameters of BP-CELM are incrementally updated efficiently during the learning step of networks. Simulation results show that BP-CELM can greatly enhance the learning effectiveness, reduce the number of hidden nodes.

References

1. Huang, G.B., Zhu, Q.Y., Siew, C.K.: Extreme learning machine: theory and applications. Neurocomputing **70**(1), 489–501 (2006)
2. Huang, G.B., Chen, L., Siew, C.K.: Universal approximation using incremental constructive feedforward networks with random hidden nodes[J]. IEEE Trans. Neural Netw. **17**(4), 879–892 (2006)
3. Yang, Y.M., Wu, Q.M.J., Wang, Y.N., Zeeshan, K.M., Lin, X.F., Yuan, X.: Data partition learning with multiple extreme learning machines. IEEE Trans. Cybern. **45**(8), 1463–1475 (2015)
4. Cao, J., Lin, Z.: Extreme learning machines on high dimensional and large data applications: a survey. Math. Probl. Eng. **2015**, 1–13, ID 103796 (2015)
5. Tang, Y.G., Han, Z.Z., Liu, F.C., Guan, X.P.: Identification and control of nonlinear system based on Laguerre-ELM Wiener model. Commun. Nonlinear Sci. Numer. Simul. **38**, 192–205 (2016)
6. Yang, Y.M., Wang, Y.N., Yuan, X.F., Chen, Y.H., Tan, L.: Neural network-based self-learning control for power transmission line deicing robot. Neural Comput. Appl. **22**(5), 969–986 (2013)
7. Huang, G., Song, S., Gupta, J.N.D., Wu, C.: Semi-supervised and unsupervised extreme learning machines. IEEE Trans. Cybern. **44**(12), 2405–2417 (2014)
8. Yang, Y.M., Wang, Y.N., Yuan, X.F.: Bidirectional extreme learning machine for regression problem and its learning effectiveness. IEEE Trans. Neural Netw. **23**(9), 1498–1505 (2012)
9. Zhu, W.T., Miao, J., Qing, L.Y.: Constrained extreme learning machines: a study on classification cases. Comput. Sci. (2015). arXiv:1501.06115
10. Krishnasamy, G., Paramesran, R.: Hessian semi-supervised extreme learning machine. Neurocomputing **207**, 560–567 (2016)
11. Liao, S.Z., Feng, C.: Meta-ELM: ELM with ELM hidden nodes. Neurocomputing **128**, 81–87 (2014)
12. Zhu, Q.Y., Qin, A.K., Suganthan, P.N., Huang, G.B.: Evolutionary extreme learning machine. Pattern Recognit. **38**(10), 1759–1763 (2005)
13. Barron, A.R.: Universal approximation bounds for superpositions of a sigmoidal function. IEEE Trans. Inf. Theory **39**(3), 930–945 (1993)
14. Huang, G.B., Chen, L.: Convex incremental extreme learning machine. Neurocomputing **70** (16–18), 3056–3062 (2007)
15. Yang, Y.M., Wu, Q.M.J.: Extreme learning machine with subnetwork hidden nodes for regression and classification. IEEE Trans. Cybern. **210**(3), 1–14 (2015)
16. Denis, S., Serre, M.D.: Theory and Applications. Springer, New York (2002)

Data Fusion Using OPELM for Low-Cost Sensors in AUV

Jia Guo, Bo He, Pengfei Lv, Tianhong Yan and Amaury Lendasse

Abstract With mobility, security, intelligence and other advantages, autonomous underwater vehicle (AUV) becomes an indispensable instrument in the complex underwater environment. Owing to the independence of external signal (such as GPS) which is restricted or invalid in the water, inertial navigation system (INS) has become the most suitable navigation and positioning system for Underwater Vehicles. However, as the excessive reliance of sensor data, the precision of INS can be affected by sensor data especially heading angle data from low-cost sensor such as attitude and heading reference system (AHRS) and digital compass. Therefore, how to fuse low-cost sensor information to get more accurate data becomes the key to improve navigation accuracy. Based on the original Extreme Learning Machine (ELM) algorithm, the Optimally Pruned Extreme Learning Machine (OPELM) algorithm is presented as a more robust and general methodology in 2010, which make it possible to realize data fusion by using a more reliable network. In this paper, we proposed a method of data fusion which using Optimally-Pruned Extreme Learning Machine (OPELM) to improve the accuracy of heading angle from AHRS and digital compass. Our method has already been demonstrated by a range of real datasets, and it outperforms current available Kalman Filtering algorithms in efficiency.

J. Guo · B. He (✉) · P. Lv
School of Information Science and Engineering, Ocean University of China,
238 Songling Road, Qingdao 266100, China
e-mail: bhe@ouc.edu.cn

T. Yan (✉)
School of Mechanical and Electrical Engineering, China Jiliang University,
258 Xueyuan Street, Xiasha High-Edu Park, Hangzhou 310018, China
e-mail: thyan@163.com

A. Lendasse
Department of Mechanical and Industrial Engineering and the Iowa Informatics Initiative,
The University of Iowa, Iowa City, IA 52242-1527, USA

J. Cao et al. (eds.), *Proceedings of ELM-2016*, Proceedings in Adaptation,
Learning and Optimization 9, DOI 10.1007/978-3-319-57421-9_22

Keywords AUV · Data fusion · OPELM · Kalman filtering · Neural network

1 Introduction

Autonomous underwater vehicle (AUV) is an indispensable instrument in the complex underwater environment such as the exploration of seabed sediment, due to its flexibility and autonomy. Unlike unmanned vehicles and unmanned aerial vehicles, GPS which employs as a significant navigation data, is limited or even unavailable in water. However, simultaneous localization and mapping (SLAM) [1–3] can map surroundings in real time and acquire estimated position information simultaneously. Therefore, SLAM is increasingly attracting global attention and becoming a research hot topic in unknown environment mobile robot's navigation and positioning areas [4–6], which can provide feasible solutions for the realization of autonomous navigation.

Traditionally, there is a lot of sensors installed in AUV, including the AHRS, digital compass, pressure sensor, GPS and the doppler velocity log (DVL) which are mainly used for navigation and positioning. As the characteristics of AHRS and digital compass are low-cost, small and low power consumption, they are widely used in motor vehicles and unmanned aerial vehicles. AHRS contains a plurality of axial sensor, which can provide heading, pitch and roll information for AUV. However, the process of surge, acceleration, deceleration and even other factors will inevitably bring angle error. Once the angle especially heading angle is not accurate, navigation accuracy would not be guaranteed. Even though most of AHRS can be inputted external Global Navigation Satellite System (GNSS) signal to correct angle error in the process of moving. Nevertheless, GPS is invalid in the water so that the compensation of angle is ineffective. And when referred to digital compass, it is vulnerable to interference of ferromagnetic substance. Thus, using a single sensor data cannot meet the demand for navigation accuracy.

Data fusion is the process of integration of multiple data and knowledge representing the same real-world object into a consistent, accurate, and useful representation. The goal of data fusion is to combine relevant information from two or more data sources into a single one that provides a more accurate description than any of the individual data sources [7]. At present, there are several methods of data fusion. The mature information fusion methods mainly include Kalman Filtering [8], Bayesian approach [9, 10], fuzzy method [11, 12], Dempster-Shafter [13], and Neural Network [14, 15]. Although Bayesian has the axiomatic foundation and plain mathematical properties, the main difficulty of Bayesian is that it is hard to establish an accurate description of probability distribution especially when the original data is given by the low-cost sensor. Different from Bayesian, fuzzy method not only can deal with problem which owns inexact description, but also can merge information adaptively. However, it only applies to the circumstance which has little and qualitative information. Data fusion based on Dempster-Shafter has an advantage in processing large amounts of data, while it often become invalid for independence data. Kalman Filtering is the most successful data fusion method in multi-sensor system by far. But low prediction accuracy, filter divergence result from cumulative prediction error are the biggest problem existing in the Kalman Filtering. The cumulative prediction error comes from the inaccurate state equation of integrated navigation system [16, 17]. The application of Neural Network in information fusion does not has a long history. Neural network which has high-speed parallel computing power and strong adaptive learning ability can achieve optimal signal processing, while the uncertainly of learning error limits the application of neural networks. In 2010, Yoan Miche et al. presented the OPELM algorithm [18]. The OPELM methodology is not only considerably faster than the Multilayer Perception (MLP) [19] and the Least-Squares Support Vector Machine (LS-SVM) [20], but also has excellent performance in terms of robustness and generality. OPELM makes it possible to realize data fusion by using a more robust and general network.

After comprehensive research of advantages and disadvantages of the above algorithms, firstly, we used OPELM to fuse heading data from AHRS and digital compass, then the heading angle correction model was generated. Moreover, the new heading angle obtained by the model give a more precise direction for AUV. At last, combined with Kalman Filtering, we can get more accurate location information of AUV.

The remainder of the paper is organized as follows: In Sect. 2, OPELM algorithm is reviewed. Data fusion using OPELM will be presented in Sect. 3. In Sect. 4, experiments with different datasets will be carried out to verify the performance of the proposed algorithm. Finally, we draw the main conclusions of this work.

2 Review of OPELM

The OPELM algorithm is presented as a more robust methodology, which is based on the original Extreme Learning Machine (ELM) algorithm. ELM was proposed by Huang et al. [21] and the main novelty introduced by ELM randomly chooses the input weights and biases of the hidden nodes instead of learning these parameters.

The output of Signal-Hidden Layer Feed-forward Neural Networks (SLFNs) with N hidden nodes can be presented as:

$$f_n(x) = \sum_{i=1}^{n} \beta_i G(\omega_i, b_i, x).x \in R^n, \omega_i \in R^n, \beta_i \in R^n. \tag{1}$$

where $G(\omega_i, b_i, x)$ is the ith output of hidden layer neurons corresponding to the input x. $\beta = [\beta_{i1}, \beta_{i2}, \ldots, \beta_{im}]^T$ is the connecting link between the ith hidden layer neurons and output neurons weight vector.

For N arbitrary input sample $(x_i, t_i) \in R^n \times R^m$, where $x_i = [x_{i1}, x_{i2}, \ldots, x_{in}] \in R^n$ and $t_i = [t_{i1}, t_{i2}, \ldots, t_{im}] \in R^m$, given N hidden layer neurons and activation function $G(\omega_i, b_i, x)$, β_i, ω_i and b_i can be found out to make SLFNs close to the N samples with zero error.

$$\sum_{j=1}^{N} \beta_j G(\omega_j, b_j, x_i) = t_i \quad i = 1, 2, \ldots, N. \tag{2}$$

To simplify, the above Eq. (2) can be written equivalently as:

$$\mathrm{H}\beta = \mathrm{T}\ . \tag{3}$$

where

$$\mathrm{H}(\omega_1, \ldots, \omega_N, b_1, \ldots, b_N, x_1, \ldots, x_N) = \begin{bmatrix} G(\omega_1, b_1, x_1) & \cdots & G(\omega_N, b_N, x_1) \\ \vdots & \ddots & \vdots \\ G(\omega_1, b_1, x_N) & \cdots & G(\omega_N, b_N, x_N) \end{bmatrix}_{N \times N}$$

$$\beta = [\beta_1^T, \ldots, \beta_N^T]_{N \times m}^T\, T = [t_1^T, \ldots, t_N^T]_{N \times m}^T. \tag{4}$$

In the formula (4), H is the hidden output matrix of the neural network, the *i*th column of H is the *i*th hidden layer neurons output corresponding to input $x_1, x_2, \ldots, x_n$. The training process of SLFNs is equivalent to find the least squares solution of linear system $\mathrm{H}\beta = \mathrm{T}$. It has been proved that the global optimal output weights can be written as:

$$\hat{\beta} = H^* R. \tag{5}$$

where H^* is Moore-Penrose inverse matrix of the hidden layer output matrix H.

To remove the useless neurons of the hidden layer, OPELM algorithm ranks the best neurons by multi-response sparse regression (MRSR) algorithm. As an extension of least angle regression (LARS), MRSR provides a ranking of the kernels. Finally, the leave-one-out (LOO) validation method is used to select the actual best number of neurons. The most specific details of the MRSR algorithm and LOO algorithm can be found from two reported papers [22, 23] respectively.

The whole structure of OPELM is shown in Fig. 1.

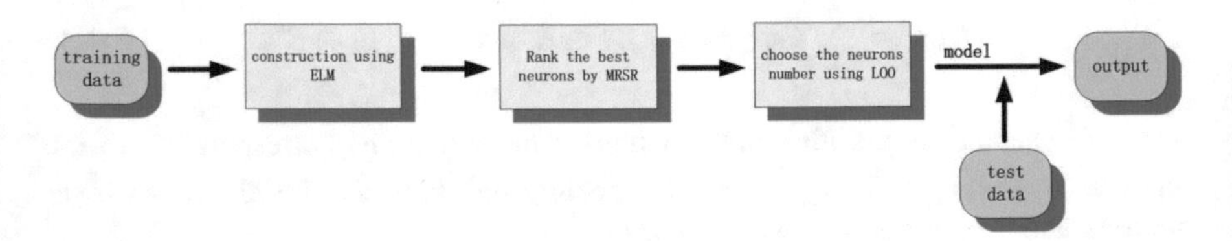

Fig. 1 The whole structure of OPELM

3 Data Fusion Using OPELM

3.1 The Selection of Input Data

The datasets used in this paper were collected by Sailfish AUV at Menlou Reservoir (Yantai, China). The essential parameters of the Sailfish autonomous underwater vehicle are in Table 1. Figure 2 displays the location of sensors in Sailfish AUV. And the deployment figure of AUV is shown in Fig. 3.

In this system, we adopted TCM5 and AHRS 100 to measure the angle of AUV. The TCM5 combines 3-axis of PNI Corporation's patented Magneto-Inductive (MI) magnetic sensors and a 3-axis MEMS accelerometer in a single module. Despite it has small size, light weight, no error accumulation and many other advantages, the local geomagnetic field where location TCM5 used is vulnerable to the influences of various ferromagnetic and electromagnetic distortion. So, there is often an error between measured angle and the actual direction. And this would affect the accuracy of measurement. Unlike TCM5 using magnetic sensor, AHRS 100 integrates three MEMS gyroscope. Based on Coriolis effect, MEMS gyroscope measures the angular velocity through the change of capacitance. The change of angle is integral by the angular velocity. Therefore, either AHRS angle or AHRS angular velocity can be chosen as input. In this paper, we chose AHRS angle for data fusion. Among three Euler angles, the heading angle has the greatest effect on the track, so we only fuse the heading angles from AHRS 100 and TCM5. Variables corresponding to the input-output are as follows. And according to the proportion of 3:1, the dataset is typically separated into training set and testing set in offline mode Table 2 shows input variables.

Table 1 Essential parameters of the Sailfish autonomous underwater vehicle (AUV)

Parameter	Length	Width	Tonnage	Payload	Maximal speed	Endurance
Value	2.3 m	210 mm	72 kg	4 kg	5knot	10 h

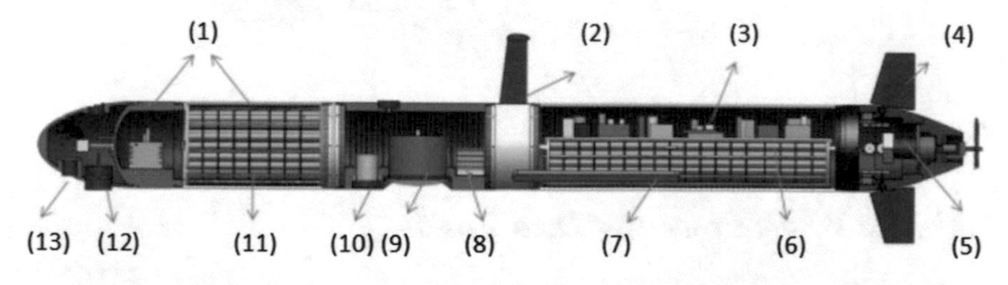

Fig. 2 Sailfish AUV structure diagram. (*1*) Shell; (*2*) antenna; (*3*) electronic control system; (*4*) rudder; (*5*) propeller; (*6*) lithium battery; (*7*) side scan sonar; (*8*) navigation system; (*9*) DVL; (*10*) charge coupled device; (*11*) replaceable lithium battery; (*12*) load dump; (*13*) depth gauge

Fig. 3 Launch AUV in the sea trial

Table 2 Input variables

Input			
GPS_Heading	TCM_Heading	AHRS_Acc_X	AHRS_Heading

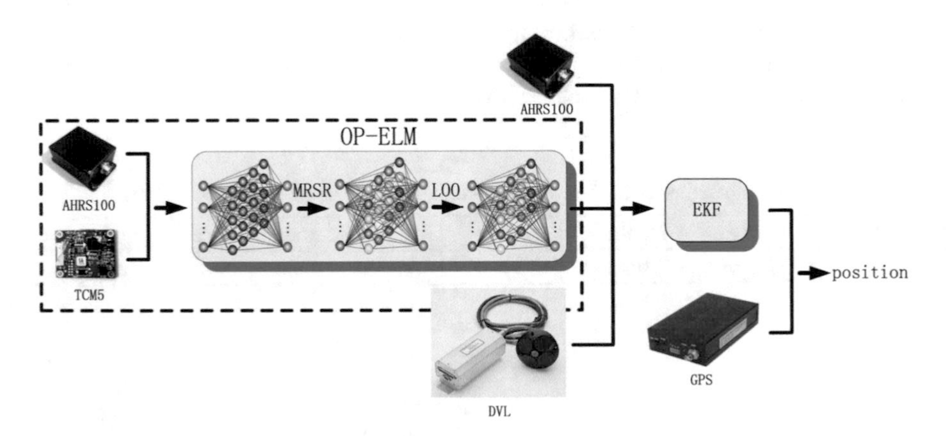

Fig. 4 The whole process of data fusion

3.2 The Whole Process of Data Fusion

The whole process of Data Fusion is as follows (Fig. 4).

The key of data fusion in this system is marked with black dashed box. First, we use OPELM to learn and fuse heading angle information form AHRS 100 and TCM5, in order to obtain a more accurate value of the heading angle. Afterwards, based on the output heading angle, the velocity from DVL and other angles and acceleration information from AHRS 100, we take advantage of Kalman Filtering algorithm to generate the rough track of AUV. Finally, if GPS is effective, AUV would get accurate position information by using GPS to modify the rough track; on the contrary, if GPS is ineffective, AUV would get position information from Kalman Filtering.

3.3 Performance Evaluation

The performance of the Data Fusion using OPELM is measured by the root mean square error (RMSE).

$$E_{track} = \sqrt{\frac{\sum_{n=1}^{n=N}\left(\widehat{Y}_{i_{track}} - Y_{i_{track\_true}}\right)^2}{N}}. \qquad (6)$$

where E_{track} is the track error of the Data Fusing using OPELM. $\widehat{Y}_{i_{track}}$ are the output track produced by output heading angle, while $Y_{i_{track\_true}}$ is the actual track. And N is the total number of samples.

3.4 Online Data Fusion Using OPELM

Data fusion using OPELM is not only suitable for offline simulation data processing, but also for fusing data in real-time. Firstly, through the AUV sailing in the designated area, we will get the heading angle model which can be used to fuse AHRS 100 and TCM5 heading angle for AUV. Then the fused heading angle and other sensor information are sent into Kalman Filtering to get optimal position estimation.

4 Experimental Results

The proposed method of Data Fusion has been evaluated on a range of datasets. We compared with other algorithms. Due to heading angle fused by ELM algorithm has poor stability and its chaotic angle data interfere the display of other data, we just compare ELM with other algorithms in the first dataset as shown in Fig. 5.

Figure 6 shows the heading angle which gotten from four different methods (OPELM, TCM, AHRS and GPS). Figure 7 reveals the heading angle error between GPS and other methods, respectively. As can be seen clearly, the heading

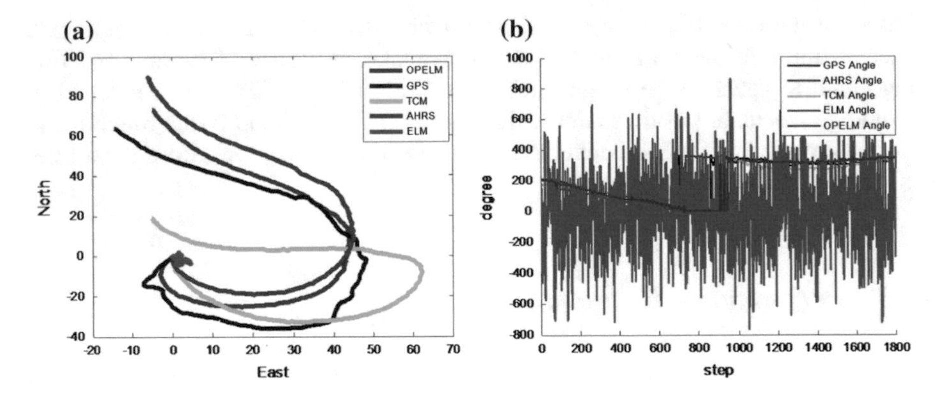

Fig. 5 The comparison of trial **a** and heading angle **b** by different algorithm including ELM

Fig. 6 The comparison of heading angle **a** and **b** by different algorithm except ELM algorithm

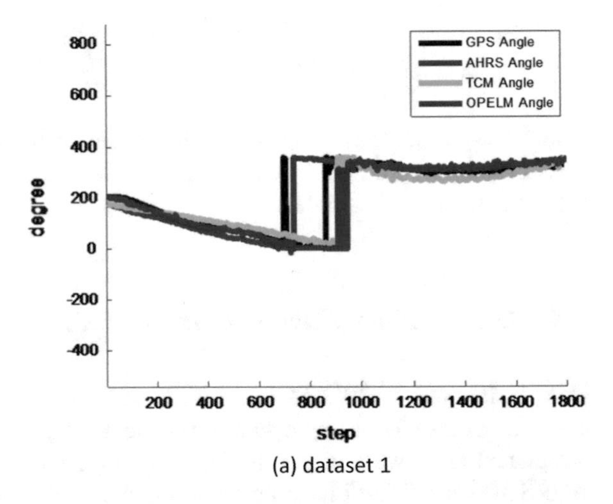

(a) dataset 1

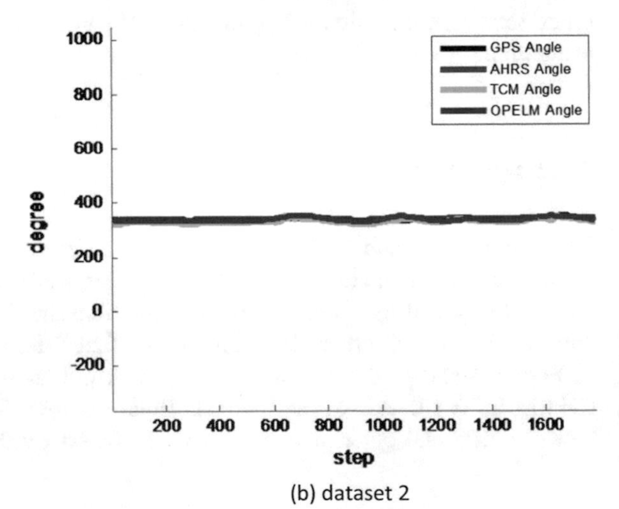

(b) dataset 2

Fig. 7 The heading angle error of different methods

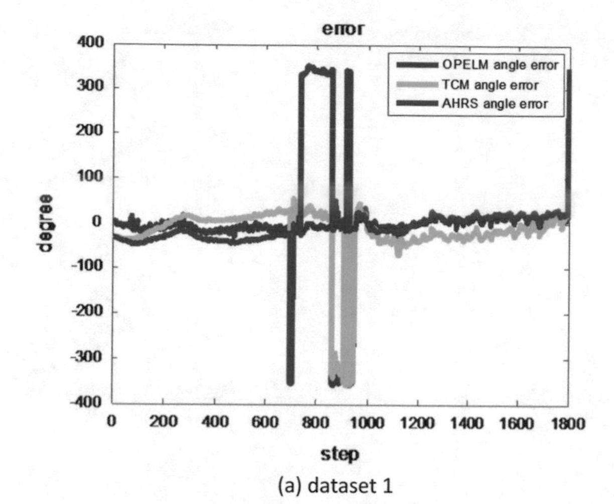

(a) dataset 1

(b) dataset 2

angle values in Fig. 6b are almost unchanged and revealed to be nearly a straight line. Correspondingly, the heading angle error of raw sensor data which are produced by AHRS 100, TCM5 and fused data are displayed in Fig. 7. Although it seems that AHRS 100 data has better stability than other data, the heading angle error from AHRS 100 is almost whole above zero degrees. This indicates that there is always a positive error existing in attitude and heading reference system, which may bring indelible positive error. In addition to a negative heading angle error similar to AHRS 100, TCM5 still has a larger margin of error. Compared with AHRS 100 and TCM5, the heading angle data fused by OPELM have a relatively stability error which also can be offset. Even if the heading angle constantly changes as shown in Fig. 7, the heading angle which is generated by the fusion still has good performance than others (Fig. 8).

Fig. 8 The comparison of trial by different method

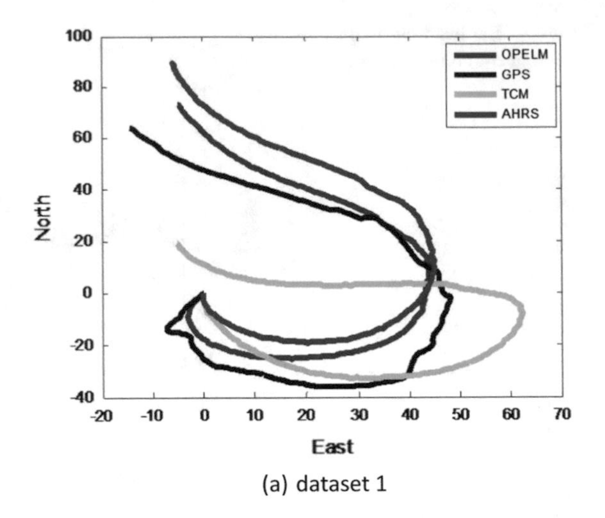

(a) dataset 1

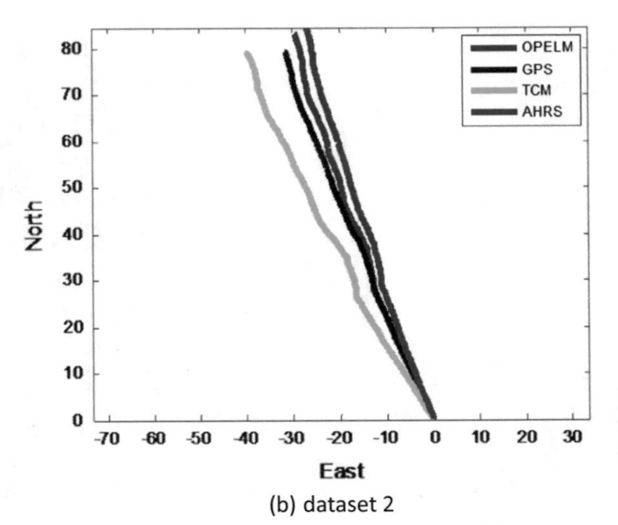

(b) dataset 2

As shown in Fig. 8, the trajectories are corresponding to heading angle from different methods, respectively. No matter the heading angle is a constant or is changing in real time, the trajectory of OPELM is more close to the true path (GPS) than the green or blue trajectory which using a single angle data. And the RMSE of different trajectory is explained in Fig. 9. When compared with other trajectory, the track error of the Data Fusing using OPELM is far less than other methods in general. It is proved by experimental data that the data fusing using OPELM is effective.

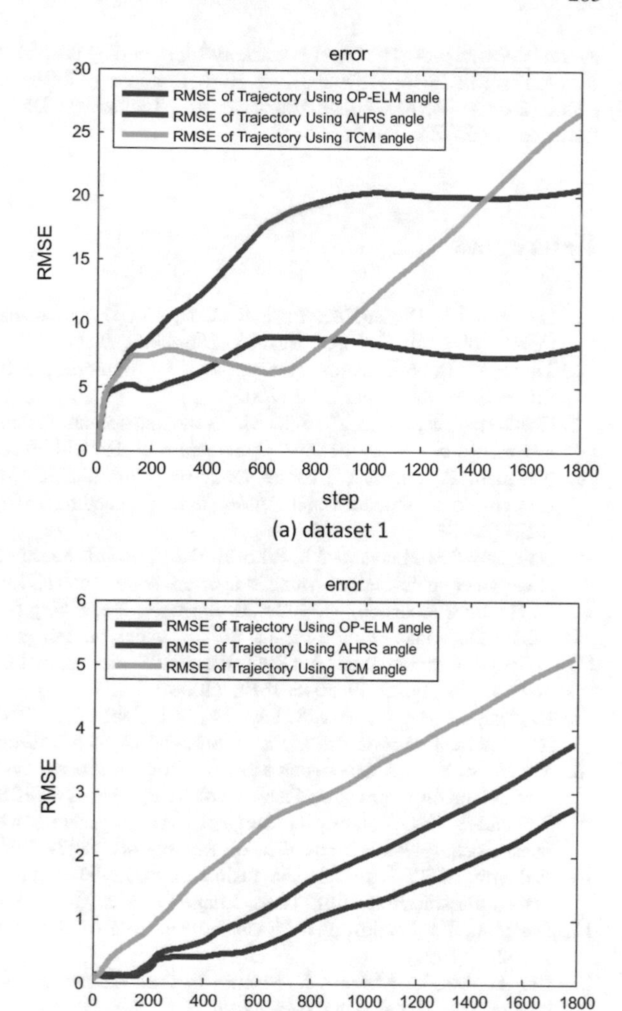

Fig. 9 the RMSE of different trajectory

5 Conclusion

In this paper, we proposed a data fusion method using OPELM to improve the navigation accuracy for AUV. Compared to the directly using of low-cost sensors (such as TCM5 or AHRS 100) data to execute Kalman Filtering, the proposed data fusion exhibit better accuracy, which RMSE improved at least 53%, or even as high as 64.7% in the complex conditions. The experimental results verify the validity and efficiency of the proposed method for AUV.

Acknowledgements The paper is partially supported by the Natural Science Foundation of China (41176076, 31202036, 51379198), the High Technology Research and Development Program of China (2014AA093400, 2014AA093410), and Technology Development Program of Shandong Province (2013GHY11507).

References

1. Leonard, J.J., Durrant-Whyte, H.F., Cox, I.J.: Dynamic map building for an autonomous mobile robot. Int. J. Robot. Res. **11**, 286–298 (1992)
2. Thrun S.: Robotic mapping: a survey. In: Exploring Artificial Intelligence in the New Millennium, vol. 1, pp. 1–35 (2002)
3. Cheeseman, P., Smith, R., Self, M.: A stochastic map for uncertain spatial relationships. In: Proceeings of 4th International Symposium on Robotic Research, pp. 467–474 (1987)
4. Newman, P., Leonard, J.: Pure range-only sub-sea SLAM. In: 2003 IEEE International Conference on Robotics and Automation, Proceedings. ICRA'03, vol. 2, pp. 1921–1926. IEEE (2003)
5. Newman, P.M., Leonard, J.J., Rikoski, R.J.: Towards constant-time SLAM on an autonomous underwater vehicle using synthetic aperture sonar. In: The Eleventh International Symposium on Robotics Research, pp. 409-420. Springer, Heidelberg (2005)
6. Ribas, D., Ridao, P., Neira, J.: SLAM using an imaging sonar for partially structured underwater environments. In: 2006 IEEE/RSJ International Conference on Intelligent Robots and Systems, pp. 5040–5045. IEEE (2006)
7. Haghighat, M., Abdel-Mottaleb, M., Alhalabi, W.: Discriminant correlation analysis: real-time feature level fusion for multimodal biometric recognition (2016)
8. Vershinin, Y.A.: A data fusion algorithm for multisensor systems. In: The Fifth International Conference on Information Fusion, vol. 1, pp. 341–345. IEEE (2002)
9. Fasbender, D., Radoux, J., Bogaert, P.: Bayesian data fusion for adaptable image pansharpening. IEEE Trans. Geosci. Remote **46**, 1847–1857 (2008)
10. Williams, D.P.: Bayesian data fusion of multiview synthetic aperture sonar imagery for seabed classification. IEEE Trans. Image Process. **18**, 1239–1254 (2009)
11. Ramponi, F.R.G.: Fuzzy methods for multisensor data fusion. IEEE Instrum. Meas. **43**, 288–294 (1994)
12. Liu, X., Ma, L., Mathew, J.: Machinery fault diagnosis based on fuzzy measure and fuzzy integral data fusion techniques. Mech. Syst. Sig. Process. **23**, 690–700 (2009)
13. Xu, C., Geng, W., Pan, Y.: Review of dempster-shafer method for data fusion. Acta Electron. Sinca **29**, 393–396 (2001)
14. Wan, W., Fraser, D.: Multisource data fusion with multiple self-organizing maps. IEEE Trans. Geosci. Remote **37**, 1344–1349 (1999)
15. Jimenez, L.O., Morales-Morell, A., Creus, A.: Classification of hyperdimensional data based on feature and decision fusion approaches using projection pursuit, majority voting, and neural networks. IEEE Trans. Geosci. Remote **37**, 1360–1366 (1999)
16. Abdel-Hamid, W., Noureldin, A., El-Sheimy, N.: Adaptive fuzzy prediction of low-cost inertial-based positioning errors. IEEE Tans. Fuzzy Syst. **15**, 519–529 (2007)
17. Noureldin, A., Karamat, T.B., Eberts, M.D., et al.: Performance enhancement of MEMS-based INS/GPS integration for low-cost navigation applications. IEEE Tans. Veh. Technol. **58**, 1077–1096 (2009)
18. Miche, Y., Sorjamaa, A., Bas, P., et al.: OPELM: optimally pruned extreme learning machine. IEEE T. Neural Netw. **21**, 158–162 (2010)
19. Haykin, S.: Neural Networks: A Comprehensive Foundation, 2nd edu. Prentice Hall, Englewood Cliffs (1998)

20. Suykens, J., Gestel, T.V., Brabanter, J.D., Moor, B.D., Vanderwalle, J.: Least-Squares Support-Vector Machines. World Scientific, Singapore (2002)
21. Huang, G.B., Zhu, Q.Y., Siew, C.K.: Extreme learning machine: theory and applications. Neurocomputing. **70**, 489–501 (2006)
22. Simil, T., Tikka, J.: Multiresponse sparse regression with application to multidimensional scaling. In: International Conference on Artificial Neural Networks, pp. 97–102. Springer (2005)
23. Feng, D., Chen, F., Xu, W.: Efficient leave-one-out strategy for supervised feature selection. TsingHua Sci Technol. **18**, 629–635 (2013)

Printed in the United States
By Bookmasters